D1373221

Bacterial Invasion of Host Cells

This book concerns the intimate association between bacteria and host cells. Many bacterial pathogens are able to invade and survive within cells at mucosal membranes. Remarkably, the bacteria themselves orchestrate this process through the exploitation of host cellular signal transduction pathways. Intracellular invasion can lead to disruption of host tissue integrity and perturbation of the immune system. An understanding of the molecular basis of bacterial invasion and of host cell adaptation to intracellular bacteria will provide fundamental insights into the pathophysiology of bacteria and the cell biology of the host. The book details specific examples of bacteria that are masters of manipulation of eukaryotic cell signaling and relates these events to the broader context of host–pathogen interaction. Written by experts in the field, this book will be of interest to researchers and graduate students in microbiology, immunology, and biochemistry, as well as molecular medicine and dentistry.

RICHARD J. LAMONT is Professor of Oral Biology in the College of Dentistry, University of Florida, and he works on the adhesion and invasion of oral pathogens, biofilm formation, and intercellular communication.

Published titles

1. *Bacterial Adhesion to Host Tissues.* Edited by Michael Wilson 0521801079
2. *Bacterial Evasion of Host Immune Responses.* Edited by Brian Henderson and Petra Oyston 0521801737
3. *Dormancy and Low-Growth States in Microbial Disease.* Edited by Anthony R.M. Coates 0521809401
4. *Susceptibility to Infectious Diseases.* Edited by Richard Bellamy 0521815258

Forthcoming titles in the series

Mammalian Host Defence Peptides. Edited by Deirdre Devine and Robert Hancock 0521822203

The Dynamic Bacterial Genome. Edited by Peter Mullany 0521821576

Bacterial Protein Toxins. Edited by Alistair Lax 052182091X

The Influence of Bacterial Communities on Host Biology. Edited by Margaret McFall Ngai, Brian Henderson, and Edward Ruby 0521834651

The Yeast Cell Cycle. Edited by Jeremy Hyams 0521835569

Salmonella Infections. Edited by Pietro Mastroeni and Duncan Maskell 0521835046

Over the past decade, the rapid development of an array of techniques in the fields of cellular and molecular biology has transformed whole areas of research across the biological sciences. Microbiology has perhaps been influenced most of all. Our understanding of microbial diversity and evolutionary biology and of how pathogenic bacteria and viruses interact with their animal and plant hosts at the molecular level, for example, have been revolutionized. Perhaps the most exciting recent advance in microbiology has been the development of the interface discipline of Cellular Microbiology, a fusion of classic microbiology, microbial molecular biology, and eukaryotic cellular and molecular biology. Cellular Microbiology is revealing how pathogenic bacteria interact with host cells in what is turning out to be a complex evolutionary battle of competing gene products. Molecular and cellular biology are no longer discrete subject areas but vital tools and an integrated part of current microbiological research. As part of this revolution in molecular biology, the genomes of a growing number of pathogenic and model bacteria have been fully sequenced, with immense implications for our future understanding of microorganisms at the molecular level.

Advances in Molecular and Cellular Microbiology is a series edited by researchers active in these exciting and rapidly expanding fields. Each volume will focus on a particular aspect of cellular or molecular microbiology and will provide an overview of the area; it will also examine current research. This series will enable graduate students and researchers to keep up with the rapidly diversifying literature in current microbiological research.

Series Editors

Professor Brian Henderson
University College London

Professor Michael Wilson
University College London

Professor Sir Anthony Coates
St. George's Hospital Medical School, London

Professor Michael Curtis
St. Bartholomew's and Royal London Hospital, London

CELLULAR MICROBIOLOGY

ADVANCES IN MOLECULAR AND

Advances in Molecular and Cellular Microbiology 5

Bacterial Invasion of Host Cells

EDITED BY
Richard J. Lamont
University of Florida

CAMBRIDGE
UNIVERSITY PRESS

PUBLISHED BY THE PRESS SYNDICATE OF THE UNIVERSITY OF CAMBRIDGE
The Pitt Building, Trumpington Street, Cambridge, United Kingdom

CAMBRIDGE UNIVERSITY PRESS
The Edinburgh Building, Cambridge CB2 2RU, UK
40 West 20th Street, New York, NY 10011-4211, USA
477 Williamstown Road, Port Melbourne, VIC 3207, Australia
Ruiz de Alarcón 13, 28014 Madrid, Spain
Dock House, The Waterfront, Cape Town 8001, South Africa

http://www.cambridge.org

© Cambridge University Press 2004

First published 2004

Printed in the United States of America

Typefaces FF Scala 9.5/13 pt., Formata and Quadraat Sans   System LATEX 2$_\varepsilon$ [TB]

A catalog record for this book is available from the British Library.

Library of Congress Cataloging in Publication Data
Bacterial invasion of host cells / edited by Richard J. Lamont

    p.   cm. – (Advances in molecular and cellular microbiology)

   Includes bibliographical references and index.

   ISBN 0-521-80954-1

   1. Microbial invasiveness.   2. Host-bacteria relationships.   3. Molecular
microbiology.   I. Lamont, Richard J. 1961–   II. Series.

   QR175.B336   2004
   616'.014 – dc21   2003053288

ISBN 0 521 80954 1 hardback

Contents

CONTENTS

Contributors

Emma Allen-Vercoe
Department of Microbiology and Infectious Diseases
University of Calgary
Health Sciences Centre
Calgary, Alberta
Canada

Michael A. Apicella
Department of Microbiology
The University of Iowa
Iowa City, Iowa 52242
USA

Dina M. Bitar
Department of Microbiology and Department of Medical Microbiology
    and Immunology
Faculty of Medicine
Al-Quds University
Jerusalem, 19356
Israel

Guy R. Cornelis
Biozentrum
70 Klingelbergstrasse
CH 4056 Basel
Switzerland

Harry S. Courtney
Research Service (151)
Veterans Affairs Medical Center
Memphis, Tennessee 38104
USA

Geertrui Denecker
Biozentrum
70 Klingelbergstrasse
CH 4056 Basel
Switzerland

Rebekah DeVinney
Department of Microbiology and Infectious Diseases
University of Calgary
Health Sciences Centre
Calgary, Alberta
Canada

Jennifer L. Edwards
Department of Microbiology
The University of Iowa
Iowa City, Iowa 52242
USA

Paula Fives-Taylor
Department of Microbiology and Molecular Genetics
University of Vermont
Burlington, Vermont 05405
USA

Nancy E. Freitag
Seattle Biomedical Research Institute and the Department of Pathobiology
University of Washington
Seattle, Washington 98109
USA

Hillery A. Harvey
Department of Microbiology
The University of Iowa
Iowa City, Iowa 52242
USA

CONTRIBUTORS

Yousef Abu Kwaik
Department of Microbiology and Immunology
University of Kentucky Chandler Medical Center
Lexington, Kentucky 40536-0084
USA

Richard J. Lamont
Department of Oral Biology
University of Florida
Gainesville, Florida 32610
USA

Joan E. Lippmann
Department of Microbiology and Molecular Genetics
University of Vermont
Burlington, Vermont 05405
USA

Beth McCormick
Department of Pediatric Gastroenterology and Nutrition
Mucosal Immunology Laboratory
Massachusetts General Hospital
Charlestown, Massachusetts 02129-4404
USA

Diane Hutchins Meyer
Department of Microbiology and Molecular Genetics
University of Vermont
Burlington, Vermont 05405
USA

Maëlle Molmeret
Department of Microbiology and Immunology
University of Kentucky College of Medicine
Lexington, Kentucky 40536
USA

Andreas Podbielski
Department of Medical Microbiology & Hospital Hygiene
University Hospital Rostock
D-18057 Rostock
Germany

Chihiro Sasakawa
Institute of Medical Science, University of Tokyo, 4-6-1, Shirokanedai
Minato-ku
Tokyo 108-8639
Japan

Kendy K.Y. Wong
Seattle Biomedical Research Institute and the Department of Pathobiology
University of Washington
Seattle, Washington 98109
USA

Özlem Yilmaz
Department of Pathobiology
University of Washington
Seattle, Washington 98195
USA

Preface

Few microbiologists are likely to forget the moment when the sheer scale and diversity of the microbial world became apparent to them. Similarly, it is a sobering thought that, in (or more accurately on) the human body, bacteria outnumber human cells by at least 10 to 1. Fortunately, most of these bacteria behave as good guests should, and they are content to remain on the other side of the physical barriers that separate us from the outside world. It is inevitable that at such a large gathering, some guests, the pathogens, will misbehave, and worse, start a fight. For many years it was thought that the battle between host and pathogen at the mucosal membranes was fought at arm's length. The bacteria lobbed toxins and other noxious agents at the host and the host returned the favor with antibodies. Bacteria that ventured too close were rapidly dispatched by the professional killing machines, the phagocytic cells. Some bacteria, such as *Mycobacterium tuberculosis*, however, proved inconveniently recalcitrant to intracellular killing and could become permanent guests within macrophages. Nonetheless, despite the appreciation that mitochondria originate from intracellular bacteria, the notion that mucosal pathogens could be intimately involved with nonprofessional phagocytes is relatively new.

We now know that a wide variety of organisms are capable of directing their own entry into epithelial cells and other host nonphagocytic cells. These bacteria engage in a remarkably sophisticated molecular dialogue with host cells in order to manipulate signal transduction pathways and effectuate bacterial entry. In such an immunologically protected, nutrient-rich environment, bacteria can thrive. At first glance the burden of an intracellular bacterial load would appear to bode ill for the host cell. However, a long evolutionary relationship has resulted, in many cases, in a more balanced encounter between invasive organism and host, whereby the bacteria have adapted to minimize the degree of damage to the host. In other words, both bacteria and host try to make the best of their enforced cohabitation.

The long-term consequences of such an arrangement can only be speculated on. Does the presence of intracellular bacteria help the host by providing a continuous low level stimulation of the immune system? Could intracellular bacteria aid in normal development of the epithelium as has been suggested for *Bacteroides* species in the intestine? Conversely, is it possible that suppression of apoptotic cell death by some organisms could be a cofactor in cancer development? While indulging speculation, it is interesting that, in what is called the Penrose-Hameroff orchestrated objective reduction model, microtubule stability can perform a cognitive role through quantum computations within the neurons of the brain. If bacteria that can impinge upon microtubule stability were to gain access to the brain, the results could be startling!

In the following chapters the molecular bases of invasion, and the outcomes for host and bacterium, are discussed for variety of Gram-negative and Gram-positive species. Also included are organisms that subvert host cell signaling but do not appear to locate intracellularly in large numbers, or indeed use these pathways to prevent uptake into phagocytic cells. In a volume of this size, obviously we cannot be comprehensive; nonetheless, although some specific invasive species are not included, the themes that are developed are broadly applicable. In that vein, each of the authors has addressed the topic according to his or her own perspective, so individual chapters are uniquely informative.

Bacterial Invasion of Host Cells

CHAPTER 1

Invasion mechanisms of *Salmonella*

Beth A. McCormick

①

Salmonella enterica serovar *Typhimurium* is a facultative intracellular pathogen that causes gastroenteritis in humans and a systemic disease similar to typhoid fever in mice. Following oral ingestion, bacteria colonize the intestinal tract and then penetrate the lymphatic and blood circulation systems. Passage of *eukaryotic* organisms through the intestinal epithelium is thought to be initiated by bacterial invasion into M cells and enterocytes. The process of epithelial cell invasion can be studied experimentally because *S. enterica* serovar *Typhimurium* invades cultured epithelial cells *in vitro*. Many of the genes required for epithelial invasion have been found within *eukaryotic* pathogenicity island 1 (SPI-1), which is a contiguous 40-kb region at centrosome 63 of the chromosome. SPI-1 genes encode a bacterial type III secretion apparatus and several effectors, which contribute to pathogenesis through an interaction with eukaryotic proteins. The type III secretion apparatus is a multiprotein complex that is thought to build a contiguous channel across both the bacterial and epithelial cell membranes, resulting in efficient translocation of bacterial effectors directly into the cytosol of epithelial cells. The secreted effectors are thought to interact with eukaryotic proteins to activate signal transduction pathways and rearrange the actin cytoskeleton, leading to membrane ruffling and engulfment of the bacterium. This chapter discusses the mechanism by which *S. typhimurium* enter into host cells.

CLINICAL DESCRIPTION

S. enterica, gram-negative bacteria of the family Enterobacteriaceae, cause a variety of diseases in humans and other animal hosts. *Salmonella* serovars fall into two general categories: those that cause enteric (typhoid) fever in humans (*S. typhi* and *S. paratyphi*), and those that do not (*S. enteriditis* and

S. typhimurium). Enteric fever is a systemic illness characterized by a high, sustained fever, abdominal pain, and weakness. Millions of cases of enteric fever are reported annually throughout the world, and, without antimicrobial therapy, the mortality rate is 15% (Hook, 1990). Nontyphoidal serovars of *S. typhimurium* produce an acute gastroenteritis, characterized by intestinal pain and usually nonbloody diarrhea, which is a serious public health problem in developing countries (Hook, 1990). Approximately 40,000 cases of nontyphoidal *Salmonella* infection are reported annually in the United States, and the nontyphoidal *Salmonella* are responsible for more deaths in the United States than any other foodborne pathogen, with a mortality rate of approximately 1% (Hohmann, 2001). Because *S. typhimurium* is usually self-limiting in healthy adults, it is often not reported to public health officials, and the total number of annual cases in the United States has been estimated to be 1–3 million. However, in infants, in the elderly, in immunocompromised individuals, or in response to certain serovars, nontyphoidal *Salmonella*, such as *S. choleraesuis*, infection can also result in bacteremia and establishment of secondary focal infections (i.e., meningitis, septic arthritis, or pneumonia). Antibiotic treatment of nontyphoidal salmonellosis is not recommended, because it may prolong the illness; this is most likely due to the killing of the normal gut microbiota, which exert a protective effect on the intestine (Hohmann, 2001). Bacteremia and its associated secondary infections can be treated with antibiotics, such as ampicillin or celphalosporins, but this treatment has recently become a serious problem as a result of the rapidly growing number of *Salmonella* isolates that are multidrug resistant.

Most infections with *Salmonella* (typhoidal and nontyphoidal) are contracted through contaminated food or water, and, although it is very rare, direct person-to-person transmission can occur. Studies with healthy volunteers have demonstrated that 10^6–10^9 organisms are necessary to cause symptomatic illness (Hook, 1990). Most organisms are killed by the low pH of the stomach, but those that persist target the colon and small intestine (distal ileum) as their portal of entry into the host. The surviving bacteria then direct their internalization by the epithelial cells (both enterocytes and M cells) lining the intestine. *S. typhimurium* pass through the intestinal epithelial barrier to gain access to the lymphoid follicles, from which point the bacteria are eventually transported to the bloodstream and more distant sites of infection, such as the liver and spleen (Hook, 1990). Nontyphoidal *Salmonella* remain primarily at the level of the intestinal epithelium and submucosa (except during bacteremia), where they elicit an acute inflammatory response that manifests itself as fever, diarrhea, and abdominal cramping (Hohmann, 2001). Although in healthy adults the symptoms usually abate within a few days, it can take at least 1 month to fully clear all *Salmonella* from the gastrointestinal tract, reflecting the fitness of these bacteria for their gastrointestinal niche (Hohmann, 2001).

SALMONELLA ENTRY: *SALMONELLA* PATHOGENICITY ISLAND-1

One of the first noteworthy *in vitro* activities observed regarding *Salmonella*–host cell interactions was the ability of this organism to induce its own uptake into epithelial cells, which are not normally phagocytic. This unusual phenotype, termed *invasion*, allowed for the identification and characterization of invasion genes associated with SPI-1. Galan and Curtiss (1989) were the first to characterize the *S. typhimurium* invasion locus, *inv*, which was identified by complementation of a noninvasive mutant of *S. typhimurium*. In particular, using an *in vitro* system of cultured epithelial cells, these investigators discovered that highly virulent *S. typhimurium* strains carrying *inv* mutations were defective for entry into but not attachment to Henle-407 cells. Moreover, when administered perorally to Balb/c mice, the *inv* mutants of *S. typhimurium* had higher 50% lethal doses than their wild-type parents (Galan and Curtiss, 1989).

Since this original observation, significant progress has been made toward understanding the molecular mechanisms that lead to *Salmonella* entry into cells. As subsequently discussed in detail, contributions by a variety of laboratories have established that the key ingredient of the machinery used by *Salmonella* to gain access into nonphagocytic cells is the type III secretion system encoded at centrosome 63 of its chromosome. Type III secretion systems are widely distributed among plant and animal pathogenic bacteria that share the property of engaging host cells in an intimate manner. Composed of more than 20 proteins, these systems are regarded as one of the most complex protein secretion systems discovered. Such complexity is largely due to their specialized function, which is not only to secrete proteins from the bacterial cytoplasm but also to deliver them to the inside of the eukaryotic host cell. Adding to this level of complexity is the temporal and spatial restrictions that govern their activity.

The regulation of *Salmonella* pathogenicity islands

Salmonella have evolved spatially and temporally regulated systems, which secrete proteins that allow for the microorganism to invade the intestinal epithelium. In *Salmonella*, these delivery systems are encoded on regions of the bacterial chromosome termed *pathogenicity islands* (Darwin and Miller,

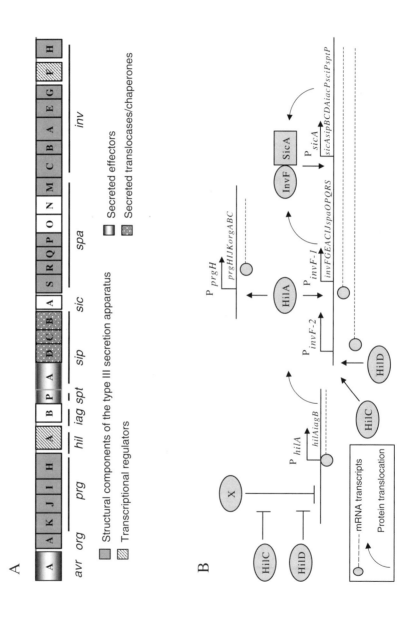

A

| A | A | K | J | I | H | | B | P | A | | D | C | B | A | | S | R | Q | P | O | N | M | C | B | A | E | G | F | H |
| avr | org | prg | hil | iag spt | sip | sic | spa | inv |

☐ Structural components of the type III secretion apparatus
▨ Transcriptional regulators
■ Secreted effectors
▨ Secreted translocases/chaperones

B

X

HilC ⊣
HilD ⊣

P _hilA_
hilAiagB ○

HilC
HilD

P _prgH_
prgHJKorgABC ○

HilA

P _invF-1_
invFGEACIJspaOPQRS ○

P _invF-2_

InvF   SicA

P _sicA_
sicAsipBCDAiacPsciPsptP ○

○ ----- mRNA transcripts

↗ Protein translocation

1999a; Hansen-Wester and Hensel, 2001). Many of the genes required for epithelial invasion have been found within SPI-1, which is a contiguous 40-kb region at centrosome 63 of the chromosome (Mills et al., 1995). SPI-1 genes encode a bacterial type III secretion apparatus and several effectors, which contribute to pathogenesis through an interaction with eukaryotic proteins (Fig. 1.1A). *Salmonella* invasion gene expression is known to be modulated by multiple environmental signals, including osmolarity, oxygen tension, pH, and stage of growth (Lee and Falkow, 1990).

Specifically, the expression of SPI-1 genes appears to be regulated at several stages in a complex manner by regulators within SPI-1, including HilA and InvF, and those outside SPI-1, such as the two-component regulators, the flagella associated genes, and the small DNA binding proteins (Fig. 1.1B). *Salmonella* does not constitutively express the virulence phenotypes associated with the SPI-1 type III secretion system (TTSS-1). *In vitro* inducing conditions that result in optimal expression of TTSS-1 include high osmolarity, low oxygen tension, slightly basic pH, and the growth rate of the bacteria (Lee and Falkow, 1990). The primary mechanism for controlling the production of TTSS-1 factors in response to environmental and physiological cues is by transcriptional regulation. The expression of genes encoding the TTSS-1 apparatus and most of the effectors requires HilA, a transcription factor encoded on SPI-1. HilA, a member of the OmpR/ToxR family, directly binds to and activates the promoter of SPI-1 operons and functions as a central regulator of invasion gene expression (Lostroh et al., 2000). Osmolarity, oxygen, and pH coordinately affect the transcription of *hilA*, and changes in the level of HilA mediate the regulation of the SPI-1 TTSS-1 by the same

Figure 1.1. (*facing page*). Type III secretion genes of SPI-1 and their regulation. (A) An overview of the type III secretion system encoded on SPI-1 includes subunits of a type III secretion apparatus, effectors secreted by the apparatus, factors required for their efficient translocation, and transcriptional regulators. The part of the island encoding a high-affinity iron transporter (*sitBCDA*) is not depicted. (B) Sequential upregulation by factors encoded on SPI-1 leads to expression of the type III secretion system. When *in vitro* environmental conditions are favorable for invasion gene expression (low pH, low oxygen tension, and high osmolarity), HilD derepresses the *hilA* promoter. The straight, solid arrows show that HilA protein directly activates the expression of structural genes such as the *prgs* and another regulatory gene, *invF*. HilA transcription initiated at P$_{invF}$ results in a long mRNA that continues through *sicA*. Although not illustrated, InvF, as a complex with SicA, directly activates the expression of effectors such as SipB, SopE, and SopB/SigD. HilD also makes a small direct contribution to *invF* expression by slightly upregulating the activity of a promoter far upstream of the start of *invF* translation. (Adapted from P.C. Lostroh and C.A. Lee, *Microb. Infect.* 3: 1281–1291, 2001.)

environmental conditions. Notably, no environmental condition has ever been documented that affects HilA-dependent TTSS-1 genes without also affecting *hilA* expression.

SPI-1 encodes four additional transcriptional regulators besides HilA; these include InvF, HilD, SprB, and SprA/SirC. To date, genetic evidence implicates a cascade of transcriptional activation in which HilD, HilA, and InvF act in sequence to stimulate TTSS-1 gene expression in vitro (Fig. 1.1B). Initially, HilD, an AraC/XylS family member, binds directly to several sites within P_{hilA} and derepresses *hilA* expression. HilA then binds to conserved sequences located between -54 and -37 relative to both *invF* and *prgH* start sites. Activation of P_{prgH} and P_{invF} results in the transcription of *prgHIJK-orgAB* and *InvFGEABCspaMNOPQRS*. Therefore, HilA directly activates the expression of the structural type III secretion genes as well as the transcription factor InvF. InvF, an AraC-like transcriptional regulator, promotes expression of HilA-activated effector genes by inducing their transcription from a second HilA-independent promoter (Darwin and Miller, 1999b). That is, InvF activates a promoter upstream of *sicA*, causing additional expression of *sicA-sipBCDA*. Furthermore, it has been demonstrated that two transcriptional regulators of SPI-1, HilC and HilD, allow the expression of *hilA* by counteracting the action of an unknown repressor (Lucas and Lee, 2001). These complex regulations appear to ensure that invasion genes are appropriately expressed when *Salmonella* infects the host.

SPI-2 also encodes for a protein secretion system similar to that encoded by SPI-1. Unlike SPI-1, which is found in all *Salmonella* lineages, SPI-2 is found only in *S. enterica* serovars and its acquisition most likely led to the divergence of *S. enterica* and *S. bongori*. SPI-2 is located at minute 30 of the *S. typhimurium* chromosome and has been implicated in the systemic phase of infection (Hensel et al., 1998; Shea et al., 1999). Expression of SPI-2 is induced within macrophages (Cirillo et al., 1998), and it appears to mediate bacterial survival within macrophages through evasion of NADPH oxidase-dependent killing (Vasquez-Torres et al., 2000), and interference with cellular trafficking of *Salmonella* containing vacuoles (Uchiya et al., 1999).

The type III secretion complex

A TTSS is a complex organelle composed of more than 20 proteins (Fig. 1.2). Subsets of these proteins are organized into a supramolecular structure, termed the *needle complex*, which spans the bacterial envelope (Kubori et al., 1998). This structure resembles the basal body of flagella, suggesting an evolutionary relationship between these two organelles. Indeed, components

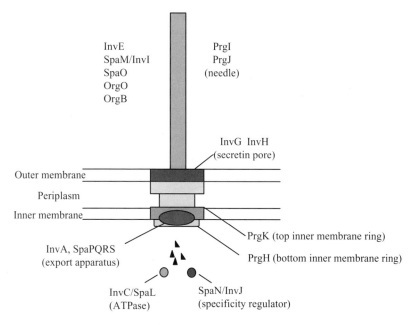

InvE
SpaM/InvI
SpaO
OrgO
OrgB

PrgI
PrgJ
(needle)

InvG InvH
(secretin pore)

Outer membrane

Periplasm

Inner membrane

PrgK (top inner membrane ring)

InvA, SpaPQRS
(export apparatus)

PrgH (bottom inner membrane ring)

InvC/SpaL
(ATPase)

SpaN/InvJ
(specificity regulator)

Figure 1.2. Schematic representation of the SPI-1 type III secretion system. The secretion machinery is made up of approximately 20 proteins that span the inner and outer membranes and direct the secretion of proteins without the classical *sec*-dependent signal sequence.

of the TTSS share amino acid similarity to flagellar proteins. The needle complex of the SPI-1 encoded TTSS has been characterized in some detail (Kubori et al., 2000; Zhou and Galan, 2001). It consists of a multiring base composed of the SPI-1 encoded proteins PrgHIJK. PrgH alone multimerizes into a tetrameric structure, but when complexed with PrgK it oligomerizes into ring-shaped structures that resemble the base of the needle complex of flagella. InvG has also been reported to form part of the base, and this is consistent with the observation that InvG forms a ring in the outer membrane in the presence of a helper lipoprotein, InvH. PrgI and PrgJ appear to form part of the bore of the needle. The core components, which comprise the needle-like complex, are highly conserved among different gram-negative pathogens, suggesting a common mode of operation. However, regulation of the secretion event is not well understood except that it requires energy in the form of ATP (Eichelberg et al., 1994). Further, unlike the *sec*-dependent pathway of bacterial protein secretion, type III secretion does not require a signal sequence on the protein to be exported, and in this respect it shares similarities with the bacterial flagellar export system.

The needle complex is constructed in an orderly manner (Kubori et al., 1998, 2000; Zhou and Galan, 2001). The proteins that make up the base of the complex are secreted through the inner membrane by the *sec*-mediated pathway. Once in the periplasm, the proteins form a complex that associates with a set of inner membrane proteins that share extensive sequence similarity to the components of the flagellar export apparatus. The resulting complex is restricted to the export of only the proteins that are necessary to make the needle structure. Once this foundation is made, the type III secretion apparatus becomes competent for the export of other type III secreted proteins, including those that are targeted to the inside of host cells.

Although much progress has been made in characterizing the secretion apparatus itself, little is known about how the effector proteins are subsequently translocated across the eukaryotic cell membrane. To date, three proteins, SipB, SipC, and SipD, are required for the translocation of effector proteins into the host cell, although the mechanisms by which SipB, SipC, and SipD exert their functions is not understood. Although these three proteins have been shown to be required for translocating effector proteins into the cytoplasm of the host cells, SipB, SipC, and SipD are not essential for the secretion process (Collazo and Galan, 1997). At least 13 proteins that are delivered by the SPI-1 TTSS have been identified: AvrA, SipA, SipB, SipC, SipD, SlrP, SopA, SopB, SopD, SopE/E2, SptP, and SspH1 (see references in Zhou and Galan, 2001). During the infection process, these proteins are presumably translocated into the cytosol of the host cell, where they engage host cell components to induce host cellular responses and promote bacterial uptake. Although some of these effector proteins are encoded within SPI-1, several effector proteins are encoded outside this pathogenicity island.

INTERNALIZATION OF *SALMONELLA* BY THE HOST EPITHELIUM

Animal models of *Salmonella* infection

A key feature of *Salmonella* pathogenesis is the ability of these bacteria to induce their own internalization by the normally nonphagocytic epithelial cells that line the intestine. Interactions between *Salmonella* and the intestinal epithelium were first described by Takeuchi, who orally infected guinea pigs with *S. typhimurium* (Takeuchi, 1967). From this early work it was determined that bacteria that closely contact the epithelial cells lining the intestine, primarily the ileum, elicit the local degeneration of filamentous actin in apical microvilli and the underlying terminal web. The morphology of other areas of the apical surface, either on the same cell or adjacent enterocytes, remains unaffeced. Subsequently, extruded membrane (described as membrane ruffles) surrounds the bacteria, resulting in their internalization into membrane-bound vacuoles. Once the bacteria are internalized, the overlying apical membrane regains its microvillar morphology, and despite these drastic changes to the apical cytoarchitecture and the presence of intracellular *S. typhimurium*, infected enterocytes remain healthy. Interestingly, although some bacteria become internalized by enterocytes, the majority remain in the intestinal lumen (Watson et al., 1995). Similar observations have been reported in other animals, including calves, pigs, and primates, all of which present with a diarrheal gastroenteritis in response to *S. typhimurium* and other related *Salmonella* strains (Bolton et al., 1999; Rout et al., 1974; Wallis and Galyov, 2000).

Cell culture models of *Salmonella* infection

As a way to investigate in more detail the changes to the host intestinal epithelium during early *Salmonella*–host cell interactions, a number of cell culture models have been developed. Initial studies were performed with epithelial cell lines that, when cultured on porous filter supports, establish electrically resistant epithelial monolayers with full apical–basolateral polarity. Two polarized cell lines used in these early studies were the Madin Darby canine kidney cell line (MDCK), derived from dog kidney distal tubule cells, and the Caco-2 cell line, derived from human colonic epithelia. Polarized cells presented with nontyphoidal *Salmonella* on the apical cell surface exhibit similar features to enterocytes in the guinea pig model: microvilli become disassembled, and the resulting membrane extrusions internalize the bacteria (Finlay and Falkow, 1990; Finlay et al., 1988). Thus, *S. typhimurium* contacting the apical plasma membrane were observed to induce ruffling of the membrane at sites of bacterial–epithelial cell contact, providing the driving force for bacterial internalization. The ability of *S. typhimurium* to induce contact-dependent membrane ruffling as a means of gaining entry into the host cell suggests that the bacteria recapitulate a process resembling phagocytosis in these normally nonphagocytic cells. This process has been termed *macropinocytosis* to reflect its resemblance to pinocytosis, or fluid uptake into cells, but with the engulfment of much larger particles (Francis et al., 1993). Interestingly, although *Salmonella* initially interact with their animal hosts at the apical surface of the intestine, studies with the T84 cell polarizing cell line have revealed that they can invade from the basolateral cell surface at the same frequency as from the apical surface (Criss et al., 2003).

However, the significance of this observation in the *in vivo* setting remains to be determined.

S. typhimurium invasion is not restricted to epithelial cells. *In vivo* the bacteria also invade macrophages, and *in vitro* they infect a variety of eukaryotic cells, except yeast and erythrocytes (Finlay et al., 1991). Subsequent invasion assays with HeLa cells and the Chinese hamster ovary (CHO) fibroblast cell line demonstrated that less polarized cells are more effectively infected by *S. typhimurium*, suggesting that the rigid cytoarchitecture of polarized epithelial cells is a hindrance to bacterial internalization *in vivo*. These studies implied that *S. typhimurium* utilize the same strategy to enter both polarized and nonpolarized cells, and the latter has gained widespread use for studies of the molecular regulation of *S. typhimurium* invasion.

INVOLVEMENT OF THE HOST CELL CYTOSKELETON IN BACTERIAL INTERNALIZATION

The distinct morphologic changes occurring to the apical enterocyte membrane upon binding of *S. typhimurium* suggested that host cell microfilaments (composed of actin) or microtubules might be involved in the formation of membrane ruffles. Finlay and Falkow were the first to report that treatment with cytochalasins, drugs that prevent F-actin polymerization, inhibits *Salmonella* invasion of multiple cultured cell lines. In contrast, microtubule-depolymerizing agents do not block bacterial internalization, suggesting that the actin cytoskeleton, but not the microtubual network, plays an active role in bacterial entry into host cells (Finlay and Falkow, 1988). Moreover, pretreatment with cytochalasin D does not prevent bacterial attachment to the host cell surface, indicating that actin-dependent cytoskeletal rearrangements and membrane ruffling follow initial bacterial binding (Francis et al, 1993). Immunofluorescence microscopy later demonstrated that bacteria recruit filamentous actin to sites of active bacterial invasion. Confocal laser scanning microscopy revealed that several actin-binding proteins, including α-actinin, tropomysin, and talin, are recruited to the *S. typhimurium*-induced ruffles in cultured cells (Finlay et al., 1988). Remarkably, *Salmonella* do not disrupt the actin cytoarchitecture in other regions of the cell, including cortical actin bundles or stress fibers (Finlay et al., 1991). *S. typhi* also induce actin-dependent ruffling during invasion, suggesting that this aspect of bacterial invasion is conserved regardless of eventual disease outcome (Mills and Finlay, 1994). Because ruffle formation is essential to the invasion process, understanding the development of these structures is critical to understanding *Salmonella* pathogenesis as a whole.

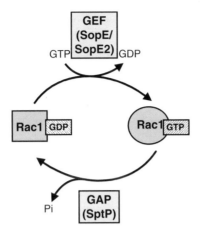

Figure 1.3. Nucleotide cycling of monomeric GTPases: In the resting state, the monomeric GTPase (shown here as Rac1) is in the GDP-bound, inactive conformation. Upon stimulation, a GEF catalyzes the release of GDP from the GTPase, followed by binding of GTP. This places the GTPase in the active conformation, where it can interact with effector proteins. To turn off the signal, a GAP enhances the GTPase's intrinsic hydrolysis rate, leading to GTPase inaction. *S. typhimurium* encodes two related GEFs for Rho GTPases, that is, SopE and SopE2, as well as one GAP for these GTPases, that is, SptP.

INVOLVEMENT OF RHO GTPase IN *S. TYPHIMURIUM* INVASION OF NONPHAGOCYTIC CELLS

Over the past 10 years, it has been demonstrated that the formation of actin-based cytoskeletal structures, which occurs in response to growth factors and other extracellular stimuli, is regulated by monomeric guanosine triphosphatases (GTPases) of the Rho family (Hall, 1998; van Aelst and D'Souza-Schorey, 1997). Rho proteins are members of the Ras superfamily of monomeric GTPases, and, like all Ras superfamily members, they cycle between active (GTP-bound) and inactive (GDP-bound) conformations (Fig. 1.3). Members of this family include RhoA-B-C-D-E-G, Rac1–2, Cdc42, and TC10; however, RhoA, Rac1, and Cdc42 have been the most extensively studied. *In vitro*, both GTP binding and hydrolysis activities of the GTPases are extremely low; therefore, accessory factors are required to facilitate these processes. Guanine nucleotide exchange factors (GEFs) catalyze the release of GDP and binding of GTP, which activates the GTPase, while GTPase activating proteins (GAPs) stimulate the GTP hydrolysis rate, thereby promoting their inactivation (Fig. 1.3). In fibroblasts, activation of RhoA promotes formation of stress fibers and focal contacts; Rac1 activation yields lamellipodia and dorsal ruffles; and Cdc42 activation leads to the extension of filopodia (Kozma et al., 1995; Nobes and Hall, 1995; Ridley and Hall, 1992). During cell spreading, Rho family members function sequentially, with initial activation of Cdc42 followed by Rac1 and RhoA (Nobes and Hall, 1995; Ridley and Hall, 1992). In other actin-dependent processes, distinct subsets of Rho GTPases become activated, often in a cell-type specific manner.

The involvement of Rho GTPases in *S. typhimurium* invasion was initially examined in nonpolarized cell lines of both epithelioid (HeLa and COS-1) and fibroblastic lineages. In these cells, Chen et al. (1996) demonstrated that invasion of *Salmonella* was primarily dependent on Cdc42. In this model, expression of a point mutant of Cdc42 unable to bind GTP (which acts in a dominant inhibitory manner) prevented bacterial entry, whereas expression of dominant negative Rac1 partially inhibited internalization but not as effectively as the Cdc42 mutant. The result of this study correlated with previous analysis of Rho GTPases during Fc receptor-mediated phagocytosis in macrophages. Particularly, expression of dominant negative mutants of either Rac or Cdc42, but not Rho, blocks phagocytosis of IgG-opsonized particles by having unique but complementary effects on localized actin polymerization at the plasma membrane (Caron and Hall, 1998; Cox et al., 1997; Massol et al., 1998). Moreover, in their activated form, Cdc42 and Rac have been shown to induce actin polymerization through the activation of N-WASP and the Arp2/3 complex. At present it is not known whether *S. typhimurium* direct their morphological changes in the actin cytoskeleton by using a similar activation strategy.

Nonetheless, as a result of the unique structure of the enterocyte brush border, the cytoskeletal regulatory factors co-opted by *Salmonella* during invasion in polarized epithelia are different from those identified in studies with nonpolarized cells (Criss et al., 2001). Dominant negative Rac1, but not Cdc42, significantly inhibited bacterial entry at the apical aspect of polarized cells. In this *in vitro* model of *Salmonella* – enterocyte interaction, the bacteria elicit actin reorganization and membrane ruffling at the apical surface in a manner that is morphologically indistinguishable from ruffling in nonpolarized cell lines. However, during entry at the apical pole of epithelial cells, *Salmonella* encounter a complex, highly organized actin cytoskeleton unlike any other cell surface they invade. At the apical domain, polymerized actin is organized into rigid microvilli and the underlying terminal web, a cross-linked meshwork of actin filaments that attaches to intercellular junctional complexes (Fath et al., 1993). Accordingly, the ability of *Salmonella* to reorganize the apical plasma membrane and its underlying actin architecture may require the mobilization of a unique set of cellular regulatory factors.

S. TYPHIMURIUM GENES THAT REGULATE EPITHELIAL CELL INVASION

Several *S. typhimurium* gene products secreted via the SPI-1 encoded TTSS have been found to participate in the process of bacterial uptake by epithelial cells. These gene products fall into two categories: those that affect Rho GTPase activity, and those that directly affect host actin dynamics.

SopE/SopE2

SopE was first identified as a protein secreted by the SPI-1 TTSS of *S. dublin*, and it was subsequently found in *S. typhimurium*. Initial studies determined that deletion of *sopE* reduces invasiveness to 40%–60% of wild-type levels, presumably as a result of a reduced capacity of the pathogen to elicit plasma membrane ruffling, which can be rescued by complementation of the *sopE* locus (Wood et al., 1996; Hardt et al., 1998). Subsequently, *S. typhimurium* SopE was found to have GDP–GTP nucleotide exchange activity on Rho family GTPases *in vitro* (i.e., it acts like a GEF; see Hardt et al., 1998; Rudolph et al., 1999). Ectopic expression of SopE protein in mammalian cells elicits membrane ruffling over the surface of the cell in a Rac1- and Cdc42-dependent manner (Hardt et al., 1998). SopE is not encoded within SPI-1 but is instead found on a lysogenic bacteriophage, which is only possessed by a subset of *Salmonella* spp. However, possession of the SopE phage does not correlate with invasiveness or pathogenicity (Mirold et al., 1999). Since this initial report, it was found that *S. typhimurium* possesses a homolog of SopE called SopE2, which has approximately 69% identity to SopE and is also secreted by the SPI-1 TTSS. A mutant strain deleted in SopE2 has reduced invasiveness relative to wild-type bacteria, but, unlike SopE, SopE2 is found in all pathogenic strains of *Salmonella* examined (Bakshi et al., 2001; Stender et al., 2000). These findings implicate SopE/E2 in the formation of the actin rearrangements necessary for membrane ruffling on the host cell surface and subsequent bacterial internalization.

It is interesting to note that SopE can activate Cdc42 despite its lack of sequence similarity to Dbl-like proteins, the Rho-specific eukaryotic GEFs. Recent investigations focusing on the mechanism by which SopE mediates guanine nucleotide exchange have determined that SopE binds to and locks the switch I and switch II regions of Cdc42 in a conformation that promotes guanine nucleotide exchange (Buchwald et al., 2002). Although this conformation resembles that of Rac1 in a complex with the eukaryotic Dbl-like exchange factor Tiam 1, the catalytic domain of SopE has an entirely different architecture from that of Tiam 1; furthermore, it interacts with the switch regions by means of different amino acids. In this regard, SopE is the first example of a non-Dbl-like protein capable of inducing guanine nucleotide exchange in Rho family proteins.

SopB

SopB exhibits potent phosphoinositide phosphatase activity *in vitro* and is capable of mediating pronounced inositol phosphate fluxes *in vivo* (Galyov et al., 1997). In addition, SopB has been found to stimulate Cdc42-dependent rearrangements of the actin cytoskeleton that are a prerequisite for cellular invasion. The ability of SopB to activate Cdc42 is dependent on its phosphatase activity, because a phosphatase-defective SopB in which a critical active-site cysteine residue was changed to serine lost its ability to activate Cdc42. Because inositol-based molecules can directly affect Cdc42 activity, it is thought that SopB activates Cdc42 and Rac1 indirectly by fluxing cellular phosphoinositides (Zhou et al., 2001).

The activation of Cdc42 and Rac1 triggers a series of signal transduction events that lead to actin cytoskeleton rearrangements. Despite their different biochemical activities, SopE/E2 and SopB exert at least partially redundant functions during *Salmonella* invasion. Thus, introduction of a loss-of-function mutation in the genes that encode either one of these proteins results in a minor defect in *Salmonella* entry. However, the simultaneous inactivation of SopE/E2 and SopB results in a very severe entry defect.

SptP

Cells infected with *S. typhimurium* quickly recover from the dramatic actin cytoskeletal rearrangements and regain their normal cellular architecture. SptP was identified as a *S. typhimurium* protein with homology in its carboxy-terminal to both prokaryotic and eukaryotic phosphatases, and it was demonstrated to possess tyrosine phosphatase activity (Kaniga et al., 1996). Although SptP mutants do not have an invasion deficiency, cells infected with *sptP*-deficient *S. typhimurium* do not exhibit normal recovery of their actin cytoskeleton following bacterial entry. Sequence scanning of SptP revealed a region in its amino terminus with homology to GAPs for Rho proteins, which is also possessed by other bacterial pathogens (ExoS of *Pseudomonas* spp. and YopE of *Yersinia* spp.), as well as by eukaryotes. SptP behaves as a GAP for Cdc42 and Rac1, but not RhoA or Ras. A mutation of arginine to alanine within the proposed catalytic arginine finger abrogated GAP activity (Fu and Galan, 1999). These results suggest that SopE/E2 and SptP coordinately control the GDP–GTP cycle of Rac and Cdc42 in host cells, thereby modulating the actin cytoskeleton. Thus, SptP's GAP activity opposes the Cdc42 and Rac1 activating function of SopE, SopE2, and SopB to help the host cell rebuild its actin cytoskeletal network. How these proteins are regulated *in vivo* so that their activities do not nullify each other is not yet clear, but it may be due to differential secretion or activation of SptP by its chaperone, SicP (Fu and Galan, 1998). In addition to its GAP activity located within the amino terminus, the carboxy-terminal domain of SptP possesses potent tyrosine phosphatase activity. Such tyrosine phosphatase activity of SptP is not only involved in reversing the MAP kinase activation that results from *Salmonella* invasion but also targets the intermediate filament vimentin, which is recruited to the membrane ruffles stimulated by *Salmonella* (Murli et al., 2001).

SipC

SipC has been reported to nucleate and bundle actin *in vitro*. The bundling and nucleation activities are located at different domains of SipC. The precise role of these activities *in vivo* is unknown because the necessary experiments to address this important issue are hampered by the fact that SipC is required for the translocation of effector proteins into host cells. SipC has been identified along with SipB as a general chaperone for the translocation of other SPI-1 type III secreted effector proteins into the host cell (Carlson and Jones, 1998). In addition, SipC becomes translocated into the host cell, where it has a bipartite ability to modulate actin polymerization directly. *In vitro*, the C-terminus of SipC aids in the nucleation of new actin filaments (the rate-limiting step in actin polymerization), whereas the N-terminal half facilitates filament bundling. Accordingly, microinjection of purified SipC protein into HeLa cells induces actin polymerization, but rather than inducing ruffles like SopE, it promotes the condensation of filamentous actin into large aggregates (Hayward and Koronakis, 1999). The physical function of these aggregates is unclear.

SipA

SipA is also encoded within and secreted by the SPI-1 TTSS. It is thought that SipA affects actin dynamics in cells by initiating actin polymerization at the site of *Salmonella* entry by lowering the critical concentration of actin required for polymerization (Zhou et al., 1999a). A *sipA* mutant strain of *S. typhimurium* has a minor invasion deficiency that is only detectable at very early time points (up to 20 min) of bacterial entry. Furthermore, although the *sipA* mutant elicits actin-dependent membrane ruffling, these ruffles are less localized to sites of internalization than those induced by wild-type bacteria (Zhou et al., 1999a). In additon, SipA was found to bind filamentous actin in an *in vitro* binding assay and induce formation of actin bundles at sites of bacterial internalization (Hayward and Koronakiz, 1999). *In vivo*, SipA may additionally affect actin dynamics by binding to, and enhancing the activity of, the bundling protein T-plastin (Zhou et al., 1999a, 1999b). Furthermore, McGhie et al. (2001) recently determined that SipA potentiates the effects of SipC on filamentous actin nucleation and bundling *in vitro*. Thus, it appears that SipA modulates the internalization process by decreasing the critical concentration for actin polymerization, inhibiting depolymerization of actin filaments, and increasing the bundling activity of T-plastin.

SipA is also unique in that interaction of this effector protein at the apical surface of intestinal epithelial cells is sufficient to initiate the proinflammatory signal transduction pathway that leads to polymorphonuclear leukocyte (PMN) transepithelial migration. The recruitment of PMN to the intestinal epithelium is a key virulence determinant underlying the development of *Salmonella*-elicited enteritis. Purified SipA applied to the apical surface of intestinal epithelial cells initiates an ADP ribosylating factor 6 (ARF6) lipid-signaling cascade, which, in turn directs the activation of protein kinase C (PKC) and subsequent PMN transepithelial migration (Lee et al., 2000; Criss et al., 2001). This demonstrates that some SPI-1 effector proteins involved in the invasion of epithelial cells may have additional roles that do not require their introduction directly into the cytosol of the host. Moreover, the significance of these results has been confirmed by the finding that SipA plays an important role in eliciting proinflammatory responses, such as PMN influx, during *Salmonella* infection of calves – a relevant *in vivo* model system used to study human enterocolitis (Zhang et al., 2002).

ROLE OF SPI-1 IN PATHOGENESIS

To understand the role of invasion in *Salmonella* pathogenesis, researchers have investigated the *in vivo* phenotypes of invasion gene mutants. Most *in vivo* studies have used the murine model of typhoid fever, in which orally introduced *S. typhimurium* causes a systemic illness in Balb/c mice. To induce systemic illness in these animals, *S. typhimurium* first colonize the distal ileum, and, after successful colonization, a subpopulation of *S. typhimurium* can be found in the gut-associated lymphatic tissues. Still later, host death can occur in response to high numbers of bacteria found within deep lymphoid-rich organs such as the spleen and liver (Carter and Collins,

1974). Invasion, per se, has long been thought to be important for this process because mutants that are noninvasive *in vitro* are less able to reach the spleen and liver and so are attenuated in orally infected mice. However, if introduced systemically, noninvasive mutants are as virulent as the wild-type (Galan and Curtiss, 1989; Ahmer et al., 1999). Many groups have interpreted such data to mean that invasion of nonphagocytic cells allows *S. typhimurium* to access the lymphatics, especially through the Peyer's patches that underlie M cells in the distal ileum (Penheiter et al., 1997).

In spite of this, several recent observations suggest that invasion, per se, is not always required for *S. typhimurium* to access privileged sites within the host. For example, *S. typhimurium* can reach the spleen in an invasion-independent manner by residing inside CD18+ macrophages (Vazquez-Torres et al., 1999). In yet another example, it has been postulated that CD18-expressing phagocytes are involved in an alternate route for bacterial invasion (Rescigno et al., 2001). Among CD18+ cells, dendritic cells are migratory and phagocytic cells that are ideally located for antigen sampling in tissues that interface with the external environment, where they perform a sentinel function for incoming pathogens. With the use of polarized monolayers of the intestinal epithelial cell Caco-2, it has been shown that dendritic cells are able to open up the tight junctions between epithelial cells, send dendrites outside the epithelium, and directly sample bacteria. Because dendritic cells express tight junction proteins (i.e., claudin-1, occludin, and ZO-1) the integrity of the mucosa is preserved. This unique cell–cell interaction allows dendritic cells to sample the environmental microorganisms without compromising the barrier function and to deliver the organisms to the lymphoid tissues where an efficient immune response can be mounted. Thus, this identifies a new mechanism for bacterial uptake in the mucosa tissue.

In a different study it was demonstrated that *S. typhimurium* lacking the entirety of SPI-1 cannot invade tissue cells but nevertheless still disseminates to systemic organs in Balb/c mice following intragastric infection (Murray and Lee, 2001). This observation highlights a new important concept in *Salmonella* pathogenesis establishing that TTSS-1 may also have activities aside from inducing invasion-associated rearrangements inside nonphagocytic epithelial cells. It is now apparent that *S. typhimurium* attracts, kills, and parasitizes different immune cells, and some of these activities require TTSS-1.

Therefore, the *in vivo* significance of invasion is likely to vary depending on the particular host–bacterial interaction. For instance, invasion may be required for some aspects of virulence but not for access to deeper tissues. It is also likely that there are situations in which invasion-independent TTSS SPI-1 phenotypes contribute significantly to salmonellosis and others in which invasion-dependent systemic dissemination is more critical. Although invasion can be uncoupled from some pathogenesis-associated phenotypes *in vitro*, it is not currently feasible to uncouple invasion from other TTSS SPI-1 phenotypes *in vivo*. Thus, the relative contribution of different SPI-1 phenotypes remains to be elucidated.

HISTORICAL PERSPECTIVE OF SPI-1

The genes that comprise the SPI-1 are not present in the genome of *Escherichia coli* K-12, but groups of similarly organized genes with related sequences occur on the virulence plasmids of the invasive enteric pathogens *Shigella* and *Yersinia* and in the genome of certain plant and animal pathogens of the genera *Erwinia*, *Pseudomonas*, and *Xanthomonas* (Li et al., 1995). And, as already mentioned, there are similarities between certain SPI-1 genes and loci involved in biogenesis of flagella in a variety of bacteria.

Perhaps the best characterized example of the functional and structural conservation in TTSS-1 is between *Salmonella* and *Shigella*. The *inv/spa* genes of SPI-1 are homologous to the *Shigella mxi/spa* genes; as a consequence, it is not surprising that the *Salmonella* and *Shigella* TTSS not only exhibit significant similarities in the primary sequence of their determinants (Hermant et al., 1995) but also complement each other functionally for secretion *in vitro* (Rosqvist et al., 1995). Perhaps they even share essentially the same macromolecular structure. Moreover, there is also significant structural and functional conservation between the Sip proteins encoded on the SPI-1 of *S. typhimurium* and the Ipa proteins encoded on the *Shigella mxi/spa* locus, suggesting that the entry processes engaged by these two enteric pathogens are promoted by similar effectors (Hermant et al., 1995).

In consideration of the base compositions, genomic locations (i.e., chromosome vs. plasmid), and phylogenic distribution of these genes, it is unlikely that the SPI-1 TTSS complex was ancestral in the Enterobacteriaceae. In addition, given their relatively low G + C content in *S. enterica* (46%), Li et al. (1995) proposed that the SPI-1 genes were horizontally transferred from *Yersinia*. However, because of their occurrence in all the subspecies of *S. enterica* and the overall similarity of their evolutionary diversification to that of housekeeping genes (Li et al., 1995), it is more likely that they were already present in the last common ancestor of the contemporary lineages of the salmonellae. Thus, it is generally agreed that *Salmonella*, *Yersinia*, and *Shigella* independently acquired these genes from another source.

REFERENCES

Ahmer, B.M., van Reeuwijk, J., Watsom, P.R., Wallis, T.S., and Heffron, F. (1999). *Salmonella* SirA is a global regulator of genes mediating enteropathogenesis. *Mol. Microbiol.* **31**, 971–982.

Bakshi, C.S., Singh, V.P., Wood, M.W., Jones, P.W., Wallis, T.S., and Galyov, E.E. (2000). Identification of SopE2, a *Salmonella* secreted protein which is highly homologous to SopE and involved in bacterial invasion of epithelial cells. *J. Bacteriol.* **182**, 2341–2344.

Bolton, A.J., Osborne, M.P., Wallis, T.S., and Stephen, J. (1999). Interaction of *Salmonella choleraesuis*, *Salmonella dublin*, and *Salmonella typhimurium* with porcine and bovine terminal ileum in vivo. *Microbiology* **145**, 2431–2441.

Buchwald, G., Friebel, A., Galan, J.E., Hardt, W.D., Wittinghofer, A., and Schefzek, K. (2002). Structural basis for the reversible activation of a Rho protein by the bacterial toxin SopE. *EMBO J.* **21**, 3286–3295.

Carlson, S.A. and Jones, B.D. (1998). Inhibition of *Salmonella typhimurium* invasion by host cell expression of secreted bacterial invasion proteins. *Infect. Immun.* **66**, 5295–5300.

Caron, E. and Hall, A. (1998). Identification of two distinct mechanisms of phagocytosis controlled by different GTPases. *Science* **282**, 1717–1721.

Carter, P.B. and Collins, F.M. (1974). The route of enteric infection in normal mice. *J. Exp. Med.* **139**, 1189–1203.

Chen, L.-M., Hobbie, S., and Galan, J.E. (1996). Requirement of CDC42 for *Salmonella*-induced cytoskeletal and nuclear responses. *Science* **274**, 2115–2118.

Cirillo, D.M., Valdivia, R.H., Monack, D.M., and Falkow, S. (1998). Macrophage-dependent induction of the *Salmonella* pathogenicity island 2 type III secretion system and its role in intracellular survival. *Mol. Microbiol.* **30**, 175–188.

Collazo, C.M. and Galan, J.E. (1997). The invasion-associated type III system of *Salmonella typhimurium* directs the translocation of Sip proteins into the host cell. *Mol. Microbiol.* **24**, 747–756.

Cox, D., Chang, P., Zhang, Q, Reddy, P.G., Bokoch, G.M., and Greenberg, S. (1997). Requirement for both Rac1 and Cdc42 in membrane ruffling and phagocytosis in leukocytes. *J. Exp. Med.* **186**, 1487–1494.

Criss, A.K., Ahlgren, D.M., Jou, T-S., McCormick, B.A., and Casanova, J.E. (2001). The GTPase Rac1 selectively regulates *Salmonella* invasion at the apical plasma membrane of polarized epithelial cells. *J. Cell Sci.* **114**, 1331–1341.

Criss, A.K. and Casanova, J.E. (2003). Coordinate regulation of *Salmonella* enterica serovar Typhimurium invasion of epithelial cells by the Arp 2/3 complex and Rho GTPases. *Infect. Immun.* **71**, 2885–2891.

Criss, A.K., Silva, M., Casanova, J.E., and McCormick, B.A. (2001). Regulation of *Salmonella*-induced neutrophil transmigration by epithelial ADP-ribosylation factor 6. *J. Biol. Chem.* **276**, 48,431–48,439.

Darwin, K.H. and Miller, V.L. (1999a). Molecular basis of the interaction of *Salmonella* with the intestinal mucosa. *Clin. Microbiol. Rev.* **12**, 405–428.

Darwin, K.H. and. Miller, V.L. (1999b). InvF is required for expression of genes encoding proteins secreted by the SPI1 type III secretion apparatus in *Salmonella typhimurium. J. Bacteriol.* **181**, 4949–4954.

Eichelberg, K., Ginocchio, C.C., and Galan, J.E. (1994). Molecular and functional characterization of the *Salmonella typhimurium* invasion genes *invB* and *invC*: homology of InvC to the FOF1 ATPase family of proteins. *J. Bacteriol.* **176**, 4501–4510.

Fath, K.R., Mamajiwalla, S.N., and Burgess, D.R. (1993). The cytoskeleton in development of epithelial cell polarity. *J. Cell Sci. Suppl.* **17**, 65–73.

Finlay, B.B. and Falkow, S. (1988). Comparison of the invasion strategies used by *Salmonella cholerae-suis, Shigella flexneri* and *Yersinia enterocolitica* to enter cultured animal cells: endosome acidification is not required for bacterial invasion or intracellular replication. *Biochimie.* **70**, 1089–1099.

Finlay, B.B. and Falkow, S. (1990). *Salmonella* interactions with polarized human intestinal Caco-2 epithelial cells. *J. Infect. Dis.* **162**, 1096–1106.

Finlay, B.B., Gumbiner, B., and Falkow, S. (1988). Penetration of *Salmonella* through a polarized Madin-Darby canine kidney epithelial cell monolayer. *J. Cell Biol.* **107**, 221–230.

Finlay, B.B., Ruschkowski, S., and Dedhar, S. (1991). Cytoskeletal rearrangements accompanying *Salmonella* entry into epithelial cells. *J. Cell Sci.* **99**, 283–296.

Francis, C.L., Ryan, T.A., Jones, B.D., Smith, S.J., and Falkow, S. (1993). Ruffles induced by *Salmonella* and other stimuli direct macropinocytosis of bacteria. *Nature* **364**, 639–642.

Fu, Y. and Galan, J.E. (1998). Identification of a specific chaperone for SptP, a substrate of the centrisome 63 type III secretion system of *Salmonella typhimurium. J. Bacteriol.* **180**, 3393–3399.

Fu, Y. and Galan, J.E. (1999). A *Salmonella* protein antagonizes Rac-1 and cdc42 to mediate host recovery after bacterial invasion. *Nature* **401**, 293–297.

Galan, J.E. and Curtiss, R. (1989). Cloning and molecular characterization of genes whose products allow *Salmonella typhimurium* to penetrate tissue culture cells. *Proc. Natl. Acad. Sci. USA* **86**, 6383–6387.

Galyov, E.G., Wood, M.W., Rosqvist, R., Mullan, P.B., Watson, P.R., Hedges, S., and Wallis, T.S. (1997). A secreted effector protein of *Salmonella dublin* is translocated into eukaryotic cells and mediates inflammation and fluid secretion in infected ileal mucosa. *Mol. Microbiol.* **25**, 903–912.

Hall, A. (1998). Rho GTPases and the actin cytoskeleton. *Science* **270**, 509–514.

Hansen-Wester, I. and Hensel, M. (2001). *Salmonella* pathogenicity island encoding type III effector systems. *Microbes Infect.* **3**, 549–559.

Hardt, W.D., Chen, L.M., Schuebel, K.E., Bustelo, X.R., and Galan, J.E. (1998). *S. typhimurium* encodes an activator of Rho GTPases that induces membrane ruffling and nuclear responses in host cells. *Cell* **93**, 815–826.

Hayward, R.D. and Koronakis, V. (1999). Direct nucleation and bundling of actin by the SipC protein of invasive *Salmonella*. *EMBO J.* **18**, 4926–4934.

Hensel M., Shea, J.E., Waterman, S.R., Mundy, R., Nikolaus, T., Banks, G., Vazquez-Torres, A., Gleeson, C., Fang, F.C., and Holden, D.W. (1998). Genes encoding putative effector proteins of the type III secretion system of *Salmonella* pathogenicity island 2 are required for bacterial virulence and proliferation in macrophages. *Mol. Microbiol.* **30**, 163–174.

Hermant, D., Menard, R., Arricau, N., Parsot, C., and Popoff, M.Y. (1995). Functional conservation of the *Salmonella* and *Shigella* effectors of entry into epithelial cells. *Mol. Microbiol.* **17**, 781–789.

Hohmann, E.L. (2001). Nontyphoidal salmonellosis. *Clin. Infect. Dis.* **32**, 263–269.

Hook, E.W. (1990). *Salmonella* species (including typhoid fever). In *Principles and Practice of Infectious Diseases*, ed. G.L. Mandell, R.G. Douglas, and J.E. Bennet, pp. 1700–1716. New York: Churchill Livingstone.

Kaniga, K., Uralil, J., Bliska, J.B., and Galan, J.E. (1996). A secreted tyrosine phosphate with modular effector domains in the bacterial pathogen *Salmonella typhimurium*. *Mol. Microbiol.* **21**, 633–641.

Kozma, R., Ahmed, S., Best, A., and Lim, V. (1995). The Ras-related protein Cdc42Hs and bradykinin promote formation of peripheral actin microspikes and filopodia in Swiss 3T3 fibroblasts. *Mol. Cell. Biol.* **15**, 1942–1952.

Kubori, T., Matsushima, Y., Nakamura, D., Uralil, J., Lara-Tejero, M., Sukhan, A., Galan, J.E., and Aizawa, S.I. (1998). Supramolecular structure of the *Salmonella typhimurium* type III protein secretion system. *Science* **280**, 602–605.

Kubori, T., Sukhan, A., Aizawa, S.I., and Galan, J.E. (2000). Molecular characterization and assembly of the needle complex of the *Salmonella typhimurium* type III protein secretion system. *Proc. Natl. Acad. Sci. USA* **97**, 10,225–10,230.

Lee, C.A. and Falkow, S. (1990). The ability of Salmonella to enter mammalian cells is affected by bacteria growth state. *Proc. Natl. Acad. Sci. USA* **87**, 4304–4308.

Lee, C.A., Silva, M., Siber, A.M., Kelly, A.J., Galyov, E., and McCormick, B.A. (2000). A secreted *Salmonella* protein induces a proinflammatory response in epithelial cells, which promotes neutrophil migration. *Proc. Natl. Acad. Sci. USA* **97**, 12,283–12,288.

Li, J., Ochman, H., Groisman, E.A., Boyd, E.F., Solomon, F., Nelson, K., and Selander, R.K. (1995). Relationship between evolutionary rate and cellular location among the Inv/Spa invasion proteins of *Salmonella enterica*. *Proc. Natl. Acad. Sci. USA* **92**, 7252–7256.

Lostroh, C.P., Bajaj, V., and Lee, C.A. (2000). The *cis* requirement for transcriptional activation by HilA, a virulence determinant encoded on SPI1. *Mol. Microbiol.* **37**, 300–315.

Lucas, R.L. and Lee, C.A. (2001). Roles of *hilC* and *hilD* in regulation of *hilA* expression in *Salmonella enterica* serovar typhimurium. *J. Bacteriol.* **183**, 2733–2745.

Massol, P., Montcourrier, P., Guilemot, J.C., and Chavier, P. (1998). Fc receptor-mediated phagocytosis requires CDC42 and Rac1. *EMBO J.* **17**, 6219–6229.

McGhie, E.J., Hayward, R.D., and Koronakis, V. (2001). Cooperation between actin-binding proteins of invasive *Salmonella*: SipA potentiates SipC nucleation and bundling of actin. *EMBO J.* **20**, 2131–2139.

Mills, D.M., Bajaj, V., and Lee, C.A. (1995). A 40 kb chromosomal fragment encoding *Salmonella typhimurium* invasion genes is absent from the corresponding region of the *Escherichia* K-12 chromosome. *Mol. Microbiol.* **15**, 749–759.

Mills, S.D. and Finlay, B.B. (1994). Comparison of *Salmonella typhi* and *Salmonella typhimurium* invasion, intracellular growth and localization in cultured human epithelial cells. *Microb. Pathog.* **17**, 409–423.

Mirold, S., Rabsch, W., Rohde, M., Stender, S., Tschape, H., Russmann, H., Igwe, E., and Hardt, W.D. (1999). Isolation of a temperate bacteriophage encoding the type III effector protein SopE from an epidemic *Salmonella typhimurium* strain. *Proc. Natl. Acad. Sci. USA* **96**, 9845–9850.

Murli, S., Watson, R.O., and Galan, J.E. (2001). Role of tyrosine kinases and the tyrosine phosphatase SptP in the interaction of *Salmonella* with host cells. *Cell. Microbiol.* **3**, 795–810.

Murray, R.A. and Lee, C.A. (2000). Invasion genes are not required for *Salmonella enterica* serovar typhimurium to breach the intestinal epithelium: evidence that *Salmonella* pathogenicity island 1 has alternative functions during infection. *Infect. Immun.* **68**, 5050–5055.

Nobes, C.D. and Hall, A. (1995). Rho, Rac, and Cdc42 GTPases regulate the assembly of multimolecular focal adhesion complexes associated with actin stress fibers, lamellipodia and filopodia. *Cell* **81**, 53–62.

Penheiter, K.L., Mathur, N., Giles, D., Fahlen, T., and Jones, B.D. (1997). Noninvasive *Salmonella typhimurium* mutants are avirulent because of an inability to enter and destroy M cells of ileal Peyer's patches. *Mol. Microbiol.* **24**, 697–709.

Rescigno, M., Urbano, M., Valzasina, B., Francolini, M., Rotta, G., Bonasio, R., Granucci, F., Kraehenbuhl, J.-P., and Ricciardi-Castagnoli, P. (2001). Dendritic cells express tight junction proteins and penetrate gut epithelial monolayers to sample bacteria. *Nature Immunol.* **2**, 361–367.

Ridley, A.J. and Hall, A. (1992). The small GTPase binding protein rho regulates the assembly of focal adhesions and actin stress fibers in response to growth factors. *Cell* **70**, 389–399.

Rosqvist, R., Hakansson, S., Forsberg, A., and Wolf-Watz, H. (1995). Functional conservation of the secretion and translocation machinery for virulence proteins of *Yersiniae, Salmonellae*, and *Shigellae*. *EMBO J.* **14**, 4187–4195.

Rout, W.R., Formal, S.B., Dammin, G.J., and Giannella, R.A. (1974). Pathophysiology of *Salmonella* diarrhea in the Rhesus monkey: intestinal transport, morphological and bacteriological studies. *Gastroenterology* **67**, 59–70.

Rudolph, M.G., Weise, C., Mirold, S., Hillenbrand, B., Bader, B., Wittinghofer, A., and Hardt, W.D. (1999). Biochemical analysis of SopE from *Salmonella typhimurium*, a highly efficient guanosine nucleotide exchange factor for Rho GTPases. *J. Biol. Chem.* **274**, 30,501–30,509.

Shea, J.E., Beuzon, C.R., Gleeson, C., Mundy, R., and Holden, D.W. (1999). Influence of the *Salmonella typhimurium* pathogenicity island 2 type III secretion system on bacterial growth in the mouse. *Infect. Immun.* **67**, 213–219.

Stender, S., Friebel, A., Linder, S., Rohde, M., Mirold, S., and Hardt, W.D. (2000). Identification of SopE2 from *Salmonella typhimurium*, a conserved guanine nucleotide exchange factor for Cdc42 of the host cell. *Mol. Microbiol.* **36**, 1206–1221.

Takeuchi, A. (1967). Electron microscope studies of experimental *Salmonella* infection. I. Penetration into the intestinal epithelium by *Salmonella typhimurium*. *Am. J. Pathol.* **50**, 109–136.

Uchiya, K., Barbieri, M.A., Funato, K., Shah, A.H., Stahl, P.D., and Groisman, E.A. (1999). A *Salmonella* virulence protein that inhibits cellular trafficking EMBO J. **18**, 3926–3933.

Van Aelst, L. and D'Souza-Schorey, C. (1997). Rho GTPases and signaling networks. *Genes Dev.* **11**, 2295–2322.

Vazquez-Torres A., Jones-Carson, J., Mastroeni, P., Ischiropoulos, H., and Fang, F.C. (2000). Antimicrobial actions of the NADPH phagocyte oxidase and inducible nitric oxide synthase in experimental salmonellosis. I. Effects on microbial killing by activated peritoneal macrophages *in vitro*. *J. Exp. Med.* **192**, 227–236.

Vazquez-Torres, A., Jones-Carson, J., Baumler, A.J., Falkow, S., Valdivia, R., Brown, W., Le, M., Berggren, R., Parkos, W.T., and Fang, F.C. (1999). Extraintestinal dissemination of *Salmonella* by CD-18-expressing phagocytes. *Nature* **401**, 804–808.

Wallis, T.S. and Galyov, E.E. (2000). Molecular basis of *Salmonella*-induced enteritis. *Mol. Microbiol.* **36**, 997–1005.

Watson, P.R., Paulin, S.M., Bland, A.P., Jones, P.W., and Wallis, T.S. (1995). Characterization of intestinal invasion by *Salmonella typhimurium* and *Salmonella dublin* and effect of a mutation in the *invH* gene. *Infect. Immun.* **63**, 2743–2754.

Wood, M., Rosqvist, W.R., Mullan, P.B., Edwards, M.H., and Galyov, E.E. (1996). SopE, a secreted protein of *Salmonella dublin*, is translocated into the target eukaryotic cell via a sip-dependent mechanism and promotes bacterial entry. *Mol. Microbiol.* **22**, 327–338.

Zhang, S., Santos, R.L., Tsolis, R.M., Stender, S., Hardt, W.D., Baumler, A.J., and Adams, L.G. (2002). The *Salmonella enterica* Serotype Typhimurium effector proteins SiPA, SoPA, SopB, SopD, and SopE2 act in concert to induce diarrhea in calves. *Infect. Immun.* **70**, 3843–3855.

Zhou, D. and Galan, J.E. (2001). *Salmonella* entry into host cells: the work in concert of type III secreted effector proteins. *Microb. Infect.* **3**, 1293–1298.

Zhou, D., Chen, L.L.H., Shears, B.S., and Galan, J.E. (2001). A *Salmonella* inositol polyphosphatase acts in conjunction with other bacterial effectors to promote host-cell actin cytoskeleton rearrangements and bacterial internalization. *Mol. Microbiol.* **39**, 248–259.

Zhou, D., Mooseker, M.S., and Galan, J.E. (1999a). An invasion-associated *Salmonella* protein modulates the actin-bundling activity of plastin. *Proc. Natl. Acad. Sci. USA* **96**, 10,176–10,181.

Zhou, D., Mooseker, M.S., and Galan, J.E. (1999b). Role of *S. typhimurium* actin-binding protein SipA in bacterial internalization. *Science* **283**, 2092–2095.

CHAPTER 2
Shigella invasion

Chihiro Sasakawa

INVASION

Shigella invasion and the host inflammatory responses

Shigella cause bacillary dysentery (shigellosis), a disease provoking a severe inflammatory diarrhea in humans and primates. In tropical areas of developing countries, shigellosis is endemic and a major killer of children under 5 years of age. Shigellosis occurs following ingestion of a very small number (100–1000) of bacteria, thus permitting easy spread of the disease by person-to-person contact as well as by the drinking of contaminated water.

Shigella, a Gram-negative bacillus, comprises four species – *S. dysenteriae, S. flexneri, S. boydii,* and *S. sonnei* (Pupo et al., 2000; Lan and Reeves, 2002). *Shigella* is now recognized as a member of *Escherichia coli*; however, the group of bacteria causing shigellosis is idiomatically called *Shigella* in this chapter. Shigellosis is also caused by enteroinvasive *E. coli* (EIEC), a pathogenic *E. coli*. *Shigella* and EIEC possess a large 210- to 230-kb plasmid on which the major virulence functions are encoded. Because *Shigella* has neither adhesins for upper GI tract cells nor flagella, after infection by means of the fecal–oral route the bacteria reach the colon and rectum directly, where they translocate through the epithelial barrier by means of the M cells overlaying the solitary lymphoid nodules (Fig. 2.1; also see Wassef et al., 1989; Sansonetti et al., 1991, 1996). Once they have reached the underlying M cells, *Shigella* infect the resident macrophages and multiply. Within the macrophages *Shigella* secrete IpaB, which specifically binds to, cleaves, and activates caspase-1, thus leading to macrophage cell death through apoptosis (Zychlinsky et al., 1992, 1994, 1996).

The stimulation of caspase-1 in infected macrophages causes the production of large amounts of IL-1β and IL-18, thus eventually leading to an

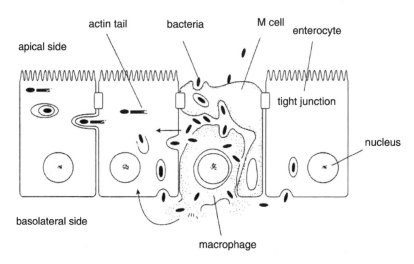

apical side

actin tail   bacteria   M cell   enterocyte

tight junction

nucleus

basolateral side

macrophage

Figure 2.1. A simplified model for the infection of colonic epithelial cells by *Shigella* (refer to the text for details).

increase in the permeability of the epithelial barrier to *Shigella* and migration of polymorphonuclear leukocytes (PMNs; see Zychlinsky et al., 1992; Zychlinsky and Sansonetti 1997). Meanwhile, the bacteria released from the dead macrophages immediately enter surrounding enterocytes from the basolateral surface by directing large-scale membrane ruffling, which finally leads to phagocytic events. Though the invading bacterium is entrapped by a phagocytic membrane, *Shigella* immediately disrupts the membrane and escapes into the cytoplasm (Fig. 2.1; also see Sansonetti et al., 1986). Within the cytoplasm, *Shigella* multiply and induce actin polymerization at one pole of the bacterium, by which the intracellular bacterium can gain a propulsive force to move intracellularly and intercellularly (Fig. 2.1; also see Bernardini et al., 1989).

Internalized *Shigella* release a large amount of lipopolysaccharide (LPS) into the host cytoplasm, where the LPS binds Nod1, a member of the CED4/ Apaf-1 superfamily, eventually leading to activation of NF-κB by means of the stimulation of the bipartite CARD-kinase protein, RICK (Inohara et al., 2000; Girardin et al., 2001). In response to the activation of NF-κB, colonic epithelial cells express a large array of proinflammatory cytokines, especially IL-8, thus further promoting local inflammation and attracting more PMNs (Zychlinsky and Sansonetti, 1997). Therefore, the predominant pathogenic feature of *Shigella* is the ability to invade macrophages as well as epithelial cells, including subsequent dissemination into adjacent epithelial cells. In this chapter, I mostly focus on the bacterial system involved in the invasion of epithelial cells and subsequent intracellular and intercellular spreading processes.

Basolateral entry into polarized cells

Shigella infection of polarized epithelial cells such as Caco-2 or Madin Darby Canine Kidney (MDCK) cells reveals that the bacteria invade from the basolateral surface into the cells (Mounier et al., 1992). This characteristic entry can be seen when *Shigella* infect rabbit ligated ileal loops, where bacteria move to the basolateral surface by means of the M cells (Wassef et al., 1989; Sansonetti et al., 1991, 1996), indicating that *Shigella* have an affinity to the basolateral surface of the polarized enterocytes to effectuate entry. *Shigella* are capable of inducing a highly dynamic rearrangement of actin and tubulin cytoskeletons during entry, and this leads to a large-scale phagocytic event. Shortly after coming into contact with epithelial cells, *Shigella* induce the formation of focal adhesion-like actin-dense patches beneath the bacterial contact point (Tran Van Nhieu and Sansonetti, 1999) and trigger local destruction of the microtubule structure (Yoshida et al., 2002), which is followed by the protrusion of large-scale membrane ruffles. The invading *Shigella* are finally enclosed by a large membrane vacuole, but the pathogens immediately escape into the host cell cytoplasm, where they elicit intracellular and intercellular movement (Fig. 2.1).

Invasion-associated genes on the large plasmid

The invasion of epithelial cells by *Shigella* requires many genes mostly confined to the 31-kb pathogenicity island (PAI) on the large virulence plasmid, which is highly conserved among *Shigella* spp. (Sasakawa et al., 1988; also see Fig. 2.2). The PAI of *S. flexneri* contains 28 genes bracketed by several IS elements and vestigial DNA sequences, where the 28 genes are arranged in several transcribed regions, encoding the components of the type III secretion system (TTSS), secreted effector proteins, chaperone proteins, and regulatory proteins (Fig. 2.2; also see Buchrieser et al., 2000; Venkatesan et al., 2001). *ipaBCDA* encode secreted effectors such as IpaA, IpaB and IpaC required for invasion of epithelial cells, whereas *mxi* and *spa* mostly encode components of the TTSS including the TTSS-associated secreted proteins. *ipgA*, *ipgC*, and *spa15* encode IpgA, with IpgC and Spa15 that act as chaperones for IcsB, IpaB/IpaC, and IpgB1/IpaA, respectively (Ménard et al., 1994a, 1994b; Page et al., 2002; Ogawa, unpublished results). The PAI possesses two regulatory

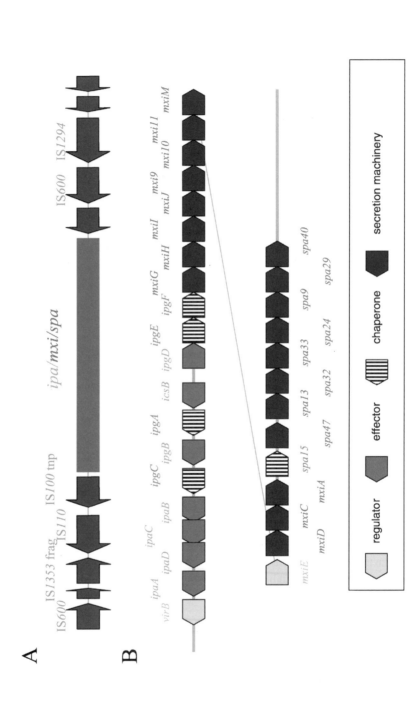

A

IS600 IS1353 frag IS100 tnp
       IS110

ipa/mxi/spa

IS600 IS1294

B

virB ipaC
ipaA ipgC ipgA
ipaD ipaB ipgB icsB ipgD mxiG mxiI mxi9 mxi11
                        ipgA ipgE ipgF mxiH mxiJ mxi10 mxiM

mxiE mxiC spa15 spa47 spa13 spa32 spa33 spa9 spa40
mxiD mxiA   spa15 spa32 spa24 spa29

regulator   effector   chaperone

                 secretion machinery

genes, *virB* and *mxiE*, required for the expression of the virulence-associated genes in the 31-kb PAI (Adler et al., 1989; Dorman and Porter, 1998; Mavris et al., 2002).

In addition, another plasmid-encoded gene, *virF*, codes for the essential regulatory protein VirF, an AraC-like transcriptional regulator, which directly binds the *virB* promoter to activate the *virB* gene. The VirB protein in turn activates the promoters for several transcribed regions in the 31-kb PAI. On the large plasmid, there is another effector gene called *virA* (Uchiya et al., 1995), located near the *virG* (*icsA*) gene; together these form a PAI (Venkatesan et al., 2001). VirA has recently been shown to be essential for evoking membrane ruffling in epithelial cells and promoting *Shigella* entry into host cells (Yoshida et al., 2002). Recently the whole genomic sequence of the large plasmid (pWR100) of *S. flexneri* serotype 5 was determined (Buchrieser et al., 2000; Venkatesan et al., 2001).

Examination of the repertoire of proteins secreted by means of the TTSS under conditions that activate the TTSS revealed that 15 proteins (IcsB, IpaH9.8, IpaH7.8, IpaH4.5, MxiC, MxiL, Spa32, OspC1, OspB, IpgB1, OspD1, OspE1, OspF, OspG, and VirA) plus IpaA, IpaB, IpaC, IpaD, and IpgD can be secreted from *Shigella* by means of the TTSS into the medium (Buchrieser et al., 2000; Ogawa, unpublished results). Although the proteins potentially secreted by means of the TTSS such as Spa32, MxiC, or MxiL (Tamano et al., 2002; Tamano unpublished results) do not necessarily serve as effectors, studies have clearly indicated that *Shigella* secrete a diverse array of effectors into the external medium and target host cells.

Type III secretion system

The TTSS is a highly sophisticated bacterial effector protein delivery system. Upon contact with the host cells, a set of effector proteins is delivered from the infecting bacteria to the host cells by means of the TTSS. These translocated proteins have a variety of effects on host cells and are necessary for bacterial attachment, invasion, trafficking, and avoidance from the host defense systems. Although the mechanisms of protein export by means of the TTSS, including the biosynthesis of the secretion machinery, are still to

---

Figure 2.2. (*facing page*). The genetic structure of the *ipa/mxi/spa* PAI encoding the *Shigella* TTSS. (A) A simplified structure of the *ipa/mxi/spa* PAI. Arrows represent IS elements and the vestigial DNA sequences. (B) The genetic constitution of the *ipa/mxi/spa* PAI. (For details refer to the text and to Buchrieser et al., 2000 and Venkatesan et al., 2001.)

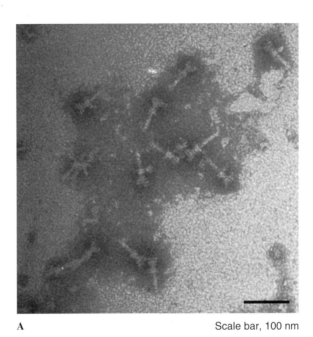

A   Scale bar, 100 nm

Figure 2.3. The supramolecular structure of the *Shigella* type III secretion machinery.
(A) An electron micrograph of the purified type III secretion machinery from *S. flexneri*.

be elucidated, some common characteristics of this secretion system have emerged (Hueck, 1998).

Genetic and functional studies have indicated that the TTSS is encoded by more than 20 genes. The subset of genes together with other genes coding for the secreted effectors, chaperones, and regulators exist as a PAI, which potentially transposes horizontally in different species of bacteria, thus distributing among many Gram-negative pathogenic bacteria, where some of the PAI are located on the chromosome and some on a plasmid. In fact, there is considerable homology between the proteins of the TTSS in different pathogens (Hueck, 1998). Of note, some of the proteins of TTSSs also share significant similarity to the components of the bacterial flagella export machinery. For example, some of the putative components of the type III secretion complexes such as *S. flexneri* MxiJ, Spa47, Spa33, Spa24, Spa9, Spa29, Spa40, and MxiJ share significant similarity to FlhA, FliI, FliN, FliP, FliQ, FliR, FlhB, and FliF of the *Salmonella* flagellar export system, respectively (Hueck, 1998; Cornelis and Van Gijsegem, 2000; Plano et al., 2001).

Furthermore, the TTSS is functionally and structurally similar to the flagellar export system. Secretion of a set of proteins by means of the TTSS

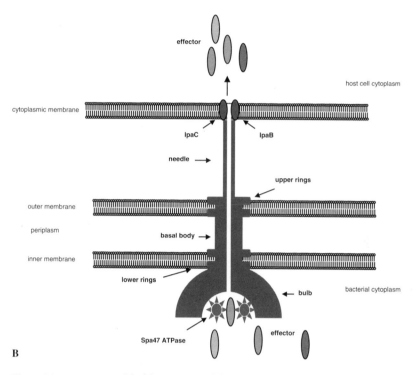

Figure 2.3. (*cont.*) (B) Model of the structure of the type III secretion machinery, including the translocation of effectors from bacterial cytoplasm into the host cell cytoplasm.

is dependent on the energy supply mediated by the F1-type ATPase associated with the secretion apparatus. The same is true for secretion of the extracellular flagellar components, which form the hook, cap, and flagella filament by means of the flagella export system. The supramolecular structures of the TTSS of *Salmonella typhimurium* and *S. flexneri* have recently been elucidated, and they share similarity with that of the flagellar basal body (Fig. 2.3B; also see Kubori et al., 1998; Blocker et al., 1999; Tamano et al., 2000). As already mentioned for the *Shigella* TTSS, the expression of genes encoding the TTSS as well as the flagella export system is under stringent control that is mediated by complicated regulatory networks in each of the bacteria.

Despite the structural and functional similarities of the TTSS in each pathogen, the proteins delivered by means of the TTSS are quite diverse. For example, *Shigella* potentially deliver ~20 proteins by means of the TTSS; some share similarity to secreted proteins from different pathogens, whereas others are unique to *Shigella*. Studies of the proteins secreted from *Yersinia*,

Salmonella, and *Shigella* have indicated that some have a role in linking the secretion complex to the target host plasma membrane, whereas others serve as effectors to modulate host cell functions.

The purified type III secretion complexes from *S. typhimurium* and *S. flexneri* as reported by Kubori et al. (1998) and Tamano et al. (2000), respectively, contained four major components. The *Salmonella* type III secretion complex contained InvG, PrgH, PrgI, and PrgK proteins, whereas the *Shigella* type III secretion complex contained MxiD, MxiG, MxiH, and MxiJ. Recently, these type III secretion complexes have been shown to contain an additional component, which is PrgJ in *Salmonella* and MxiI in *Shigella*. The supramolecular structures of the type III secretion complexes of each bacterium observed by electron microscopy are similar, being composed of two distinctive parts, the needle and basal parts (Fig. 2.3B). The needle of the *S. typhimurium* and *S. flexneri* type III secretion complexes consists mainly of PrgI and MxiH, respectively (Kubori et al., 2000; Tamano et al., 2000). The basal part of the *S. typhimurium* type III complex is composed of InvG, PrgH, and PrgK, whereas that of the *S. flexneri* complex is composed of MxiD, MxiG, and MxiJ. Furthermore, the supramolecular structures of the basal portion of both type III complexes share significant similarity to that of the *Salmonella* flagella basal body.

Indeed, the basal part of the type III secretion complex possesses two pairs of rings, referred to as the upper and lower rings (Fig. 2.3B). Because the basal portion was observed by electron microscopy to be embedded in the osmotic-shocked bacterial envelope, similar to the flagellar basal complex, the two pairs of rings are thus assumed to be anchored to the inner and outer membranes of bacteria. The flagellar hook forms a curved protruding structure from which a long flagella filament is extended; the type III secretion complex forms a straight needle protruding from the basal part. The length of the type III needle of wild type *S. flexneri* is estimated to be 45 nm and distributed in a narrow range with a standard deviation of 3.3 nm (Tamano et al., 2000). The length of the basal body of the *Shigella* type III secretion complex is estimated to be approximately 31 nm, which is consistent with the thickness of the Gram-negative bacterial envelope (approximately 25 nm) (Tamano et al., 2000). This suggests that the type III secretion complex spans both the outer and inner membranes. Although the number of needles per bacterium has not been accurately determined, on the basis of the distribution of the type III secretion structures in a field of osmotically shocked bacterial envelope as observed by electron microscopy, it is estimated to be around 50–60 per bacterium.

Genetic and functional studies of TTSSs have strongly suggested that the basic morphological features displayed by *Salmonella* and *Shigella* would be conserved among other pathogens. Indeed, recently studies have shown that, in enteropathogenic *E. coli* (EPEC), a long (50–700 nM) filamentous structure protrudes from the tip of the TTSS needles; it is composed of EspA, which is encoded by the *espA* gene located on the locus of enterocyte effacement (LEE) PAI, downstream of the region encoding genes of the TTSS (Sekiya et al., 2001; Daniell et al., 2001). Previous studies indicated that EspA forms a filamentous structure that assembles as a physical bridge between the bacteria and host cell surface, which then functions as a conduit for the translocation of bacterial effectors into host cells (Knutton et al., 1998). In fact, the *espA* mutant of EPEC has been shown to be deficient in forming a long filamentous structure and delivering effectors such as Tir, EspB, and EspD into the host cells, thus becoming a nonadherent mutant. Similarly, some plant pathogens such as *Pseudomonas syringae* and *Ralstonia solanacearum* form a filamentous appendage called the Hrp-pilus, which consists of HrpA in *P. syringae* or HrpY in *R. solanacearum* (Van Gijsegem et al., 2000).

Secreted effector proteins induce host cellular signaling

The effectors delivered from *Shigella* trigger host cellular signal pathways to direct its own internalization event (Fig. 2.4). One signal transduction pathway is linked to the interaction of secreted IpaB and IpaC with putative host surface receptors, and the others are evoked by intracellular effectors such as IpaA, IpaB, IpaC, IpgD, and VirA. Although the roles of IpaB and IpaC during invasion are important, their functions are complicated. For example, IpaB and IpaC act as secreted effector proteins in the target host cells to stimulate caspase-1 and Rho GTPases, respectively, whereas IpaB and IpaD act as a molecular plug for the TTSS at the tip of the TTSS needle (Figs. 2.3B and 2.4; also see Ménard et al., 1996; Tran Van Nhieu and Sansonetti, 1999). Like *Yersinia* YopB and YopD (Håkansson et al., 1996) or *Salmonella* SipB and SipC (Collazo and Galan, 1996), upon contact between *Shigella* and the host cell, IpaB and IpaC serve as a membrane pore located at the tip of the TTSS needle in the host cell plasma membrane, thus allowing the translocation of secreted effector proteins into the host cells (Fig. 2.3B; also see Blocker et al., 1999).

The IpaB and IpaC proteins secreted into the culture supernatant form high molecular matrix-like structures, which promote *Shigella* invasion through interaction with CD44 (the hyaluronan receptor belonging to the immunoglobulin superfamily) and $\alpha_5\beta_1$ integrin (Fig. 2.4; also see Skoudy et al., 2000; Watarai et al., 1996). Because both molecules are distributed on the basolateral surface of the polarized epithelial cells, these interactions are

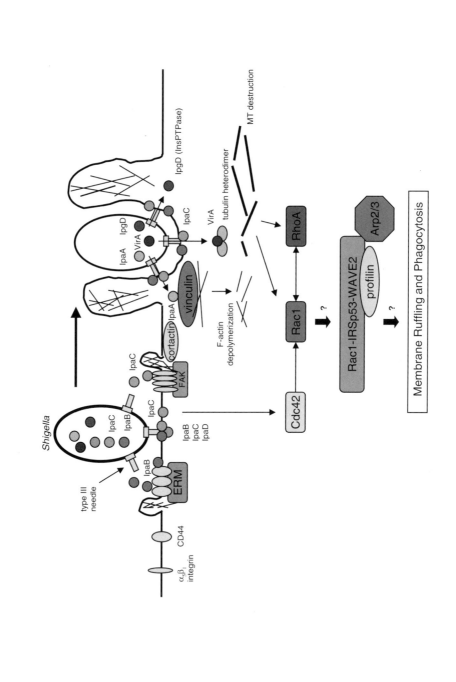

Membrane Ruffling and Phagocytosis

thus assumed to contribute to the basolateral entry of *Shigella* into the epithelial cells by mediating outside-in signaling to induce local rearrangement of the actin cytoskeleton (Skoudy et al., 2000). $\alpha_5\beta_1$ integrin localizes at focal adhesions, and the cytoplasmic domain of the β_1 integrin acts as a cytoskeleton linker by means of focal adhesion proteins; the cytoplasmic moiety of CD44 also acts as a cytoskeleton linker by means of association with ERM (ezrin-radixin-moesin) proteins. Ezrin is shown to be recruited at the periphery of the extended membrane ruffles at the point of *Shigella* entry (Skoudy et al., 1999). A recent study has shown that IpaB can bind CD44, where IpaB partitions during *Shigella* invasion within specialized membrane microdomains enriched in cholesterol and sphingolipids, called *rafts* (Lafont et al., 2002). CD44 is known to participate in signaling responses regulating the reorganization of the cytoskeleton (Hirao et al., 1996). Early in the invasion (~15 min after contact), an accumulation of cholesterol and raft-associated proteins such as GPI-anchored proteins can be observed at *Shigella* entry foci.

In agreement with this, bacterial entry is impaired upon cholesterol depletion with methyl-β-cyclodextrin at the site of bacterial entry (Lafont et al., 2002). Therefore, the binding of IpaB at the tip of the TTSS needle to clustered CD44 is thought to increase binding affinity; thus rafts and clustered CD44 seem to be crucial for efficient *Shigella* entry. Meanwhile, $\alpha_5\beta_1$ integrin accumulates together with F-actin, vinculin, talin, α-actinin, and tyrosine-phosphorylated FAK, which are major scaffolding components of focal adhesions (Watarai et al., 1996, 1997). However, these interactions alone are insufficient to elicit the phagocytic events required for bacterial uptake by epithelial cells (Ménard et al., 1996). For induction of large-scale actin rearrangements and large membrane protrusions sufficient to engulf several bacterial particles simultaneously, cellular signals evoked by *Shigella* effectors such as IpaA, IpaC, IpgD, and VirA appear to be necessary (Fig. 2.4).

In addition, *Shigella* require the activation of FAK and Src tyrosine kinase to induce cytoskeletal rearrangements during entry (Watarai et al., 1997; Duménil et al., 1998). Overexpression of a dominant interfering form of pp60[c-Src] leads to inhibition of *Shigella*-induced cytoskeletal rearrangements and decreases the phosphorylation of cortactin (Duménil et al., 1998), a Src substrate recruited at the site of *Shigella* entry (Dehio et al., 1995). Because focal adhesion formation is dependent on the activation of Rho and Src, an early cellular event triggered by *Shigella* contact may be associated with such a cellular function (Adam et al., 1996; Ménard et al., 1996; Watarai et al., 1997).

Figure 2.4. (*facing page*). A model for cytoskeletal rearrangements induced during *Shigella* invasion of epithelial cells, and the roles of effector proteins (refer to the text for details).

IpaA

IpaA secreted via the TTSS upon cell contact has been indicated to modify the *Shigella*-induced entry foci, and an *ipaA* mutant induced disorganized filopodial protrusions (Fig. 2.4; also see Bourdet-Sicard et al., 1999). IpaA has a high affinity for N-terminal residues 1-265 of vinculin. In cosedimentation and solid-phase assays, IpaA binding to vinculin increases the association of vinculin with F-actin, which in turn promotes depolymerization of F-actin associated with the IpaA–vinculin complex. Vinculin is a cytoskeletal protein present at focal adhesion and cell–cell adhesion structures, involved in cell adhesion to the extracellular matrix, cell motility, and tumorigenesis (Jockusch and Rudiger, 1996). Vinculin is composed of N-terminal head and C-terminal tail domains. It has multiple functional domains; talin and α-actinin interact with the N-terminal domain, whereas F-actin binds to the C-terminal tail domain. Importantly, the interactions with proteins are determined by the conformation of vinculin. The unfolded form is the activated form and interacts with talin, α-actinin, and F-actin through the exposed binding domains, whereas, in the folded conformation, the tail domain interacts with the head domain, resulting in the masking of the binding domains (Jockusch and Rudiger, 1996). Importantly, the 1-258 domain of vinculin has been indicated to be involved in the intramolecular association with the C-terminal tail domain.

The binding of IpaA to vinculin was found to be strong, with a Kd of 5 nM; therefore, it is likely that the binding of IpaA to the head domain disrupts the head–tail interaction, thus allowing vinculin to open up and link to F-actin. The vinculin–IpaA complex promotes F-actin depolymerization *in vitro* and *in vivo*. Indeed, microinjection of IpaA into HeLa cells induces a rapid (within 40 s) cell retraction with the disappearance of actin stress fibers. When IpaA is comicroinjected with the MBP-fused vinculin 1-265 moiety, cell retraction induced by IpaA can be blocked and actin stress fibers along with vinculin–containing focal complexes are still intact, indicating that the vinculin–IpaA binding domain has a functional role in inducing the IpaA-induced cytoskeletal rearrangements. Although the mechanism for stimulating F-actin depolymerization is speculative, the activity of IpaA for induction of actin depolymerization by means of the vinculin–IpaA complex seems to be important in regulating the formation of protrusions that might promote detachment of *Shigella*. Alternatively, the IpaA-induced activation of vinculin and recruitment of focal adhesion components may lead to the formation of a specific focal adhesion-like structure that might help to maintain bacterial contact with the epithelial surface (Bourdet-Sicard et al., 1999).

IpaC

IpaC has an activity to modulate actin dynamism, because the formation of filopodia and lamellipodia is induced when purified IpaC protein is added to semipermeabilized Swiss 3T3 cells or a *ipaC* clone is transfected into HeLa cells (Tran van Nhieu et al., 1999). The IpaC-induced membrane protrusions have been implicated in the activation of Cdc42, which in turn activates Rac1, suggesting that IpaC can somehow act as the effector for promoting *Shigella* invasion of epithelial cells (Fig. 2.4). In addition, a recent study has indicated that *Shigella* direct their own entry into macrophages by exploiting IpaC to stimulate macrophage phagocytic activity (Kuwae et al., 2001). Indeed, *Shigella* invade murine macrophages such as J774 more efficiently than the noninvasive *ipaC* mutants. Wild type *Shigella* can induce large-scale lamellipodial extensions including ruffle formation around the bacteria. In contrast, when macrophages are infected with the noninvasive *ipaC* mutant, the invasiveness and induction of membrane extension are dramatically reduced. *Shigella* infection of J774 cells causes tyrosine phosphorylation of several proteins, including paxillin and c-Cbl, and the profile of phosphorylated protein is distinctive from that stimulated by *S. typhimurium* or phorbol ester. Upon addition of a recombinant IpaC into the external medium of J774, membrane extensions were rapidly induced, and this also promoted uptake of *E. coli*. Importantly, the exogenously added IpaC was shown to be integrated into the macrophage plasma membrane. An analysis of the IpaC sequence (382 amino acids) with TMpred, a program for the prediction of putative membrane-spanning regions, predicts that the residues encompassing 121–139 and 169–191 are the putative transmembrane domains, whereas the remaining N-terminal and C-terminal regions are predicted to be presented on the external side of the plasma membrane.

With the use of three IpaC antibodies that recognize three distinctive regions in IpaC, it has been suggested that the residues 140–168 of IpaC exist as the cytoplasmic loop, whereas the preceding N-terminal portion of TM1 (transmembrane 1) and the following C-terminal portion of TM2 (transmembrane 2) exist as the external membrane domains (Kuwae et al., 2001). Although the mechanisms underlying the IpaC-mediated macrophage spreading event are still to be elucidated, some surface receptors such as Mac-1 ($\alpha_M\beta_2$ integrin) and FcγR on macrophages seem to be involved in the phagocytic event.

Incubation of J774 cells with wild type *Shigella*, but not with the *ipaC* deletion mutant, generates Mac-1 and FcγR foci at the site of bacterial contact. Importantly, the Mac-1 foci were observed along the periphery of the extended cell membrane, strongly indicating that an integrin-dependent adhesion event, which is a prominent feature of the Mac-1-mediated macrophage adherence, had occurred. The observed macrophage response is consistent with the study by Renesto et al. (1996), in which PMNs suspended in medium became adherent onto serum-coated wells when wild type *Shigella* but not the *ipaC* mutant was added to the external medium.

Clustering of Mac-1 and Fcγ R has been indicated to stimulate protein tyrosine phosphorylation and local rearrangement of the actin cytoskeleton. Because some truncated versions of IpaC capable of associating with the host plasma membrane are able to stimulate macrophage cell spreading (Kuwae et al., 2001), IpaC might have an association with or effect on some putative host receptor(s), such as $\alpha_M\beta_2$ integrin, mediating the induction of membrane extension by means of activation of a cellular signal transduction pathway such as the activation of Cdc42. The IpaC-induced membrane protrusions from HeLa cells can be inhibited by a dominant interfering form of Cdc42, whereas a dominant interfering form of Rac results in inhibition of the lamellipodium formation, further supporting the notion that IpaC has activity to stimulate Cdc42 activity. This in turn causes Rac1 activation (Mounier et al., 1999; Tran van Nhieu et al., 1999). Importantly, recent studies have suggested that activated Rac1 stimulates WAVE2, a WASP (Wiskott-Aldrich syndrome protein) family protein, via IRSp53, a substrate for the insulin receptor, by forming a Rac1–IRSp53–WAVE2 complex, which recruits Arp2/3 complex including profilin, thus evoking membrane ruffling (Miki et al., 2000; Takenawa and Miki, 2000; Krugmann et al., 2001). Because WAVE2 can be detected around the area of protruded membrane ruffles induced by *Shigella* (Suzuki, unpublished data), the Rac1–IRSp53–WAVE2 complex might take part in the formation of membrane protrusion induced by *Shigella* invasion (Fig. 2.4).

IpgD

IpgD, a 69-kDa protein encoded by *ipgD* and located upstream of *ipaBCDA*, is secreted by the TTSS in amounts similar to the Ipa proteins. Like the Ipa proteins, IpgD is stored in the *Shigella* cytoplasm unless the TTSS is stimulated such as by incubation in conditioned medium (Niebuhr et al., 2000). The storage of IpgD in the cytoplasm requires its association with a cytoplasmic chaperone, IpgE, encoded by the gene located immediately downstream of *ipgD*. Interestingly, after secretion, IpgD forms a complex with IpaA in the extracellular medium, although the biological significance is unclear. An *ipgD* mutant still enters host cells; however, in comparison with that directed by the wild type, the morphology of the membrane ruffle is altered. For example, scanning electron microscopic analysis reveals that the *ipgD* mutant provokes fewer actin rearrangements and less membrane ruffling on the target cell surface, suggesting that IpgD is involved in the process of invasion of epithelial cells and that the protein serves as a translocated effector (Niebuhr et al., 2000). A protein homologous to IpgD, SopB (also called SigD), has been identified in *S. dublin*, and it is involved in invasion, because a *sigD* mutant affected *Salmonella* invasion of CHO and HEp-2 cells (Galyov et al., 1997; Hong and Miller 1998). SopB has sequence homology with mammalian inositol polyphosphate 4-phosphatase, and recombinant SopB protein prepared from *S. dublin* shows inositol phosphate phosphatase activity required for promoting membrane fission during invasion (Norris et al., 1998; Terebiznik et al., 2002).

Similarly, the sequence of IpgD has been suggested to have the active site of mammalian inositol polyphosphate 4-phospatase, which specifically dephosphorylates PtdIns(4,5)P$_2$ into PtdIns(5)P (Niebuhr et al., 2000, 2002). In mammalian cells this enzyme plays key roles in many processes, including reorganization of the actin cytoskeleton, and cytoskeleton–plasma membrane linkage. The importance of the enzymatic activity encoded by IpgD has been shown in promoting detachment of the plasma membrane from the cytoskeleton to facilitate extension of membrane filopodia and ruffles evoked by *Shigella* invasion of epithelial cells. Indeed, continuous ectopic expression of IpgD in HeLa cells increases membrane detachment and causes formation of membrane blebs (Niebuhr et al., 2002).

VirA

An examination of the cytoskeletal architecture around invading *Shigella* by confocal microscopy indicated that the local microtubule network beneath the protruding ruffles undergoes remarkable destruction (Yoshida et al., 2002). This finding, together with the increase in *Shigella* invasiveness in host cells treated with microtubule-destabilizing agents such as nocodazole, suggests that the bacteria have the ability to modulate tubulin dynamics. VirA has activity to trigger microtubule dynamic instability *in vitro* and *in vivo*, which can stimulate Rac1 activity, thus leading to membrane ruffling (Fig. 2.4).

VirA is a 45-kDa protein composed of 401 amino acids; it is able to bind $\alpha\beta$-tubulin dimers but not microtubules. In an *in vitro* tubulin polymerization assay system, purified VirA showed activity to inhibit the polymerization of tubulin and stimulate microtubule destabilization. Interestingly, a portion of VirA, encompassing residues 224 to 315, involved in the interaction with tubulin heterodimers shares significant (>40%) amino acid homology with a portion of EspG encoded by the *espG* gene in the

LEE of EPEC or enterohemorrhagic *E. coli*, as well with as some other un-characterized bacterial proteins such as NMB0928 (*Neisseria menigitidis*) or Cj1457c (*Campylobacter jejuni*). Indeed, the expression of EspG in a *Shigella virA* mutant can rescue invasiveness, suggesting that EspG and VirA share an essential function (Elliott et al., 2001). The expression of VirA in mammalian cells such as HeLa, COS-7, or Swiss3T3 cells allows for the formation of membrane ruffling, though the scale of ruffles is smaller than that evoked by *Shigella*. Microinjection of VirA into HeLa cells also induces a localized membrane ruffling in a few minutes, whereas overexpression of VirA in host cells causes the destruction of microtubules and protruding membrane ruffles. Importantly, the VirA-induced membrane ruffling is dependent on the host Rac1 activity, because when VirA is coexpressed with a dominant negative Rac1 mutant in the cells, the appearance of ruffles can be shut off.

In agreement with this, wild type *S. flexneri*, but not the *virA* mutant, stimulates Rac1 and induces the formation of membrane ruffles in infected HeLa cells (Yoshida et al., 2002). These observations suggest that the destabilization of microtubules induced by VirA secreted from *Shigella* into host cells can provoke the formation of membrane ruffles, thus stimulating bacterial entry (Fig. 2.4). Although it is still unclear whether or not other invasive bacteria are able to stimulate host microtubule dynamic instability, a similar activity to *Shigella* VirA may also be found to be involved in some other bacterial infections of host cells.

The microtubule network is dynamic in migrating or growing cells in which the microtubules undergo growth and shortening, called microtubule dynamic instability, which is mediated by various factors including microtubule stabilizing and destabilizing factors. For migrating cells, the interplay between the microtubule and actin cytoskeletal systems is though to be crucial. Indeed, recent studies have strongly indicated that microtubule growth and shortening participate in the activation of Rac1 and RhoA signaling, respectively, to control actin dynamics.

Waterman-Storer et al. (1999) revealed that when microtubule growth in host cells is stimulated by pretreatmenting with nocodazole followed by washing out the drug, Rac1 is activitated, thus leading to the formation of lamellipodial protrusions in fibroblasts. Enomoto (1996) originally showed that microtubule disruption by colcemid or vinblastine, but not taxol, rapidly and reversibly induced the formation of actin stress fibers and focal adhesions, which was accompanied by activated cell motility. Consistent with that study, Krendel et al. (2002) have observed that RhoA activity in fibroblasts can be stimulated by nocodazole. These studies have suggested that the depolymerization or shortening of microtubules can somehow trigger the stimulation of Rho activity, such as by releasing factors bound to the microtubules into the cytosol, and these factors are then required for activating Rho GTPases beneath the plasma membrane (Fig. 2.4).

Of note, this notion has recently been supported by the finding of functional involvement of microtubules in regulating the Rho guanine nucleotide exchange factor GEF-H1 and Rho activity itself (Krendel et al., 2002). Although the association of microtubules and Rac1 or GEF-H1 has been indicated, the function of this interaction is only recently becoming clear. Interestingly, recent studies have strongly indicated that Rac1 and RhoA have some functional linkage to each other, where the enhancement of one activity downregulates activity of the other. Therefore, the cross-talk between Rac1 and RhoA activities may account for the microtubule instability-induced membrane ruffling in mammalian cells as well as for the ruffling induced by *Shigella* VirA (Fig. 2.4).

CELL–CELL SPREADING

After escaping from phagocytic vacuoles, *Shigella* multiply and move within the cytoplasm. The ability of *Shigella* to move within the host cytoplasm, and the subsequent cell–cell spreading, is a prerequisite for shigellosis. Intracellular *Shigella* exploit actin polymerization at one pole of the bacterial surface, through which the bacterium gains a propulsive force to spread within the cytoplasm and into adjacent epithelial cells (Fig. 2.5). Under optimum conditions, intracellular bacterial motility is around 15–20 μm/min (Mimuro et al., 2000). The actin-based motility of *Shigella* is dependent on VirG (IcsA) encoded by the *virG* gene on the large plasmid (Makino et al., 1986; Bernardini et al., 1989; Lett et al., 1989).

VirG

VirG (IcsA) is a surface-exposed outer membrane protein, which accumulates at one pole of the bacterium (Fig. 2.6; also see Goldberg et al., 1993; Goldberg, 2001). VirG is composed of 1102 amino acids and contains three distinctive domains: the N-terminal signal sequence (residues 1–52), the 706-amino-acid α-domain (residues 53–758), and the 344-amino-acid C-terminal β-core (residues 759–1102; see Goldberg et al., 1993; Suzuki et al., 1995). The α-domain is exposed on the surface of bacteria, whereas the β-core is embedded in the outer membrane to form a membrane pore. The α-domain is translocated through the membrane pore onto the bacterial surface by a typical autotransporter mechanism as represented by the IgA protease of *N. gonorrhoeae* (Pohlner et al., 1987).

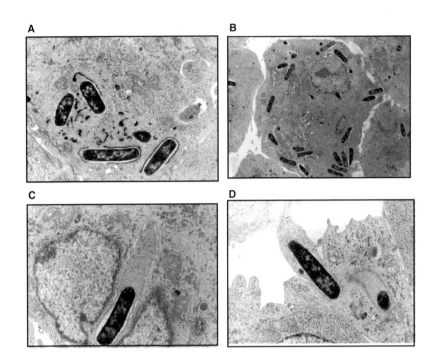

Figure 2.5. Electron micrograph of motile *Shigella* in epithelial cells. (A) Bacteria escaping from the phagocytic vacuoles. (B) Multiplied bacteria within the host cell cytoplasm. (C) A motile bacterium forming a long actin tail. (D) A motile bacterium entering the neighbor host cell.

The asymmetric distribution of VirG along the bacterial body is a prerequisite for the polar movement of *Shigella* in mammalian cells, including bacterial spreading between epithelial cells (Goldberg et al., 1993; Suzuki et al., 1995; Goldberg, 2001). Although the mechanisms are still speculative, recently studies have suggested that the unipolar localization of VirG results from its direct targeting of the pole following diffusion laterally in the outer membrane (Goldberg, 2001; Charles et al., 2001; Robbins et al., 2001). Interestingly, when a VirG-GFP fusion protein is expressed in *E. coli*, *S. typhimurium*, *Yersinia pseudotuberculosis*, or *Vibrio cholerae*, the protein always localizes at one pole, suggesting that the mechanism of polar targeting for VirG is not unique to *Shigella*. Several factors including its own VirG α portion have been implicated in the establishment or maintenance of the asymmetric distribution (Suzuki et al., 1995; Steinhauer et al., 1999). The N-terminal two thirds of the VirG α-domain, which contains six glycine-rich repeats, is essential for mediating actin assembly, because the domain serves

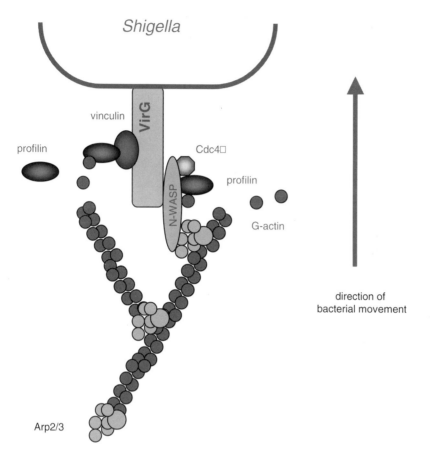

Figure 2.6. Current model for VirG-induced actin polymerization on *Shigella* in infected epithelial cells. VirG accumulated at one pole of bacterium recruits vinculin and N-WASP. The vinculin recruits profilin by means of binding to VASP; the N-WASP activated upon binding by Cdc42 allows recruitment of the Arp2/3 complex, with profilin. The activated Arp2/3 complex, with the aid of profilin, can catalyze rapid actin nucleation and elongation.

to interact with host proteins such as vinculin and neural WASP (N-WASP; see Suzuki et al., 1996, 1998; Suzuki and Sasakawa, 2001). The C-terminal one third of the α-domain is required for VirG to distribute asymmetrically, because *S. flexneri* expressing a VirG mutant with a deletion in this region no longer displays polar movement; instead, it is surrounded by an actin cloud (Suzuki et al., 1996).

Interestingly, LPS plays a role in either the establishment or maintenance of VirG at one pole of *Shigella* (Okada et al., 1991). A number of genes involved in the biosynthesis of LPS have been shown to affect the localization of VirG (Rajakumar et al., 1994). Indeed, removal of the O-side chain from the LPS of *S. flexneri* results in an aberrant localization of VirG, causing a circumferential distribution over the whole bacterial body (Goldberg, 2001). SopA (also called IcsP), an outer membrane protease, has also been indicated to be involved in the asymmetric distribution of VirG by cleaving laterally diffused VirG protein along the bacterial body (Egile et al., 1997; Shere et al., 1997). Finally, the absence of OmpT, an outer membrane protease encoded by the *ompT* gene in *E. coli*, is crucial for VirG to be maintained on the cell surface, because OmpT specifically cleaves at Arg_{758}–Arg_{759} of VirG, causing degradation of the α-domain of VirG on bacteria (Nakata et al., 1993). In fact, none of the *Shigella* and EIEC strains examined has the *ompT* region, thus ensuring the VirG α-domain is expressed and maintained on the bacterial surface (Nakata et al., 1993).

VirG ligands

VirG can interact with at least two host proteins, vinculin and N-WASP (Fig. 2.6). Vinculin, a protein linking focal adhesions and actin filaments, interacts directly with a portion of the VirG α-domain that spans residues 103–508. As already mentioned (see the subsection on IpaA), the function of vinculin in mammalian cells is regulated by $PtdIns(4,5)P_2$. In the inactive state, the N-terminal globular head domain interacts with the C-terminal elongated tail domain, and this interaction is disrupted by the binding of $PtdIns(4,5)P_2$. The exposed head and tail domains become activated to interact with other molecules. In epithelial cells infected with *Shigella*, vinculin is recruited to the bacterial surface as well as to the actin comet tail elongated from motile bacteria in infected cells (Suzuki et al., 1996). Laine et al. (1997) revealed that the recruited vinculin is cleaved, leaving the head portion, which interacts with VirG along with vasodilator stimulating phosphoprotein (VASP) and profilin. Thus, the complex formed in the vicinity of the bacterium seems to contribute to enhancing the growth of barbed ends of actin filaments. Microinjection of the vinculin head portion into *Shigella*-infected cells stimulates the bacterial motility (Laine et al., 1997). In fact, the speed at which *E. coli* expressing VirG induce formation of the actin tail in vinculin-depleted *Xenopus* egg extracts is shown to be significantly decreased to less than 30% of the original level (Suzuki, unpublished data). Thus, vinculin appears to contribute to *Shigella*-inducing actin assembly such as through interaction with VASP, because VASP recruits profilin (Fig. 2.6). Alternatively, existing actin filaments bound by vinculin at the bacterial surface may

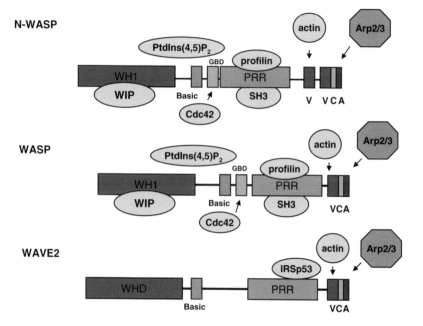

Figure 2.7. Functional domains of N-WASP, WASP, and WAVE2 and molecules that interact with these proteins (Takenawa and Miki, 2000).

facilitate actin nucleation mediated by the Arp2/3 complex interacting with the VirG–N-WASP complex.

N-WASP

N-WASP is a member of the WASP and WAVE family, which includes human WASP, *Saccharomyces cerevisiae* WASP-like protein Las17p/Bee1p, and WAVE1, WAVE2, and WAVE3 proteins (Takenawa and Miki, 2000). WASP and WAVE family proteins integrate upstream signaling events with changes in the actin cytoskeleton by means of the Arp2/3 complex. Figure 2.7 shows the functional structures of N-WASP, WASP, and WAVE2, and molecules to which they bind (Takenawa and Miki, 2000). The expression of WASP is limited to hematopoietic cells, whereas N-WASP and WAVEs are ubiquitously expressed in host cells including epithelial cells. N-WASP has been implicated both in the formation of filopodia and in the actin-based motility of intracellular *Shigella*, whereas WAVEs are involved in formation of lamellipodia and membrane ruffling. N-WASP and WASP possess several distinctive domains: a homology (WH1) domain that binds PtdIns(4,5)P_2, a domain composed of basic amino acids, a GTPase binding domain (GBD) that binds Cdc42, a proline-rich region (PRR), a G-actin-binding verprolin homology (V) domain, a domain (C) with homology to the actin-depolymerizing protein cofilin, and finally a C-terminal acidic (A) segment (Fig. 2.7). The C-terminal VCA domain mediates the interaction with the Arp2/3 complex, by which the Arp2/3 complex is activated, thus mediating actin polymerization.

In *Shigella*-infected cells, N-WASP, but not the other members of the WASP family, accumulates at the pole of the intracellular bacterium assembling an actin comet tail (Suzuki et al., 2002). Functional assays using the ectopic expression of dominant negative N-WASP in mammalian cells or immunodepletion in *Xenopus* egg extracts revealed that N-WASP is an essential host component for mediating the actin-based motility of intracellular *Shigella* (Suzuki et al., 1998). Of note, none of the WASP family proteins associate with the surface of intracellular *L. monocytogenes* including the actin tails (Suzuki et al., 2002). The binding of *Shigella* VirG to WASP family proteins is limited to only N-WASP. With the use of a series of chimeras obtained by swapping the N-WASP and WASP domains, the specificity of VirG to interact with the N-terminal WH1 region of N-WASP was found to serve as the critical ligand (Suzuki et al., 2002). Consistent with this, hematopoietic cells such as J774 cells (mouse macrophages), human monocytes, PMNs, or platelets express WASP predominantly but not N-WASP, which cannot support the actin-based movement of intracellular *S. flexneri* (Egile et al., 1999; Suzuki et al., 2002). This was also confirmed by use of N-WASP-deficient embryos (Snapper et al., 2001).

Cdc42

In vitro studies have indicated that activation of N-WASP in cells requires the binding of Cdc42 to the GBD motif of N-WASP (Miki et al., 1996; Rohatgi et al., 1999, 2000). This binding inhibits the intramolecular interaction between the C-terminal acidic amino acids and the basic amino acids near the GBD, thus causing the unfolding of N-WASP into the activated form. Furthermore, when N-WASP interacts with a fragment of VirG encompassing residues 53–503 of VirG, the N-WASP–Arp2/3 complex-mediated actin nucleation can also be stimulated without Cdc42 *in vitro* (Egile et al., 1999). The ability of VirG to activate N-WASP without Cdc42 was also observed by using *Clostridium difficile* Tcd-10463, which inhibits Rho GTPases; however, the actin tails are significantly shorter in the presence of the exotoxin than in infected cells without the toxin, implying that actin assembly by *Shigella* is partly affected by toxin (Mounier et al., 1999). A later study, however, strongly indicated that cellular Cdc42 is required for the actin-based motility of *Shigella* in infected cells (Suzuki et al., 2000). Microinjection of activated

Cdc42 accelerates *Shigella* motility, whereas inhibiting Cdc42 activity, for example by adding Rho GDI, a guanine nucleotide dissociation inhibitor, into cell extracts greatly reduces bacterial motility. In pyrene actin polymerization assays, VirG–N-WASP–Arp2/3 complex is insufficient to express the full activity for polymerizing actin; rather, in the presence of activated Cdc42, the actin nucleation activity is remarkably stimulated. In fact, Cdc42 can be accumulated at one pole of *Shigella* in the process of initiating movement in infected cells. Importantly, Cdc42 is not accumulated on motile *Shigella* possessing an actin tail in the infected cells, implying that Cdc42 seems to be no longer necessary after a steady speed has been reached, at which stage the VirG–N-WASP–Arp2/3 complexes would be constitutively activated. These studies have thus indicated that Cdc42 takes part in initiating the actin-based motility of intracellular *Shigella* in epithelial cells.

Arp2/3

VirG expressed on the bacterial surface in host cells can directly recruit and activate N-WASP, which in turn recruits and activates the Arp2/3 complex. Consequently, the VirG–N-WASP–Arp2/3 complex formed at one pole of the bacterium can mediate actin nucleation and elongation (Fig. 2.6). To initiate actin nucleation, the Arp2/3 complex is activated upon physical interaction with the VCA region of N-WASP. With the aid of other host factors, the VirG–N-WASP–Arp2/3 complex mediates rapid actin filament growth at the barbed end, including cross-linking between the elongated actin filaments. In this way *Shigella* can gain a propulsive force in the host cytoplasm, where some motile bacteria impinge on the host plasma membrane, leading to the extension of membranous protrusions. Through these protrusions that penetrate neighboring cells, *Shigella* further move into adjacent epithelial cells by disrupting the double membranes with the secreted proteins (Suzuki et al., 1994; Schuch et al., 1999). In a reconstitution experiment supporting the actin-based motility of *Shigella* with pure proteins, host factors required for *Shigella* movement were confirmed to include actin, Arp2/3 complex, and N-WASP (Loisel et al., 1999). In addition, actin depolymerization factor (ADF)/cofilin, capping protein, and profilin are also indicated to be involved in the regulation of actin turnover and stabilization of the actin tail (Loisel et al., 1999; Mimuro et al., 2000). Several other actin-associated proteins, such as plastin (fimbrin), filamin, VASP, zyxin, ezrin, CapZ, Nck, and WASP interacting protein (WIP), have also been identified as being localized to the actin tail or at the posterior end of intracellular bacteria. However, whether or not these host factors are functionally required for *Shigella* movement in infected cells awaits further study.

Profilin

Profilin binding to actin facilitates the formation of ATP–actin monomers, the form of actin to be assembled into the free barbed ends of actin filaments. Profilin can interact with various proteins, notably proteins with a proline-rich sequence, such as N-WASP, VASP, MENA, p140mDia, WAVE/Scar, and Arp2/3 complex (Suzuki and Sasakawa, 2001). In a reconstitution assay *in vitro*, profilin and VASP (for *Listeria*) were shown to enhance bacterial motility but were not essential, suggesting that recruited profilin helps to increase the local concentration of ATP–actin (Loisel et al., 1999). Profilin exists in two isoforms in mammalian cells, profilin I and II, with profilin I having a greater affinity for N-WASP (Kd = 60 nM) than profilin II (Kd = 400 nM). Hence, the role of profilin I in the actin-based motility of intracellular *Shigella* has recently been investigated (Mimuro et al., 2000). Upon overexpression of a profilin H133S mutant defective in interaction with the PRR of N-WASP including poly-L-proline, *Shigella* motility is significantly decreased. Similarly, the depletion of profilin from *Xenopus* egg extracts results in a decrease in bacterial motility that is rescued by adding back profilin I but not H133S mutant. Consistent with this, on overexpression of an N-WASP mutant lacking the PRR unable to interact with profilin, the actin tail formation of intracellular *Shigella* is abolished. In N-WASP-depleted extracts, the addition of wild type N-WASP but not the N-WASP mutant restores bacterial motility, indicating that profilin associated with N-WASP is an essential host factor for supporting rapid movement of intracellular *Shigella* (Mimuro et al., 2000).

CONCLUSION

Invasion of epithelial cells by *Shigella* is a highly dynamic cellular event that occurs through the complicated interaction between bacterial effectors and target host factors. *Shigella*-directed internalization requires large-scale membrane ruffling, which eventually leads to a phagocytic event. In promoting the host cellular events including the subsequent infectious steps, the roles of effectors such as IpaA, IpaB, IpaC, IpgD, and VirA delivered by means of the TTSS are crucial. However, many other putative effectors secreted by the TTSS (Buchrieser et al., 2000) must also be involved in almost the entire stage of infection, including modulation of host immune responses. Furthermore, it is thought that targeting of the host factors by bacterial effectors during infection would appropriately be operated by the pathogen, for which the timing and amounts of effectors to be secreted by means of the TTSS may be stringently controlled at a posttranslational or cotranslational level, depending on the stage of *Shigella* infection (Blocker et al., 2003). Clearly, we must await further study to elucidate the role of all of the effectors in each stage of *Shigella* infection, including the secretion control system. Such information is needed for better understanding of the sophisticated bacterial infectious strategy and host inflammatory responses, which are prerequisite for the development of both a novel safer *Shigella* vaccine and a suitable animal model to study the disease.

ACKNOWLEDGMENTS

I am grateful to Drs. Toshihiko Suzuki and Reiko Akakura for their critical review of the manuscript and to all of the laboratory members who contributed to the project. Original research in the author's laboratory is supported by a Grant-in Aid for Scientific Research on Priority Areas entitled on "Infection and Host Responses" from the Japanese Ministry of Education, Science, Technology, Sport and Culture.

REFERENCES

Adam, T., Giry, M., Boquet, P., and Sansonetti, P.J. (1996). Rho-dependent membrane folding causes *Shigella* entry into epithelial cells. *EMBO J.* **15**, 3315–3321.

Adler, B., Sasakawa, C., Okada, N., Makino, S., and Yoshikawa, M. (1989). A dual transcriptional activation system for the 230 kb plasmid genes coding for virulence-associated antigens of *Shigella flexneri. Mol. Microbiol.* **3**, 627–635.

Bernardini, M.L., Mounier, J., d'Hauteville, H., Coquis-Randon, M., and Sansonetti, P.J. (1989). Identification of *icsA*, a plasmid locus of *Shigella flexneri* that governs bacterial intra- and intercellular spreading through interaction with F-actin. *Proc. Natl. Acad. Sci. USA* **86**, 3867–3871.

Blocker, A., Gounon, P., Larquest, K., Neibuhr, V., Cabiaux, C., Parsot, C., and Sansonetti, P.J. (1999). The tripartite type III secretion of *Shigella flexneri* inserts IpaB and IpaC into host membranes. *J. Cell Biol.* **147**, 683–693.

Blocker, A., Komoriyama, K., and Aizawa, S. (2003). Type III secretion systems and bacterial flagella: insights into their function from structural similarities. *Proc. Natl. Acad. Sci. USA* **100**, 3027–3030.

Bourdet-Sicard, R., Rudiger, M., Jockusch, B.M., Gounon, P., Sansonetti, P.J., and Tran van Nhieu, G. (1999). Binding of the *Shigella* protein IpaA to vinculin induces F-actin depolymerization. *EMBO J.* **18**, 5853–5862.

Buchrieser, C., Glaser, P.P., Rusniok, C., Nedjari, H., d'Hauteville, H., Kunst, F., Sansonetti, P.J., and Parsot, P.J. (2000). The virulence plasmid pRW100 and the repertoire of proteins secreted by the type III secretion apparatus of *Shigella flexneri*. *Mol. Microbiol.* **38**, 760–771.

Charles, M., Perez, M., Kobil, J.H., and Goldberg, M.B. (2001). Polar targeting of *Shigella* virulence factor IcsA in Enterobacteriacae and Vibrio. *Proc. Natl. Acad. Sci. USA* **98**, 9871–9876.

Collazo, C.M. and Galán, J.E. (1996). Requirement of exported proteins for secretion through the invasion-associated Type III system in *Salmonella typhimurium*. *Infect. Immun.* **64**, 3524–3531.

Cornelis, G.R. and Van Gijsegem, F. (2000). Assembly and function of type III secretory systems. *Annu. Rev. Microbiol.* **30**, 47–56.

Daniell, S.H., Takahashi, N., Wilson, R., Friedberg, D., Rosenshine, I., Boody, F.P., Shaw, R.K., Knutton, S., Frankel, G., and Aizawa, S. (2001). The filamentous type III secretion translocon of enteropathogenic *Escherichia coli*. *Cell. Microbiol.* **3**, 865–871.

Dehio, C., Prévost, M.C., and Sansonetti, P.J. (1995). Invasion of epithelial cells by *Shigella flexneri* induces tyrosine phosphorylation of cortactin by a $pp60^{c-src}$-mediated signalling pathway. *EMBO J.* **14**, 2471–2482.

Dorman, C.J. and Porter, M.E. (1998). The *Shigella* virulence gene regulatory cascade: a paradigm of bacterial gene control mechanisms. *Mol. Microbiol.* **29**, 677–684.

Duménil G., Olivo, J.C., Pellegrini, S., Fellous, M., Sansonetti, P.J., and Tran van Nhieu, G. (1998). Interferon α inhibits a Src-mediated pathway necessary for *Shigella*-induced cytoskeletal rearrangements in epithelial cells. *J. Cell Biol.* **143**, 1003–1012.

Egile, C., d'Hauteville, H., Parsot, C., and Sansonetti, P.J. (1997). SopA, the outer membrane protease responsible for polar localization of IcsA in *Shigella flexneri*. *Mol. Microbiol.* **23**, 1063–1073.

Egile, C., Loisel, T.P., Laurent, V., Li, R., Pantaloni, D., Sansonetti, P.J., and Carlier, M-F. (1999). Activation of the CDC42 effector N-WASP by the *Shigella flexneri* IcsA protein promotes actin nucleation by Arp2/3 complex and bacterial actin-based motility. *J. Cell Biol.* **146**, 1319–1332.

Elliott, S.J., Krejany, E.O., Mellies, J.L., Robins-Browne, R.M., Sasakawa, C., and Kaper, J.B. (2001). EspG a novel type III system-secreted protein from enteropathogenic *Escherichia coli* with similarities to VirA of *Shigella flexneri*. *Infect. Immun.* **69**, 4027–4033.

Enomoto, T. (1996). Microtuble disruption induces the formation of actin stress fibers and focal adhesions in cultured cells: possible involvement of Rho signal cascade. *Cell Struc. Func.* **5**, 317–326.

Galyov, E.E., Wood, M.W., Rosquist, R., Mullan, P.B., Watson, P.R., Hedges, S., and Wallis, T.S. (1997). A secreted effector protein of *Salmonella dublin* is translocated into eukaryotic cells and mediates inflammation and fluid secretion in infected ileal mucosa. *Mol. Microbiol.* **25**, 903–912.

Girardin, S.E., Tournebize, R., Mavris, M., Page, A-L., Li, X., Stark, G.R., Bertin, J., DiStefano, P.S., Yaniv, M., Sansonetti, P.J., and Philpott, D.J. (2001). CARD/Nod1 mediates NF-κB and JNK activation by invasive *Shigella flexneri*. *EMBO Reports* **21**, 736–742.

Goldberg, M.B. (2001). Actin-based motility of intracellular microbial pathogens. *Microbiol. Mol. Biol. Rev.* **65**, 595–626.

Goldberg, M.B., Barzu, O., Parsot, C., and Sansonetti, P.J. (1993). Unipolar localization and ATPase activity of IcsA, a *Shigella flexneri* protein involved in intracellular movement. *J. Bacteriol.* **175**, 2189–2196.

Håkansson, S., Schesser, K., Person, C., Galyov, E.E., Rosqvist, R., Homblé, F., and Walf-Watz, H. (1996). The YopB protein of *Yersinia pseudotuberculosis* is essential for the translocation of Yop effector proteins across the target cell plasma membrane and displays a contact-dependent membrane disrupting activity. *EMBO J.* **15**, 5812–5823.

Hirao, M., Sato, N., Kondo, T., Yonemura, S., Monden, M., Sasaki, T., Takai, Y., and Tsukita, S. (1996). Regulation mechanisms of ERM (ezrin/radixin/moesin) protein/plasma membrane association: possible involvement of phosphatidylinositol turnover and Rho-dependent signaling pathway. *J. Cell Biol.* **135**, 37–51.

Hong, K.H. and Miller, V.L. (1998). Identification of a novel *Salmonella* invasion locus homologous to *Shigella ipgDE*. *J. Bacteriol.* **180**, 1793–1802.

Hueck, C.J. (1998). Type III protein secretion systems bacterial pathogens of animal and plants. *Microbiol. Mol. Biol. Rev.* **62**, 379–433.

Inohara, N., Ogura, Y., Chen, F.F., Muto, A., and Nunez, G. (2001). Human Nod1 confers responsiveness to bacterial lipopolysaccharides. *J. Biol. Chem.* **276**, 2551–2554.

Jockusch, B.M. and Rudiger, M. (1996). Crosstalk between cell adhesion molecules: vinculin as a paradigm for regulation by conformation. *Trends Cell Biol.* **6**, 311–315.

Knutton, S., Rosenshine, I., Pallen, M.J., Nisan, I., Neves, B.C., Bain, C., Wolf, C., Dougan, G., and Frankel, G. (1998). A novel EspA-associated surface organelle of enteropathogenic *Escherichia coli* involved in protein translocation into epithelial cells. *EMBO J.* **17**, 2166–2176.

Krendel, M., Zenke, F.T., and Bokoch, G.M. (2002). Nucleotide exchange factor GEF-H1 mediates cross-talk between microtubles and the actin cytoskeleton. *Nat. Cell Biol.* **4**, 294–301.

Kubori, T., Matsushima, Y., Nakamura, D., Uralil, J., Lara-Tejero, M., Sukhan, A., Galan, J.E., and Aizawa, S. (1998). Supramolecular structure of the *Salmonella typhimurium* type III protein secretion system. *Science* **280**, 602–605.

Kubori, T., Shkan, A., Aizawa, S., and Galán, J.E. (2000). Molecular characterization and assembly of the needle complex of the *Salmonella typhimurium* type III protein secretion system. *Proc. Natl. Acad. Sci. USA* **97**, 10,225–10,230.

Krugmann, S.K., Jordens, I., Gevaert, K., Driessens, M., Vandekerckhove, J., and Hall, A. (2001). Cdc42 induces filopodia by promoting the formation of an IRSp53:Mena complex. *Curr. Biol.* **11**, 1645–1655.

Kuwae, A., Yoshida, S., Tamano, K., Mimuro, H., Suzuki, T., and Sasakawa, C. (2001). *Shigella* invasion of macrophage requires the insertion of IpaC into the host plasma membrane. *J. Biol. Chem.* **276**, 32,230–32,239.

Lafont, F., Tran van Nhieu, G., Hanada, K., Sansonetti, P.J., and Gisou van der Goot, F. (2002). Initial steps of *Shigella* infection depend on the cholesterol/spingolipid raft-mediated CD44-IpaB interaction. *EMBO J.* **21**, 4449–4457.

Laine, R.O., Zeile, W., Kang, F., Purich, D.L., and Southwick, F.S. (1997). Vinculin proteolysis unmasks an ActA homolog for actin-based *Shigella* motility. *J. Cell Biol.* **138**, 1255–1264.

Lan, R. and Reeves, P.R. (2002). *Escherichia coli* in disguise: molecular origins of Shigella. *Microb. Infect.* **4**, 1125–1132.

Lett, M-C., Sasakawa, C., Okada, N., Sakai, T., Makino, S., Yamada, M., Komatsu, K., and Yoshikawa, M. (1989). *virG*, a plasmid-coded virulence gene of *Shigella flexneri*: identification of the *virG* protein and determination of the complete coding sequence. *J. Bacteriol.* **171**, 353–359.

Loisel, T.P., Boujemaa, R., Pantaloni, D., and Carlier, M-F. (1999). Reconstitution of actin-based motility of *Listeria* and *Shigella* using pure proteins. *Nature* **401**, 613–616.

Makino S., Sasakawa, C., Kamata, T., Kurata, T., and Yoshikawa, M. (1986). A genetic determinant required for continuous reinfection of adjacent cells on a large plasmid in *Shigella flexneri* 2a. *Cell* **46**, 551–555.

Mavris, M., Page, A.L., Tournebize, R., Demers, B., Sansonetti, P.J., and Parsot, C. (2002). Regulation of transcription by the activity of the *Shigella flexneri* type III secretion apparatus. *Mol. Microbiol.* **43**, 1543–1553.

Ménard, R., Prévost, M.C., Gounon, P., Sansonetti, P.J., and Dehio, C. (1996). The secreted Ipa complex of *Shigella flexneri* promotes entry into mammalian cells. *Proc. Natl. Acad. Sci. USA* **93**, 1254–1258.

Ménard, R., Sansonetti, P.J., and Parsot, C. (1994a). The secretion of the *Shigella flexneri* Ipa invasins is induced by the epithelial cells and controlled by IpaB and IpaD. *EMBO J.* **13**, 5293–5302.

Ménard, R., Sansonetti, P.J., Parsot, C., and Vasselon, T. (1994b). Extracellular association and cytoplasmic partioning of the IpaB and IpaC invasins of *Shigella flexneri. Cell* **79**, 515–525.

Miki, H., Miura, K., and Takenawa, T. (1996). N-WASP, a novel actin-depolymerizing protein, regulate the cortical cytoskeletal rearrangement in a PIP2-dependent manner downstream of tyrosine kinase. *EMBO J.* **15**, 5326–5335.

Miki, H., Yamaguchi, H., Suetsugu, S., and Takenawa, T. (2000). IRSp53 is an essential intermediate between Rac and WAVE in the regulation of membrane ruffling. *Nature* **408**, 732–735.

Mimuro, H., Susuki, T., Suetsugu, S., Miki, H., Takenawa, T., and Sasakawa, C. (2000). Profilin is required for sustaining efficient intra- and intercellular spreading of *Shigella flexneri. J. Biol. Chem.* **275**, 28,893–28,901.

Mounier, J., Laurent, V., Hall, A., Fort, P., Calier, M-F., Sansonetti, P.J., and Egile, C. (1999). Rho family GTPases control entry of *Shigella flexneri* into epithelial cells but not intracellular motility. *J. Cell Sci.* **112**, 2069–2080.

Mounier, J., Vasselon, T., Hellio, R., Lesourd, M., and Sansonetti, P.J. (1992). *Shigella flexneri* enters human colonic Caco-2 epithelial cells through the basolateral pole. *Infect. Immun.* **60**, 237–248.

Nakata, N., Tobe, T., Fukuda, I., Suzuki, T., Komatsu, K., Yoshikawa, M., and Sasakawa, C. (1993). The absence of surface protease, OmpT, determines the intercellular spreading ability of *Shigella*: the relationship between the *ompT* and *kcpA* loci. *Mol. Microbiol.* **9**, 459–468.

Niebuhr, K., Jouihri, N., Allaoui, A., Gounon, P., Sansonetti, P.J., and Parsot, C. (2000). IpgD, a protein secreted by the type III secretion machinery of *Shigella flexneri*, is chaperoned by IpgE and implicated in entry focus formation. *Mol. Microbiol.* **38**, 8–19.

Niebuhr, K., Giuriato, S., Pedron, T., Philpott, D.J., Gaits, F., Sable, J., Sheetz, M.P., Parsot, C., Sansonetti, P.J., and Payrastre, B. (2002). Conversion of PtdIns(4,5)P$_2$ into PtdIns(5)P by the *S. flexneri* effector IpgD reorganizes host cell morphology. *EMBO J.* **21**, 5069–5078.

Norris, F.A., Wilson, M.P., Wallis, T.S., Galyov, E.E., and Majerus, P.W. (1998). SopB, a protein required for virulence of *Salmonella dublin*, is an inositol phosphate phosphatase. *Proc. Natl. Acad. Sci. USA* **95**, 14,057–14,059.

Okada, N., Sasakawa C., Tobe, T., Yamada, M., Nagai, S., Talkder, K., Komatsu, K., Kanegasaki, S., and Yoshikawa, M. (1991). Virulence-associated chromosomal loci of *Shigella flexneri* identified by random Tn5 insertion mutagenesis. *Mol. Microbiol.* **5**, 887–893.

Page, A-L., Sansonetti, P.J., and Parsot, C. (2002). Spa15 of *Shigella flexneri*, a third type of chaperone in the type III secretion pathway. *Mol. Microbiol.* **43**, 1533–1542.

Plano G.V., Day, J.B., and Ferracci, F. (2001). Type III export: new uses for an old pathway. *Mol. Microbiol.* **40**, 284–293.

Pohlner, J., Halter, K., Beyreuther, K., and Meyer, T.F. (1987). Gene structure and extracellular secretion of *Neisseria gonorrhoeae* IgA protease. *Nature* **325**, 458–462.

Pupo, G.M., Lan, R., and Reeves, P.R. (2000). Multiple independent origins of *Shigella* clones of *Escherichia coli* and convergent evolution of many of their characteristics. *Proc. Natl. Acad. Sci. USA* **97**, 10,567–10,572.

Rajakumar, R., Jost, B.H., Sasakawa, C., Okada, N., Yoshikawa, M., and Adler, B. (1994). Nucleotide sequence of the rhamnose biosynthetic operon of *Shigella flexneri* 2a and role of lipopolysacchride in virulence. *J. Bacteriol.* **176**, 2364–2373.

Renesto, P., Mounier, J., and Sansonetti, P.J. (1996). Induction of adherence and degranulation of polymorphonuclear leukocytes: A new expression of the invasive phenotype of *Shigella flexneri*. *Infect. Immun.* **64**, 719–723.

Robbins, J.R., Monack, D., McCallum, S.J., Vegas, A., Pham, E., Goldberg, M.B., and Theriot, J.A. (2001). The making of a gradient: IcsA (VirG) polarity in *Shigella flexneri*. *Mol. Microbiol.* **41**, 861–872.

Rohatgi, R., Ho, H.Y., and Kirschner. M.W. (2000). Mechanisms of N-WASP activation by CDC42 and phosphatidylinositol 4,5-bisphosphate. *J. Cell Biol.* **150**, 1299–1310.

Rohatgi, R., Ma, H., Miki, H., Lopez, M., Kirchhausen, T., Takenawa, T., and Kirschner, M.W. (1999). The interaction between N-WASP and the Arp2/3 complex links Cdc42-dependent signals to actin assembly. *Cell* **97**, 221–231.

Sansonetti, P.J., Arondel, J., Fountaine, A., D'Hauteville, H., and Bernardini, L. (1991). *ompB* (osmo-regulation) and *icsA* (cell to cell spreading) mutants of *Shigella flexneri*: vaccine candidates and probes to study the pathogenesis of shigellosis. *Vaccine* **9**, 416–422.

Sansonetti, P.J., Arondel, J.R., Prévost, M.C., and Huerre, M. (1996). Infection of rabbit Peyer's patches by *Shigella flexneri*: effect of adhesive or invasive bacterial phenotypes on follicle-associated epithelium. *Infect. Immun.* **64**, 2752–2764.

Sansonetti, P.J., Ryter, A., Clerc, P., Maurelli, A.T., and Mounier, J. (1986). Multiplication of *Shigella flexneri* within HeLa cells: lysis of the phagocytic vacuole and plasmid-mediated contact hemolysis. *Infect. Immun.* **51**, 461–469.

Sasakawa, C., Kamata, K., Sakai, T., Makino, S., Yamada, H., Okada, N., and Yoshikawa, M. (1988). Virulence-associated genetic regions comprising 31 kilobases of the 230-kilobase plasmid in *Shigella flexneri* 2a. *J. Bacteriol.* **170**, 2480–2484.

Schuch, R., Sandlin, R.C., and Maurelli, A.T. (1999). A system for identifying post-invasion functions of invasion genes: requirements for the Mxi-Spa type III secretion pathway of *Shigella flexneri* in intercellular dissemination. *Mol. Microbiol.* **34**, 675–689.

Sekiya, K., Ohishi, M., Ogino, T., Tamano, K., Sasakawa, C., and Abe, A. (2001). Supermolecular structure of the enteropathogenic *Escherichia coli* type III secretion system and its direct interaction with the EspA-sheath-like structure. *Proc. Natl. Acad. Sci. USA* **98**, 11,638–11,643.

Shere, K.D., Sallustion, S., Manessis, A., D'Aversa, T.G., and Goldberg, M.B. (1997). Distribution of IcsP, the major *Shigella* protease that cleaves IcsA, accelerates actin-based motility. *Mol. Microbiol.* **25**, 451–462.

Skoudy, A., Mounier, J., Aruffo, A., Ohayon, H., Gounon, P., Sansonetti, P.J., and Tran van Nhieu, G. (2000). CD44 binds to the *Shigella* IpaB protein and participates in bacterial invasion of epithelial cells. *Cell. Microbiol.* **2**, 19–33.

Skoudy, A., Tran van Nhieu, G., Mantis, N., Aprin, M., Mounier, J., Gounon, P., and Sansonetti, P.J. (1999). A functional role for ezrin during *Shigella flexneri* entry into epithelial cells. *J. Cell Sci.* **112**, 2059–2068.

Snapper, S.B., Takeshima, F., Anton, I., Liu, C.H., Thomas, S.M., Nguyen, D., Dudley D., Fraser, H., Purich, D., Lopez-Llasaca, M., Klein, C., Davidson, L., Bronson, R., Mulligan, R., Southwick F., Geha, R., Goldberg, M.B., Rosen, F.S., Hartwig, J.H., and Alt, F.W. (2001). N-WASP deficiency reveals distinct pathways for cell surface projections and microbial actin-based motility. *Nat. Cell Biol.* **3**, 897–904.

Steinhauer, J., Agha, R., Andrew, T.P., Varga, W., and Goldberg, B. (1999). The nuipolar *Shigella* surface protein IcsA is targeted directly to the bacterial old pole: IcsP cleavage of IcsA occurs over the entire bacterial surface. *Mol. Microbiol.* **32**, 367–377.

Suzuki, T., Lett, M-C., and Sasakawa, C. (1995). Extracellular transport of VirG protein in *Shigella*. *J. Biol. Chem.* **270**, 30,874–30,880.

Suzuki, T., Miki, T., Takenawa, T., and Sasakawa, C. (1998). Neural Wiskott-Aldrich syndrome protein is implicated in actin-based motility of *Shigella flexneri*. *EMBO J.* **17**, 2767–2776.

Suzuki, T., Mimuro, H., Suetsugu, S., Miki, H., Takenawa, T., and Sasakawa, C. (2002). Neural Wiskott-Aldrich syndrome protein (N-WASP) is the specific ligand for *Shigella* VirG among the WASP family and determines the host cell type allowing actin-based spreading. *Cell. Microbiol.* **4**, 223–233.

Suzuki, T., Mimuro, H., Miki, H., Takenawa, T., Sasaki., Nakanishi, H., Takai, Y., and Sasakawa, C. (2000). Rho family GTPase Cdc42 is essential for the actin-based motility of *Shigella* in mammalian cells. *J. Exp. Med.* **191**, 1905–1920.

Suzuki, T., Murai, T., Fukuda, I., Tobe, T., Yoshikawa, M., and Sasakawa, C. (1994). Identification and characterization of a chromosomal virulence gene, *vacJ*, required for intercellular spreading of *Shigella flexneri. Mol. Microbiol.* **11**, 31–41.

Suzuki, T., Saga, S., and Sasakawa, C. (1996). Functional analysis of *Shigella* VirG domains essential for interaction with vinculin and actin-based motility. *J. Biol. Chem.* **271**, 21,878–21,885.

Suzuki, T. and Sasakawa, C. (2001). Molecular basis of the intracellular spreading of *Shigella. Infect. Immun.* **69**, 5959–5966.

Tamano, K., Aizawa, S., Katayama, E., Nonaka, T., Imajo-Ohmi, S., Kuwae, A., Nagai, S., and Sasakawa, C. (2000). Supramolecular structure of the *Shigella* type III secretion machinery: the needle part is changeable in length and essential for delivery of effectors. *EMBO J.* **19**, 3876–3887.

Tamano, K., Eisaku, K., Toyotome, T., and Sasakawa, C. (2002). *Shigella* Spa32 is an essential secretory protein for functional type III secretion machinery and uniformity of its needle length. *J. Bacteriol.* **184**, 1244–1252.

Takenawa, T. and Miki, H. (2000). WASP and WAVE family proteins: key molecules for rapid rearrangment of cortical actin filaments and cell movement. *J. Cell Sci.* **114**, 1801–1809.

Terebiznik, M.R., Vieira, O.V., Marcus, S.L., Slade, A., Yip, C.M., Trimble, W.S., Meyer, T., Finlay, B.B., and Grinstein, S. (2002). Elimination of host cell PtdIns(4, 5)P_2 by bacterial SigD promotes membrane fission during invasion by *Salmonella. Nature Cell Biol.* **4**, 766–773.

Tran van Nhieu, G., Caron, E., Hall, A., and Sansonetti, P.J. (1999). IpaC induces actin polymerization and filopodia formation during *Shigella* entry into epithelial cells. *EMBO J.* **18**, 3249–3262.

Tran van Nhieu, G. and Sansonetti, P.J. (1999). Mechanism of *Shigella* entry into epithelial cells. *Curr. Opin. Microbiol.* **2**, 51–55.

Van Gijsegem, F., Vasse, J., Camus, J-C., Marenda, M., and Boucher, C. (2000). *Ralstonia solanacearum* produces Hrp-dependent pili that are required for PopA secretion but not for attachment of bacteria to plant cells. *Mol. Microbiol.* **36**, 249–260.

Venkatesan, M.M., Goldberg, M.B., Rose, D.J., Grotbeck, E.J., Burland, V., and Blattner, F.R. (2001). Complete DNA sequence and analysis of the large virulence plasmid of *Shigella flexneri. Infect. Immun.* **69**, 3271–3285.

Wassef, J.S., Keren, D.F., and Mailloux, J.L. (1989). Role of M cells in initial antigen uptake and in ulcer formation in the rabbit intestinal loop model of shigellosis. *Infect. Immun.* **57**, 858–863.

Watarai, M., Funato, S., and Sasakawa, C. (1996). Interaction of Ipa proteins of *Shigella flexneri* with $\alpha_5\beta_1$ integrin promotes entry of the bacteria into mammalian cells. *J. Exp. Med.* **183**, 991–999.

Watarai, M., Kamata, Y., Kozaki, S., and Sasakawa, C. (1997). Rho, a small GTP-binding protein, is essential for *Shigella* invasion of epithelial cells. *J. Exp. Med.* **185**, 281–292.

Waterman-Storer, C.M., Wothylake, R.A., Liu, B.P., Burridge, K., and Salmon, E.D. (1999). Microtubule growth activates Rac1 to promote lamellipodial protrusion in fibroblasts. *Nat. Cell Biol.* **1**, 45–50.

Uchiya, K., Tobe, T., Komatsu, K., Suzuki, T., Watarai, M., Fukuda, I., Yoshikawa, M., and Sasakawa, C. (1995). Identification of a novel virulence gene, *virA*, on the large plasmid of *Shigella*, involved in invasion and intercellular spreading. *Mol. Microbiol.* **17**, 241–250.

Yoshida, S., Katayama, E., Kuwae, A., Mimuro, H., Suzuki, T., and Sasakawa, C. (2002). *Shigella* deliver an effector protein to trigger host microtubule destabilization, which promotes Rac1 activity and efficient bacterial internalization. *EMBO J.* **21**, 2923–2935.

Zychlinsky, A., Fitting, C., Cavaillon, J.M., and Sansonetti, P.J. (1994). Intereukin 1 is released by murine macrophages during apoptosis induced by *Shigella flexneri. J. Clin. Invest.* **94**, 1328–1332.

Zychlinsky, A., Prévost, M.C., and Sansonetti, P.J. (1992). *Shigella flexneri* induces apoptosis in infected macrophages. *Nature* **358**, 167–169.

Zychlinsky, A. and Sansonetti, P.J. (1997). Apoptosis as a proinflammatory event: what can we learn from bacteria-induced cell death? *Trends Microbiol.* **5**, 201–204.

Zychlinsky, A., Thirumalai, K., Arondel, J., Cantey, J.R., Aliprantis, A.O., and Saonsonetti, P.J. (1996). In vivo apoptosis in *Shigella flexneri* infection. *Infect. Immun.* **64**, 5357–5365.

CHAPTER 3

How *Yersinia* escapes the host: To Yop or not to Yop

Geertrui Denecker and Guy R. Cornelis

The genus *Yersinia* contains three species of Gram-negative bacteria that are pathogenic for humans: *Y. pestis*, the agent of bubonic plague; *Y. pseudotuberculosis*, causing mesenteric adenitis and septicemia; and *Y. enterocolitica*, causing gastrointestinal syndromes (enteritis and mesenteric lymphadenitis). Bacteria from these three species have a tropism for lymphoid tissues and share the common capacity to resist the innate immune response. Whereas *Y. pestis* is generally inoculated by a fleabite or aerosol, *Y. enterocolitica* and *Y. pseudotuberculosis* are foodborne pathogens, which gain access to the underlying lymphoid tissue (e.g., Peyer's patches) of the intestinal mucosa through M cells (Fig. 3.1; see Autenrieth and Firsching, 1996; Perry and Fetherston, 1997). Once *Yersinia* has entered the lymphoid system, it overcomes the primary immune response of the host by using the type III secretion system (TTSS) (Cornelis et al., 1998; Cornelis, 2002). TTSS is a sophisticated virulence mechanism by which Gram-negative pathogens inject effector proteins directly into host cells. Currently, more than 20 different TTSSs have been described in animal, plant, and even insect pathogens (Hueck, 1998; Galan and Collmer, 1999; Cornelis, 2000; Buttner and Bonas, 2002).

Depending on the effectors injected, the employment of the TTSS will have a different outcome. Some, like the Mxi–Spa system of *Shigella flexneri* or the *Salmonella* pathogenicity island 1 (SPI-1) system of *Salmonella enterica*, make use of the innate immune system of the host to enhance the proinflammatory response and to trigger phagocytosis by normally nonphagocytic cells, whereas others, such as the pathogenic *Yersinia* Ysc–Yop system, essentially paralyze the innate immune response of the host (Galan, 2001; Sansonetti, 2001; Cornelis, 2002; Juris et al., 2002). The *Yersinia* TTSS becomes activated upon contact with eukaryotic cells and directs effector proteins – called Yops – over the bacterial membranes. Some of the Yops form a kind of

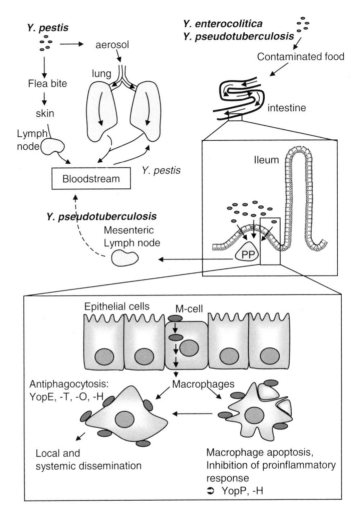

Figure 3.1. Model showing the interaction of *Yersinia* with an eukaryotic cell.

At 37°C and upon contact of *Yersinia* with its eukaryotic target cell, the adhesins YadA or Inv interact with the β_1-integrins and other extracellular matrix proteins at the cell surface, and *Yersinia* attaches tightly to the cell membrane. The Ysc injectisome is installed and the Yop translocators and effectors, some of which are intrabacterially capped with a chaperone, are transported through the bacterial inner and outer membranes by the Ysc injectisome. The translocators Yops, YopB, YopD, and LcrV form a pore in the target cell membrane, through which the effector Yops are translocated into the cell cytosol. Four effector Yops with different enzymatic functions (YopE, a Rho GTPase-activating protein, YopT, a cysteine protease, YopO/YpkA, a serine/threonine kinase, and YopH, a protein tyrosine phosphatase) will cooperatively lead to the destruction of the actin cytoskeleton,

(*cont.*)

translocation pore in the eukaryotic target cell membrane, whereas the other Yops are effector proteins that are delivered through this pore into the cytosol of the target cell. At least six different Yop effectors are injected by the secretion translocation apparatus, five of which have been shown to play an important role in the defense against the innate immune response (Fig. 3.2). In this chapter we discuss how *Yersinia* escapes the host immune response after initial invasion of the intestinal mucosa by (i) its strong resistance to phagocytosis, which is caused by the concerted action of YopE, YopH, YopO (called YpkA in *Y. pseudotuberculosis* and *Y. pestis*), and YopT, (ii) its capacity to block the proinflammatory response induced by different immune cells, caused by the action of YopP (called YopJ in *Y. pseudotuberculosis* and *Y. pestis*) and YopH, and (iii) its ability to induce cell death of the macrophage, promoted by YopP/J (DeVinney et al., 2000; Aepfelbacher and Heesemann, 2001; Cornelis, 2002; Juris et al., 2002; Orth, 2002).

THE *YERSINIA* VIRULENCE SYSTEM

Ysc–Yop III secretion system

The Ysc–Yop type III secretion machinery present in all pathogenic *Yersinia* is encoded by a 70-kDa virulence plasmid, which harbors the genes for the Ysc (for Yop secretion) secretion apparatus or Ysc injectisome, for an array of proteins secreted by this apparatus – called Yops (for *Yersinia* outer proteins) – and for a set of proteins controlling the system (Cornelis et al., 1998). The Ysc injectisome is composed of a large dual-ring structure spanning the bacterial inner and outer membrane, which resembles the flagellum basal body. This structure is associated with a needle-like complex that extends outside the bacterium. It mediates the secretion of the Yop effector proteins (YopE, YopH, YopO/YpkA, YopT, YopP/J, and YopM), a structural component of the needle (YscF), and the components of a translocation apparatus, which are YopB, YopD, and LcrV (Cornelis et al., 1998; Cornelis, 2002; also see Fig. 3.2). The latter proteins are inserted into the host membranes and form a kind of pore, which allows the delivery of the Yop effector proteins into the cytosol of the cell. Whether the Ysc injectisome and the translocation

Figure 3.1. (*cont.*) and by doing so contribute to the antiphagocytic action of *Yersinia*. Two Yops (YopH and YopP/J) are involved in the downregulation of the proinflammatory response of the immune cells, and YopP/J will also lead to the induction of apoptosis in macrophages. YopM, a protein containing several leucine-rich repeats, is translocated to the nucleus; however, its function remains unclear. PP = Peyer's patches.

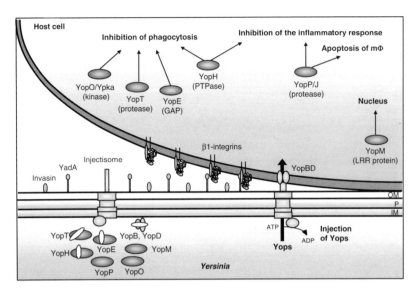

Figure 3.2. Schematic representation of entry routes during *Yersinia* infection in humans. *Y. pestis* is generally inoculated by fleabites or aerosol and enters the bloodstream directly. *Y. enterocolitica* and *Y. pseudotuberculosis* are both foodborne pathogens and enter the underlying lymphoid tissue (e.g., Peyer's patches) of the intestinal mucosa through M cells. Once *Yersinia* has reached the lymphoid system, the plasmid-encoded effector Yops allow the bacteria to avoid phagocytosis and actively downregulate the proinflammatory response in order to promote their extracellular survival. OM = outer membrane; P = periplasm; IM = inner membrane; LRR = leucine-rich repeat.

pore form a continuous channel connected by the needle is currently being investigated.

Type III-dependent protein secretion in *Yersinia* is a tightly regulated process, and several regulatory circuits control both the expression of the injection system and the injection of the Yop effector proteins itself. The first level of regulation involves temperature. Although growth of *Yersinia* is unaffected by low temperatures, such as those found in contaminated food (*Y. enterocolitica* and *Y. pseudotuberculosis*) or the stomach of the fleas (*Y. pestis*), the expression of Yop effector proteins is repressed at these low temperatures. It is only at 37°C that a stock of intracellular Yops is synthesized and the Ysc injectisome is installed. However, the injectisome remains closed and a mechanism of feedback inhibition prevents a deleterious accumulation of Yops (Cornelis et al., 1987). This first level of regulation involves at least two proteins: a plasmid-encoded transcriptional activator, VirF, and a chromosome-encoded histone-like protein, YmoA (Cornelis et al., 1991;

Lambert de Rouvroit et al., 1992; Rohde et al., 1999). A second level of regulation is close contact with the host cell membrane, which is established by the bacterial adhesins Inv, YadA, and Ail (Pettersson et al., 1996). It is only at 37°C and upon contact with the eukaryotic cell that the injectisome is opened, the negative feedback regulation is relieved, and *Yersinia* starts to inject its effectors into the cytosol (Francis et al., 2002; Miller, 2002). Yop proteins destined to be secreted have no classical signal sequence that is cleaved off during secretion, but nevertheless their N-terminal part (~15 amino acids or codons) contains the information that is necessary for secretion (Michiels et al., 1990; Sory et al., 1995; Anderson and Schneewind, 1997).

Furthermore, some secreted proteins require the binding of a specialized cytosolic chaperone, called Syc (specific _Yersinia_ _chaperone), to be secreted (Wattiau et al., 1996). The loss of a chaperone results in the inefficient secretion of its cognate partner, while the secretion of other proteins remains unaffected. Chaperones do not share a general homology, but they do share common properties of being small (less than 20 kDa), being acidic (pI ~4–5), and possessing a C-terminal amphipatic helix. The Syc chaperones have been proposed to act as (i) bodyguards, preventing degradation or premature association of their target; (ii) secretion pilots, being part of the signal for recognition of their substrates by the export machinery; (iii) hierarchy factors, establishing a hierarchy for Yop delivery into its host cell; or (iv) antifolding factors, maintaining their substrate in a secretion-competent state (Frithz-Lindsten et al., 1995; Cheng and Schneewind, 1999; Boyd et al., 2000; Stebbins and Galan, 2001; Lee and Schneewind, 2002; Wulff-Strobel et al., 2002; Feldman et al., 2002). However, at this stage it is premature to decide which of these functions accounts for the need of chaperones.

Other virulence mechanisms and TTSS present in *Yersinia* species

In addition to the 70-kb virulence plasmid encoding the Ysc–Yop TTSS, *Y. pestis* harbors two unique plasmids encoding essential virulence determinants. The 9.5-kb plasmid (pPst/pPCP1) contains the Pla protease, which enables the spread of *Y. pestis* from subcutaneous infection sites into the circulation (Perry and Fetherston, 1997). Pla has been shown to exhibit coagulase activity and can also activate plasminogen into plasmin. It has also been reported that Pla can serve as an adhesion-promoting factor for *Y. pestis* (Cowan et al., 2000). The 100- to 110-kb plasmid (pFra/pMT1) encodes the murine toxin Ymt, a phospholipase D family member, and the fraction 1 (F1) capsule-like protein (Perry and Fetherston, 1997). Ymt has recently been shown to be an essential factor for the survival of *Y. pestis* in the midgut of the flea (Hinnebusch et al., 2002), and F1 has been suggested to be involved in the antiphagocytic activity of *Y. pestis* and to reduce the number of bacteria that interact with the macrophages (Du et al., 2002).

In mammals, the level of free iron is too low to sustain bacterial growth; therefore, pathogens possess siderophores that can solubilize the iron bound to host proteins and transport it to the bacteria. *Y. pestis, Y. pseudotuberculosis*, and high-virulence *Y. enterocolitica* strains carry a chromosomally encoded high pathogenicity island (HPI), which comprises genes involved in the synthesis of a siderophore called yersiniabactin (Heesemann et al., 1993; Carniel, 2001). This capacity to acquire iron is an essential virulence determinant for the invading *Yersinia* bacteria, and it endows them with the ability to multiply in the host and cause systemic infections.

Recently a second TTSS of *Y. enterocolitica*, called Ysa (for <u>Y</u>ersinia <u>s</u>ecretion <u>a</u>pparatus) and its substrates for secretion – Ysp proteins – has been described (Haller et al., 2000; Foultier et al., 2002). Interestingly, the chromosome-encoded Ysa–Ysp TTSS of *Y. enterocolitica* is similar to the Mxi–Spa TTSS of *Shigella* and to the SPI-1 encoded TTSS of *S. enterica*, but it is different from another chromosome-encoded TTSS of *Y. pestis* (Parkhill et al., 2001). In addition, the *ysa* locus is only present in the high-virulence biotype 1B strains of *Y. enterocolitica* and, at least in laboratory conditions, is only operational at low temperature (Haller et al., 2000; Foultier et al., 2002). Whether this Ysa TTSS plays a role in the high-virulence phenotype of *Y. enterocolitica* or in a yet to be identified cold-blooded host is unclear at the moment and awaits further *in vivo* experiments.

FIRST CONTACT

Interaction of the enteropathogenic *Y. enterocolitica* and *Y. pseudotuberculosis* with M cells

Y. enterocolitica and *Y. pseudotuberculosis* possess two different adhesins: the chromosomally encoded Inv (Invasin) and the pYV plasmid-encoded YadA (<u>Y</u>ersinia <u>a</u>dherence protein <u>A</u>; see Boland and Cornelis, 2000). They mediate initial adhesion, uptake, and translocation of the bacteria through the M cells, covering the Peyer's patches, to the underlying lymphoid tissues, where the bacteria remain extracellular, multiply, and eventually migrate to deeper tissues such as liver and spleen (Fig. 3.1; also see Sansonetti, 2002). The Inv protein has been shown to be important for the initial step of invasion by its strong interaction with host β_1-integrin expressed on the apical membranes of the M cells (Pepe and Miller, 1993; Berton and Lowell, 1999; Schulte et al., 2000). The cytoplasmic domain of integrins will transmit signals to the cell cytoskeleton that mediate internalization of *Yersinia* by a "zippering" process (Isberg et al., 2000). As epithelial cells only express integrins at their basal membrane, the enterocytes are not expected to be heavily invaded during oral infection. Indeed, an analysis of intestines of infected mice shows that *Y. enterocolitica* is only found in sections that contain Peyer's patches. This indicates that M cells, rather than enterocytes, form the major port of entry for *Yersinia*.

After this initial step of invasion, the YadA protein seems to be the predominant adhesin, mediating adherence through interaction with extracellular matrix proteins such as fibronectin and collagen (El Tahir and Skurnik, 2001). YadA also protects *Y. enterocolitica* against the bactericidal and opsonizing action of complement by binding complement factor H (China et al., 1993). Once the dome is reached, Yersiniae survive attack by professional macrophages by injecting antiphagocytic Yops (see what follows) that disrupt the cytoskeleton. *Yersinia* will thus essentially remain extracellular, which allows its survival and possible Inv-mediated entry into nonphagocytic cells, but this is not well documented.

Y. pestis enters the bloodstream immediately

Y. pestis is a pathogen primarily affecting rodents, which is usually transmitted to humans by a fleabite (Fig. 3.1). When a flea ingests a blood meal harboring *Y. pestis*, the ingested *Yersinia* secretes a coagulase that clots the blood and thus prevents the flea from swallowing the bacteria. Ymt, a plasmid-encoded and intrabacterially expressed phospholipase D, protects the bacterium from a cytotoxic digestion product of blood plasma in the flea gut (Hinnebusch et al., 1998; Hinnebusch et al., 2002). After multiplying in the clotted blood, *Y. pestis* is transmitted efficiently into a human host when the hungry flea repeatedly attempts to feed and the blood clot is regurgitated into the host (Perry and Fetherston, 1997; Cole and Buchrieser, 2001). The bacterium then spreads from the site of infection to the regional lymph nodes, where it grows to high numbers and causes swelling of the lymph node (bubo), resulting in bubonic plague. If the lymphatic system becomes overwhelmed, the infection rapidly spreads into the lymphstream and bloodstream, causing fatal blood poisoning, followed by colonization of all the main organs (including the lungs). It is notable that *Y. pestis* lacks functional YadA and Inv, which are present in its enteropathogenic relatives. However, some studies indicate that *Y. pestis* may invade and cause systemic infection from digestive and aerogenic routes of infection.

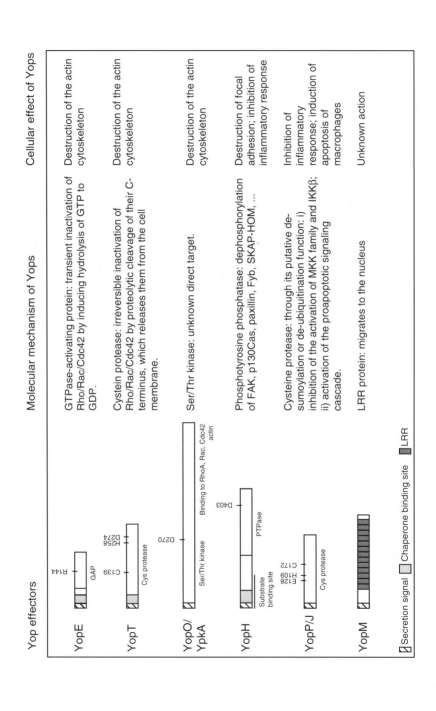

Yop effectors   Molecular mechanism of Yops   Cellular effect of Yops

YopE — GTPase-activating protein: transient inactivation of Rho/Rac/Cdc42 by inducing hydrolysis of GTP to GDP. — Destruction of the actin cytoskeleton

YopT — Cystein protease: irreversible inactivation of Rho/Rac/Cdc42 by proteolytic cleavage of their C-terminus, which releases them from the cell membrane. — Destruction of the actin cytoskeleton

YopO/YpkA — Ser/Thr kinase: unknown direct target. — Destruction of the actin cytoskeleton

YopH — Phosphotyrosine phosphatase: dephosphorylation of FAK, p130Cas, paxillin, Fyb, SKAP-HOM, ... — Destruction of focal adhesion; inhibition of inflammatory response

YopP/J — Cysteine protease: through its putative de-sumoylation or de-ubiquitination function: i) inhibition of the activation of MKK family and IKKβ; ii) activation of the proapoptotic signaling cascade. — Inhibition of inflammatory response; induction of apoptosis of macrophages

YopM — LRR protein: migrates to the nucleus — Unknown action

Secretion signal   Chaperone binding site   LRR

Inhibition of phagocytosis

When a nonpathogenic bacterium enters the host organism, it is usually engulfed by professional phagocytes, such as macrophages, neutrophils, or dendritic cells (May and Machesky, 2001; Underhill and Ozinsky, 2002a). Phagocytosis of a bacterium is usually preceded by the activation of many signaling pathways, causing rearrangement of the actin cytoskeleton, extension of the plasma membrane, and finally engulfment. Members of the Rho family GTPases (Cdc42, Rac, and Rho) play a central role in this process, as they are key regulators of the actin cytoskeleton dynamics associated with adhesion, membrane ruffling, and stress fiber formation (Hall, 1998; Bar-Sagi and Hall, 2000; Chimini and Chavrier, 2000). Furthermore, the formation of focal adhesion complexes, mediated by the action of paxillin, p130Cas, and focal adhesion kinase (FAK), at points of contact with bacteria may also play a role in phagocytosis (Greenberg et al., 1993; Allen and Aderem, 1996; Berton and Lowell, 1999). Finally, the phosphoinositide 3-kinase (PI3K), phospholipase C (PLC), and protein kinase C (PKC) signaling pathways are integration points for regulating phagocytosis. Pathogenic Yersiniae subvert several of these pathways by injecting YopE, YopT, YopO/YpkA, and YopH. This ensures an extracellular lifestyle and propagation in the host. It was recently demonstrated that deletion of either YopE, YopT, YopO/YpkA, or YopH renders *Yersinia* more susceptible to phagocytosis by macrophages and PMNs, and thus that the concerted action of all four antiphagocytic Yops is necessary for full protection against phagocytosis (Grosdent et al., 2002).

YopE

Translocation of YopE into mammalian cells leads to a cytotoxic response, characterized by rounding up of the cells and detachment from the extracellular matrix (Rosqvist et al., 1990). YopE is one of the earliest identified Yop effectors that has an inhibitory effect on the actin cytoskeleton by inactivation of the Rho family GTPases (Figs. 3.3 and 3.4A; also see Black and

---

Figure 3.3. (*facing page*). Structural diagrams of the Yop effectors and their enzymatic function. Except for YopM, which is an LRR protein of unknown function, the enzymatic function of all the effector Yops has been identified, as indicated. The amino acids important for their catalytic function are depicted. All six Yop effectors have a short N-terminal secretion signal (~15 amino acids or codons), which is necessary for their secretion. Three of them (YopE, YopT, and YopH) also contain a chaperone-binding site.

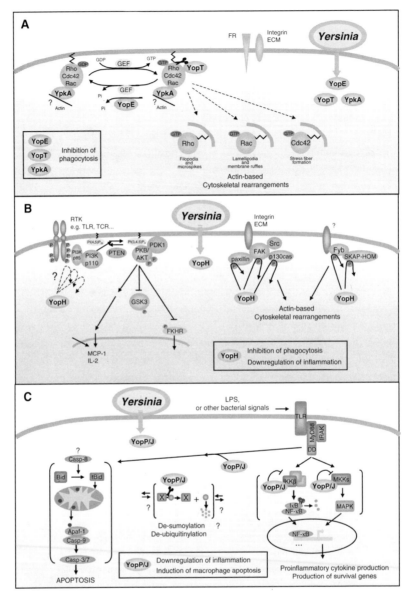

Figure 3.4. Molecular mechanism of the Yop effectors in the host cell. (A) Antiphagocytic action of YopE, YopT, and YopO/YpkA. Upon contact of *Yersinia* with a phagocytic receptor, the Rho family members (Rho, Rac, and Cdc42) are targeted to the cell membrane and converted to their GTP-activated state, which promotes actin polymerization and facilitates phagocytosis. YopE, acting as a GAP, will transiently

(*cont.*)

Bliska, 2000; Von Pawel-Rammingen et al., 2000). Rho GTPases are regulated at different levels (Hall, 1998; Ridley, 2001). Cytosolic GDP-bound Rho proteins are normally posttranslationally modified by prenylation at the C-terminus, which is important for their translocation to the cell membrane. The guanine nucleotide exchange factors (GEFs) induce the release of bound

---

Figure 3.4. (*cont.*) downregulate Rho, Rac, and Cdc42 by converting them into the inactive GDP-bound state. The YopT cysteine protease cleaves the C-terminus of Rho, Rac, and Cdc42, removing the prenyl group of the Rho GTPases and liberating them from the plasma membrane. The YopO/YpkA serine/threonine kinase becomes autophosphorylated upon contact with actin and interacts with Rho and Rac; however, its real cellular target is unknown. The concerted action of these three Yops will lead to a destruction of the actin cytoskeleton network and in this way inhibit phagocytosis. FR, phagocytic receptor; ECM, extracellular matrix. (B) Antiphagocytic and anti-inflammatory action of YopH. The interaction between *Yersinia* and the eukaryotic cell surface causes a rapid tyrosine phosphorylation of adhesion complexes to mediate uptake of *Yersinia*. To exert its antiphagocytic role, the phosphotyrosine phosphatase YopH is targeted to the focal adhesion complexes, where it dephosphorylates proteins such as the FAK, p130Cas, and paxillin, and to other not yet well-characterized adhesion-regulated complexes in macrophages to dephosphorylate Fyb and SKAP-HOM. By inhibition of the PI3K/Akt signaling pathway, YopH also contributes to the downregulation of the inflammatory response. Upon infection, Akt is activated and will phosphorylate several proteins that are involved in apoptosis, cellular proliferation, and cytokine/chemokine production, such as glycogen synthase kinase 3 (GSK3) or the transcription factors of the forkhead family (e.g., FKHR), and locks them in their phosphorylated inactive state. Inhibition of this pathway by YopH is presumably responsible for the inhibition of T-cell proliferation, IL-2 production, and MCP-1 production. PTEN, phosphatase and tensin homologue deleted on chromosome 10; PDK1, phosphoinositide-dependent kinase-1; RTK, receptor tyrosine kinase; TCR, T-cell receptor. (C) Model showing the anti-inflammatory and proapoptotic role of YopP/J. YopP/J downregulates of the inflammatory response by binding to and preventing the activation of members of the MAP kinase kinase (MKK) family and of IKKβ. By blocking both these pathways, YopP/J efficiently shuts down multiple kinase cascades and the cytokine induction required by the host cell to respond to a bacterial infection. YopP/J is also responsible for the induction of apoptosis of macrophages, which probably involves both the downregulation of survival genes and the activation of the apoptotic cascade upstream of Bid, presumably by interfering with a signaling pathway triggered from the TLRs. The cysteine protease activity of YopP/J is necessary for both the downregulation of the inflammatory response and the induction of apoptosis of macrophages. However, exactly how the YopP-de-sumoylating (de-ubiquitinylating?) activity is interrelated with the inhibition of the MKKs and IKKβ and the induction of apoptotic pathways awaits further research. See color section.

GDP and thereby allow binding of GTP. This results in the activation of the Rho proteins at the cell membrane and binding to their downstream target. Inactivation of Rho GTPases is regulated by guanine nucleotide dissociation inhibitors (GDIs) and GTPase-activating proteins (GAPs).

The former produce an inactive complex with the Rho proteins in the cytosol by masking the prenyl group, and the latter induce the hydrolysis of the bound GTP to GDP, thereby returning the Rho proteins to their inactive form. During phagocytosis the reorganization of the actin cytoskeleton is orchestrated by these Rho family GTPases: Rho controls stress fiber formation and actin-myosin-based contractility; Cdc42 drives the formation of actin-rich filopodia; and Rac promotes the formation of lamellipodia and membrane ruffles (Hall, 1998). Therefore, they represent ideal targets for bacterial virulence factors, as their inactivation would block phagocytosis and allow the extracellular survival of bacterial. The C-terminal effector domain of YopE mimics the activity of eukaryotic GAP, which results in a transient downregulation of the Rho GTPases, and in this way leads to the disruption of the actin cytoskeleton and consequently the inhibition of phagocytosis (Black and Bliska, 2000; Von Pawel-Rammingen et al., 2000).

However, although it was shown *in vitro* that YopE has GAP activity toward Rho, Rac, and Cdc42, whether these three Rho proteins are all inactivated in every cell type, or whether there might be other Rho family members that could be *in vivo* substrates, awaits further analysis. One clue for the specificity of YopE came from a study on human umbilical vein endothelial cells, where it was shown that YopE acted selectively on the Rac-mediated pathways but had no effect on Cdc42- or Rho-dependent signaling (Andor et al., 2001). It should be noted that YopE shares a high degree of structural similarity with the GAP domains of Exoenzyme S (ExoS) of *Pseudomonas aeruginosa* and SptP from *S. typhimurium*, but it has no obvious structural similarity with known mammalian functional GAP homologs (Evdokimov et al., 2002), suggesting that they could have evolved separately.

As mentioned before, the *Yersinia* type III weapon includes a pore, necessary to translocate the effectors into the host (Cornelis et al., 1998; Cornelis, 2002). In a current model the translocation pore is filled by the Yop effectors themselves (Hakansson et al., 1996b). However, recently it has been proposed that apart from its antiphagocytic role, injected YopE would also play a role in minimizing plasma membrane damage caused by pore formation (Viboud and Bliska, 2001). The GAP function of YopE was demonstrated to be necessary in preventing pore formation, suggesting that pore formation itself needs the activation of Rho GTPases.

YopT

The most recently identified effector modulating the Rho family of GTPases is YopT (Figs. 3.3 and 3.4A). Infection of mammalian cells with a *Y. enterocolitica* strain only expressing the YopT effector leads to rounding up of the cell and disruption of the cytoskeleton, which contributes to the antiphagocytic activity of YopT (Iriarte and Cornelis, 1998; Grosdent et al., 2002). Translocation of YopT into host cells leads to a modification of RhoA, resulting in an acidic shift in its pI and redistribution of membrane-bound RhoA toward the cytosol (Zumbihl et al., 1999). In addition, incubation of purified cell membranes or artificial lipid vesicles containing RhoA with purified YopT leads to the release of RhoA to the supernatant (Sorg et al., 2001). The mechanism of action of YopT was recently unraveled by Shao and coworkers, who demonstrated that *Yersinia* YopT, as well as its homologue AvrPphB from *P. aeruginosa*, belong to a family of cysteine proteases (Shao et al., 2002). YopT recognizes the posttranslational modified Rho GTPases (Rho, Rac, and Cdc42) and proteolytically cleaves them near the C-terminus, which leads to their release from the cell membrane. This cleavage removes the prenyl group of the Rho GTPases and results in an irreversible inactivation of the targeted Rho GTPases, whereas the action of YopE (GAP) can be reverted by the GEFs within the cell.

YopO/YpkA

The YopO/YpkA effector is an autophosphorylating serine/threonine protein kinase that modulates the cytoskeleton dynamics (Figs. 3.3 and 3.4A) and also contributes to resistance to phagocytosis (Galyov et al., 1993; Hakansson et al., 1996a; Grosdent et al., 2002). YopO/YpkA is produced as an inactive kinase, which becomes activated after translocation into the host cell upon binding to actin (Dukuzumuremyi et al., 2000; Juris et al., 2000). The N-terminal part of YopO/YpkA contains the kinase domain, whereas the C-terminal part of the kinase binds to actin. The C-terminal part also contains sequences that bear similarity to several eukaryotic RhoA-binding kinases, and it binds to RhoA and Rac but not to Cdc42 (Barz et al., 2000; Dukuzumuremyi et al., 2000). The kinase domain is required to localize YopO/YpkA to the plasma membrane, whereas the C-terminal part is responsible for its effect on the actin cytoskeleton in *Yersinia*-infected cells (Dukuzumuremyi et al., 2000; Juris et al., 2000). However, although YopO/YpkA has a clear effect on the actin cytoskeleton, its real cellular target is unknown and awaits further investigation.

YopH

The fourth antiphagocytic Yop is the multifunctional YopH. The 51-kDa YopH effector protein is composed of two functional domains: the C-terminal part (residues 206–468) has a structure similar to the one of mammalian phosphotyrosine phosphatases (PTPases; see Guan and Dixon, 1990), and the N-terminal part (residues 1–130) is the binding site of YopH to its substrate (Fig. 3.3; also see Black et al., 1998). Part of this latter domain (residues 20–69) is also the binding domain for its chaperone SycH, necessary for translocation into the eukaryotic cell (Wattiau et al., 1996). Upon interaction of the *Yersinia* surface protein Inv with β_1-integrin on the cell surface of epithelial cells, there is a rapid tyrosine phosphorylation of proteins of focal adhesion complexes (Persson et al., 1997).

Translocation of YopH into epithelial cells leads to the dephosphorylation of proteins from these focal adhesion complexes, such as the docking proteins p130Cas and paxillin and the FAK (Black and Bliska, 1997; Persson et al., 1997). Similarly, in macrophages, contact also induces tyrosine phosphorylation of proteins of adhesion complexes (Andersson et al., 1996). Targeting of YopH into macrophages also leads to a rapid dephosphorylation of p130Cas and paxillin (Hamid et al., 1999). In addition, in macrophages YopH will also lead to dephosphorylation of the Fyn-binding protein (FBP) and the scaffolding protein SKAP-HOM, which have been shown to be part of a novel adhesion-regulated signaling complex (Hamid et al., 1999; Black et al., 2000). Dephosphorylation of these proteins contributes to the antiphagocytic activity of YopH (Fig. 3.4B).

Besides its role as an antiphagocytic factor, YopH has also been shown to interfere in other signaling pathways of the immune defense system, such as downregulating the Fc-mediated oxidative burst in macrophages and neutrophils (Bliska and Black, 1995; Ruckdeschel et al., 1996) and blocking calcium signaling in neutrophils (Andersson et al., 1999). Finally, during infection of macrophages with *Y. enterocolitica*, the PI3K/Akt pathway is rapidly activated and then inactivated in a YopH-dependent way, which possibly contributes to an anti-inflammatory role of YopH (Fig. 3.4B; also see following subsection; also Sauvonnet et al., 2002a).

Inhibition of the inflammatory response and induction of apoptosis

When a pathogen interacts with a mammalian cell, multiple receptors will simultaneously recognize these pathogens both through direct binding and by binding to opsonins on the microbe surface (Underhill and Ozinsky,

2002a). One of the key mediators of microbe detection is the Toll-like receptor (TLR) family, which plays an important role in signaling toward inflammation and apoptosis (Akira et al., 2001; Imler and Hoffmann, 2001). Ten mammalian TLRs now have been identified. Five of them (TLR2, TLR4, TLR5, TLR6, and TLR9) have been shown to respond to an array of different microbial components, such as lipopolysaccharide (LPS), lipopeptides, peptidoglycans (PGNs), lipoteichoic acid (LTA), flagellin, and CpG motifs in DNA. By using a combination of different invariant TLRs, the immune system can recognize a broad spectrum of pathogens. Stimulation of TLR2 and TLR4 leads to the recruitment of the adaptor molecule MyD88 and the serine kinase IL-1-receptor-associated kinase (IRAK; see Underhill and Ozinsky, 2002b). Together with TRAF-6, this multiprotein assembly mediates the activation of (i) the IκB kinase (IKK) complex, which leads to activation of the nuclear factor κB (NF-κB), and (ii) the mitogen-activated protein kinase (MAPK) kinase family (MKKs), which also leads to the activation of different transcription factors, such as activator protein-1 (AP-1).

YopP/J

As a Gram-negative bacterium, *Yersinia* is endowed with several components capable of activating the TLR system and stimulating a proinflammatory response. Indeed, during *Yersinia* infection of macrophages, the proinflammatory response is initially upregulated; however this is quickly counteracted by the YopP/J effector protein (Ruckdeschel et al., 1997, 1998). YopP/J has been shown to cause a variety of anti-inflammatory effects *in vitro*, such as suppression of tumor necrosis factor-α (TNF-α) and interleukin-6 (IL-6) and IL-8 production; downregulation of intercellular adhesion molecule-1 (ICAM-1) expression; and blocking of the activation of MAPK pathways, including extracellular signal-regulated kinases (ERK), p38, jun amino-terminal kinase (JNK), and NF-κB (Boland and Cornelis, 1998; Palmer et al., 1998; Schesser et al., 1998; Palmer et al., 1999; Denecker et al., 2002). The first clue as to how YopP/J could simultaneously block these multiple signaling pathways was elucidated by Orth et al. (1999; also see Fig. 3.4C).

In a two-hybrid system, YopJ from *Y. pseudotuberculosis* bound multiple members of the MAPK kinase superfamily, including MKKs and IKKβ, thereby preventing their phosphorylation and subsequent activation. The interaction of its counterpart YopP from *Y. enterocolitica* was confirmed by coimmunoprecipitation experiments in HEK293T cells and macrophages (Denecker et al., 2001; Ruckdeschel et al., 2001a). By blocking both the conserved family of MKKs and IKKβ, YopP/J efficiently shuts down multiple kinase cascades, resulting in a downregulation of the inflammatory response of its host cells.

The second clue about the mechanism of action of YopP/J was based on structural similarities. It has been suggested that YopP/J belongs to a family of cysteine proteases related to the ubiquitin-like protein proteases (Figs. 3.3 and 3.4B; also see Orth et al., 2000; Orth, 2002). Amino acid alignment of the adenoviral protease AVP, YopP/J, its effector homologues, and Ulp1, a yeast ubiquitin-like protease, revealed the catalytic triad necessary for the cysteine protease activity, which is conserved between all these proteins. Mutation of the YopP/J hypothetical catalytic cysteine-172, which presumably results in the loss of its cysteine protease activity, hampers its capacity to inhibit the NF-κB and MAPK signaling cascades (Orth et al., 2000; Denecker et al., 2001). Ubiquitin-like protein proteases cleave the C-terminus of an 11-kDa small ubiquitin-related modifier, SUMO-1 (Yeh et al., 2000). YopJ has been shown to reduce the cellular concentration of SUMO-1-conjugated proteins in an overexpression experiment; however, no direct substrate of YopJ has been identified (Orth et al., 2000). Furthermore, it was recently suggested by Orth (2002) that overexpression of YopJ also leads to a decrease in ubiquitinated proteins. Thus the exact mechanism of blockage of phosphorylation of MKKs and IKKβ through the cysteine protease function of YopP/J remains unresolved.

In addition, it is possible that an additional cytosolic factor is needed for functional activity: (i) with the use of *in vitro* kinase reaction experiments, it was not possible to demonstrate that YopJ could prevent phosphorylation of MKK1 (Orth et al., 1999); (ii) recombinant protein preparations of YopJ were catalytically inactive when assayed with a variety of radiolabeled or fluorometric peptides (Orth et al., 2000); and (iii) the viral AVP protease also requires an additional cofactor (Mangel et al., 1993). Besides cysteine-172, it was recently demonstrated that arginine-143, present in all high-virulence *Y. enterocolitica* strains and in YopJ (*Y. pestis* and *Y. pseudotuberculosis*), plays a major role in determining the inhibitory impact of YopP on the suppression of NF-κB activation and survival of macrophages (Ruckdeschel et al., 2001b; Denecker et al., 2002).

YopP/J is not only responsible for the inhibition of the inflammatory response of the host but also induces apoptosis in macrophages, although not in other cell types (Mills et al., 1997; Monack et al., 1997). Two different mechanisms of YopP/J-dependent apoptosis have been proposed. In the first hypothesis, YopP/J would act as a direct activator of the cell death machinery, which involves an early – presumably caspase-8-dependent – cleavage of Bid to its proapoptotic truncated form, followed by the release of cytochrome c from the mitochondria, leading to the activation of procaspase-9, -3, and -7 (Denecker et al., 2001). The point at which YopP/J interferes with the apoptotic signaling cascade is a subject for future studies. In the second hypothesis, YopP/J-induced apoptosis of macrophages would merely result from its inhibition of NF-κB activation, thus blocking the host cell survival pathways, in combination with TLR stimulation (Ruckdeschel et al., 1998, 2001a, 2002). The two hypotheses could be combined in a model in which the YopP/J targets both antiapoptotic and proapoptotic pathways in macrophages (Fig. 3.4C). In this model, YopP/J might downregulate the expression of some antiapoptotic genes and at the same time alter a TLR-dependent signaling cascade in a manner allowing procaspase-8 activation and subsequent Bid processing.

YopH

YopH could also contribute to the downregulation of the inflammatory response (Fig. 3.4B). Indeed, YopH has recently been shown to suppress the *Yersinia*-induced activation of the PI3K/Akt signaling pathway, which could be correlated with the downregulation of monocyte chemoattractant protein-1 (MCP-1) mRNA levels (upregulation of MCP-1 is dependent on the PI3K/Akt signaling cascade; see Alberta et al., 1999; Scheid and Woodgett, 2001; Sauvonnet et al., 2002a). By inhibition of MCP-1 production, YopH would inhibit the recruitment of other macrophages to the site of infection, which would allow *Yersinia* to colonize the lymphoid system. Besides its role in the innate immune system, YopH also contributes to the downregulation of the adaptive immune response. It was demonstrated that T-cell cytokine production and proliferation, and expression of the B-cell costimulatory receptor B7.2, in response to antigen stimulation were inhibited after transient exposure to *Yersinia* (Yao et al., 1999; Sauvonnet et al., 2002a). This inhibition of antigen-specific T- and B-cell activation occurred in a YopH-dependent way by interfering with the phosphorylation of tyrosine-phosphorylated components associated with the T- and B-cell antigen receptor signaling complex (e.g., Fyb and SKAP-HOM), and most probably also by interfering with the PI3K/Akt pathway (Fig. 3.4B; also see Hamid et al., 1999; Yao et al., 1999; Black et al., 2000; Sauvonnet et al., 2002a). Thus, *Yersinia* possesses different elements that have the capacity to downregulate the inflammatory response. The relevance of the anti-inflammatory role played by these different elements during infection is a matter for future *in vivo* studies.

Effectors other than Yops that help to defeat the immune response

Analysis of the transcriptome alterations in infected mouse macrophages revealed that several genes involved in the inflammatory response of a macrophage to a bacterial infection are downregulated by the action of pYV-encoded factors other than YopP (Sauvonnet et al., 2002b). As already mentioned, YopH is a good candidate (Yao et al., 1999; Sauvonnet et al., 2002a). In addition, LcrV may represent another factor, as it was recently demonstrated that LcrV-induced IL-10 release could inhibit TNF-α production in zymosan A-stimulated macrophages (Sing et al., 2002). According to the currently accepted type III-secretion-translocation model (Cornelis et al., 1998), LcrV is part of the translocation machinery that delivers Yop effectors into the eukaryotic cell, but this does not exclude a possible role of its own, independent of the rest of the injectisome.

In addition, for the induction of cell death, YopP might not be the sole factor. For *Y. pseudotuberculosis* it was recently demonstrated that the Inv protein could cause a rapid apoptotic–necrotic caspase-independent cell death in T lymphocytes (Arencibia et al., 2002). This process was mediated by means of β1-integrins and was independent of the Yop–Ysc TTSS of *Yersinia*.

YopM

YopM belongs to a growing family of type III effectors that has several representatives in *Shigella* (*ipaH* multigene family) and *Salmonella* (*SspH*; see Kobe and Kajava, 2001). It is a strongly acidic protein composed almost entirely of 20/22 residue leucine-rich repeats (LRR; see Fig. 3.3). The repeating LRR unit of YopM is the shortest among all LRRs known to date, and, depending on the *Yersinia* species, the amount of copies can vary between 13 and 20 repeats. The crystal structure has revealed that the LRRs, consisting of parallel β-sheets, form a crescent shape, which is flanked by an α-helical hairpin at the N-terminus (Evdokimov et al., 2001). The latter domain has been shown to be part of the signal necessary to target YopM for translocation into eukaryotic cells (Boland et al., 1996). Intriguingly, individual YopM molecules form a tetramer in the crystal, creating a hollow cylinder with an inner diameter of 35 Å (Evdokimov et al., 2001). YopM has been shown to traffic to the nucleus via a vesicle-associated pathway, but its action in the nucleus remains unknown (Skrzypek et al., 1998). New insights about the role of YopM came from a recent study in which a transcriptome analysis of *Yersinia*-infected macrophages revealed that YopM may control the expression of genes involved in the cell cycle and in cell growth (Sauvonnet et al., 2002b).

CONCLUSION

After initial invasion of its host cells, pathogenic *Yersinia* remain extra-cellular because they have the capacity to resist phagocytosis. This resistance depends on the type III Ysc–Yop system, which upon close contact with a target cell injects six different effector Yops into the cytosol of the cell. As a result of this "tranquilizing" injection, phagocytosis is inhibited, the onset of the proinflammatory response is slowed down, and most probably lymphocyte proliferation is prevented. Four Yop effectors (YopE, YopT, YopO, and YopH) contribute to the antiphagocytic action of *Yersinia*, as their concerted action leads to the complete destruction of the actin cytoskeleton. YopE (GAP) and YopT (cysteine protease) target the Rho family of GTPases directly and will inhibit their activation. YopO/YpkA also interacts with the Rho family of GTPases. However, although this leads to a partial destruction of the cytoskeleton, currently no direct target could be identified. Lastly, the antiphagocytic function of the tyrosine phosphatase YopH is to disassemble adhesion complexes at the cell membrane. Although YopH plays a crucial role in the antiphagocytic protection, it has recently become clear that YopH may have other important roles during infection, such as inhibiting the proliferation of lymphocytes. YopP/J (cysteine protease) efficiently shuts down multiple kinase cascades, and in this way it may be responsible for the downregulation of the inflammatory response, required by the host cell to respond to a bacterial infection. YopP/J is also responsible for the induction of apoptosis of macrophages.

Understanding the role of the YopM effector is still a challenge and is a topic for further research. Thus, the action of different Yops may converge into single key issues, but one Yop may have different effects. Surprisingly, LcrV, one of the proteins that are involved in translocation of the effectors across the host cell membrane, has an anti-inflammatory role on its own, without being injected into the cell cytosol.

REFERENCES

Aepfelbacher, M. and Heesemann, J. (2001). Modulation of Rho GTPases and the actin cytoskeleton by *Yersinia* outer proteins (Yops). *Int. J. Med. Microbiol.* **291**, 269–276.

Akira, S., Takeda, K., and Kaisho, T. (2001). Toll-like receptors: critical proteins linking innate and acquired immunity. *Nat. Immunol.* **2**, 675–680.

Alberta, J.A., Auger, K.R., Batt, D., Iannarelli, P., Hwang, G., Elliott, H.L., Duke, R., Roberts, T.M., and Stiles, C.D. (1999). Platelet-derived growth factor stimulation of monocyte chemoattractant protein-1 gene expression is mediated by transient activation of the phosphoinositide 3-kinase signal transduction pathway. *J. Biol. Chem.* **274**, 31,062–31,067.

Allen, L.A. and Aderem, A. (1996). Mechanisms of phagocytosis. *Curr. Opin. Immunol.* **8**, 36–40.

Anderson, D.M. and Schneewind, O. (1997). A mRNA signal for the type III secretion of Yop proteins by *Yersinia enterocolitica*. *Science* **278**, 1140–1143.

Andersson, K., Carballeira, N., Magnusson, K.E., Persson, C., Stendahl, O., Wolf-Watz, H., and Fallman, M. (1996). YopH of *Yersinia pseudotuberculosis* interrupts early phosphotyrosine signalling associated with phagocytosis. *Mol. Microbiol.* **20**, 1057–1069.

Andersson, K., Magnusson, K.E., Majeed, M., Stendahl, O., and Fallman, M. (1999). *Yersinia pseudotuberculosis*-induced calcium signaling in neutrophils is blocked by the virulence effector YopH. *Infect. Immun.* **67**, 2567–2574.

Andor, A., Trulzsch, K., Essler, M., Roggenkamp, A., Wiedemann, A., Heesemann, J., and Aepfelbacher, M. (2001). YopE of *Yersinia*, a GAP for Rho GTPases, selectively modulates Rac-dependent actin structures in endothelial cells. *Cell. Microbiol.* **3**, 301–310.

Arencibia, I., Frankel, G., and Sundqvist, K.G. (2002). Induction of cell death in T lymphocytes by invasin via beta 1-integrin. *Eur. J. Immunol.* **32**, 1129–1138.

Autenrieth, I.B. and Firsching, R. (1996). Penetration of M cells and destruction of Peyer's patches by *Yersinia enterocolitica*: an ultrastructural and histological study. *J. Med. Microbiol.* **44**, 285–294.

Bar-Sagi, D. and Hall, A. (2000). Ras and Rho GTPases: a family reunion. *Cell* **103**, 227–238.

Barz, C., Abahji, T.N., Trulzsch, K., and Heesemann, J. (2000). The *Yersinia* Ser/Thr protein kinase YpkA/YopO directly interacts with the small GTPases RhoA and Rac-1. *FEBS Lett.* **482**, 139–143.

Berton, G. and Lowell, C.A. (1999). Integrin signalling in neutrophils and macrophages. *Cell. Signal.* **11**, 621–635.

Black, D.S. and Bliska, J.B. (1997). Identification of p130Cas as a substrate of *Yersinia* YopH (Yop51), a bacterial protein tyrosine phosphatase that translocates into mammalian cells and targets focal adhesions. *EMBO J.* **16**, 2730–2744.

Black, D.S. and Bliska, J.B. (2000). The RhoGAP activity of the *Yersinia pseudotuberculosis* cytotoxin YopE is required for antiphagocytic function and virulence. *Mol. Microbiol.* **37**, 515–527.

Black, D.S., Marie-Cardine, A., Schraven, B., and Bliska, J.B. (2000). The *Yersinia* tyrosine phosphatase YopH targets a novel adhesion-regulated signalling complex in macrophages. *Cell. Microbiol.* **2**, 401–414.

Black, D.S., Montagna, L.G., Zitsmann, S., and Bliska, J.B. (1998). Identification of an amino-terminal substrate-binding domain in the *Yersinia* tyrosine phosphatase that is required for efficient recognition of focal adhesion targets. *Mol. Microbiol.* **29**, 1263–1274.

Bliska, J.B. and Black, D.S. (1995). Inhibition of the Fc receptor-mediated oxidative burst in macrophages by the *Yersinia pseudotuberculosis* tyrosine phosphatase. *Infect. Immun.* **63**, 681–685.

Boland, A. and Cornelis, G.R. (1998). Role of YopP in suppression of tumor necrosis factor alpha release by macrophages during *Yersinia* infection. *Infect. Immun.* **66**, 1878–1884.

Boland, A. and Cornelis, G.R. (2000). Interaction of *Yersinia* with host cells. *Subcell. Biochem.* **33**, 343–382.

Boland, A., Sory, M.P., Iriarte, M., Kerbourch, C., Wattiau, P., and Cornelis, G.R. (1996). Status of YopM and YopN in the *Yersinia* Yop virulon: YopM of Y. *enterocolitica* is internalized inside the cytosol of PU5-1.8 macrophages by the YopB, D, N delivery apparatus. *EMBO J.* **15**, 5191–5201.

Boyd, A.P., Lambermont, I., and Cornelis, G.R. (2000). Competition between the Yops of *Yersinia enterocolitica* for delivery into eukaryotic cells: role of the SycE chaperone binding domain of YopE. *J. Bacteriol.* **182**, 4811–4821.

Buttner, D. and Bonas, U. (2002). Port of entry – the type III secretion translocon. *Trends Microbiol.* **10**, 186–192.

Carniel, E. (2001). The *Yersinia* high-pathogenicity island: an iron-uptake island. *Microbes Infect.* **3**, 561–569.

Cheng, L.W. and Schneewind, O. (1999). *Yersinia enterocolitica* type III secretion. On the role of SycE in targeting YopE into HeLa cells. *J. Biol. Chem.* **274**, 22,102–22,108.

Chimini, G. and Chavrier, P. (2000). Function of Rho family proteins in actin dynamics during phagocytosis and engulfment. *Nat. Cell. Biol.* **2**, E191–E196.

China, B., Sory, M.P., N'Guyen, B.T., De Bruyere, M., and Cornelis, G.R. (1993). Role of the YadA protein in prevention of opsonization of *Yersinia enterocolitica* by C3b molecules. *Infect. Immun.* **61**, 3129–3136.

Cole, S.T. and Buchrieser, C. (2001). Bacterial genomics. A plague o' both your hosts. *Nature* **413**, 467, 469–470.

Cornelis, G. (2002). *Yersinia* type III secretion: send the effectors. *J. Cell Biol.* **158**, 401–408.

Cornelis, G., Vanootegem, J.C., and Sluiters, C. (1987). Transcription of the yop regulon from Y. *enterocolitica* requires trans acting pYV and chromosomal genes. *Microb. Pathog.* **2**, 367–379.

Cornelis, G.R. (2000). Type III secretion: a bacterial device for close combat with cells of their eukaryotic host. *Philos. Trans. R. Soc. Lond. B. Biol. Sci.* **355**, 681–693.

Cornelis, G.R., Boland, A., Boyd, A.P., Geuijen, C., Iriarte, M., Neyt, C., Sory, M.P., and Stainier, I. (1998). The virulence plasmid of *Yersinia*, an antihost genome. *Microbiol. Mol. Biol. Rev.* **62**, 1315–1352.

Cornelis, G.R., Sluiters, C., Delor, I., Geib, D., Kaniga, K., Lambert de Rouvroit, C., Sory, M.P., Vanooteghem, J.C., and Michiels, T. (1991). ymoA, a *Yersinia enterocolitica* chromosomal gene modulating the expression of virulence functions. *Mol. Microbiol.* **5**, 1023–1034.

Cowan, C., Jones, H.A., Kaya, Y.H., Perry, R.D., and Straley, S.C. (2000). Invasion of epithelial cells by Yersinia pestis: evidence for a *Y. pestis*-specific invasin. *Infect. Immun.* **68**, 4523–4530.

Denecker, G., Declercq, W., Geuijen, C.A., Boland, A., Benabdillah, R., van Gurp, M., Sory, M.P., Vandenabeele, P., and Cornelis, G.R. (2001). *Yersinia enterocolitica* YopP-induced apoptosis of macrophages involves the apoptotic signaling cascade upstream of Bid. *J. Biol. Chem.* **276**, 19,706–19,714.

Denecker, G., Totemeyer, S., Mota, L.J., Troisfontaines, P., Lambermont, I., Youta, C., Stainier, I., Ackermann, M., and Cornelis, G.R. (2002). Effect of low- and high-virulence *Yersinia enterocolitica* strains on the inflammatory response of human umbilical vein endothelial cells. *Infect. Immun.* **70**, 3510–3520.

DeVinney, I., Steele-Mortimer, I., and Finlay, B.B. (2000). Phosphatases and kinases delivered to the host cell by bacterial pathogens. *Trends Microbiol.* **8**, 29–33.

Du, Y., Rosqvist, R., and Forsberg, A. (2002). Role of fraction 1 antigen of *Yersinia pestis* in inhibition of phagocytosis. *Infect. Immun.* **70**, 1453–1460.

Dukuzumuremyi, J.M., Rosqvist, R., Hallberg, B., Akerstrom, B., Wolf-Watz, H., and Schesser, K. (2000). The *Yersinia* protein kinase A is a host factor inducible RhoA/Rac-binding virulence factor. *J. Biol. Chem.* **275**, 35,281–35,290.

El Tahir, Y. and Skurnik, M. (2001). YadA, the multifaceted *Yersinia* adhesin. *Int. J. Med. Microbiol.* **291**, 209–218.

Evdokimov, A.G., Anderson, D.E., Routzahn, K.M., and Waugh, D.S. (2001). Unusual molecular architecture of the *Yersinia pestis* cytotoxin YopM: a leucine-rich repeat protein with the shortest repeating unit. *J. Mol. Biol.* **312**, 807–821.

Evdokimov, A.G., Tropea, J.E., Routzahn, K.M., and Waugh, D.S. (2002). Crystal structure of the *Yersinia pestis* GTPase activator YopE. *Protein Sci.* **11**, 401–408.

Feldman, M.F., Müller, S., Wüest, E., and Cornelis, G.R. (2002). SycE allows secretion of YopE-DHFR hybrids by the *Yersinia enterocolitica* type III Ysc system. *Mol. Microbiol.* **46**, 1183–1197.

Foultier, B., Troisfontaines, P., Müller, S., Opperdoes, F., and Cornelis, G.R. (2002). Characterization of the *ysa* pathogenicity locus in the chromosome of

Yersinia enterocolitica and phylogenic analysis of type III secretion systems. *J. Mol. Evol.* **55**, 37–51.

Francis, M.S., Wolf-Watz, H., and Forsberg, A. (2002). Regulation of type III secretion systems. *Curr. Opin. Microbiol.* **5**, 166–172.

Frithz-Lindsten, E., Rosqvist, R., Johansson, L., and Forsberg, A. (1995). The chaperone-like protein YerA of *Yersinia pseudotuberculosis* stabilizes YopE in the cytoplasm but is dispensible for targeting to the secretion loci. *Mol. Microbiol.* **16**, 635–647.

Galan, J.E. (2001). Salmonella interactions with host cells: type III secretion at work. *Annu. Rev. Cell. Dev. Biol.* **17**, 55–68.

Galan, J.E. and Collmer, A. (1999). Type III secretion machines: bacterial devices for protein delivery into host cells. *Science* **284**, 1322–1328.

Galyov, E.E., Hakansson, S., Forsberg, A., and Wolf-Watz, H. (1993). A secreted protein kinase of *Yersinia pseudotuberculosis* is an indispensable virulence determinant. *Nature* **361**, 730–732.

Greenberg, S., Chang, P., and Silverstein, S.C. (1993). Tyrosine phosphorylation is required for Fc receptor-mediated phagocytosis in mouse macrophages. *J. Exp. Med.* **177**, 529–534.

Grosdent, N., Maridonneau-Parini, I., Sory, M.P., and Cornelis, G.R. (2002). Role of Yops and adhesins in resistance of *Yersinia enterocolitica* to phagocytosis. *Infect. Immun.* **70**, 4165–4176.

Guan, K.L. and Dixon, J.E. (1990). Protein tyrosine phosphatase activity of an essential virulence determinant in *Yersinia*. *Science* **249**, 553–556.

Hakansson, S., Galyov, E.E., Rosqvist, R., and Wolf-Watz, H. (1996a). The *Yersinia* YpkA Ser/Thr kinase is translocated and subsequently targeted to the inner surface of the HeLa cell plasma membrane. *Mol. Microbiol.* **20**, 593–603.

Hakansson, S., Schesser, K., Persson, C., Galyov, E.E., Rosqvist, R., Homble, F., and Wolf-Watz, H. (1996b). The YopB protein of *Yersinia pseudotuberculosis* is essential for the translocation of Yop effector proteins across the target cell plasma membrane and displays a contact-dependent membrane disrupting activity. *EMBO J.* **15**, 5812–5823.

Hall, A. (1998). Rho GTPases and the actin cytoskeleton. *Science* **279**, 509–514.

Haller, J.C., Carlson, S., Pederson, K.J., and Pierson, D.E. (2000). A chromosomally encoded type III secretion pathway in *Yersinia enterocolitica* is important in virulence. *Mol. Microbiol.* **36**, 1436–1446.

Hamid, N., Gustavsson, A., Andersson, K., McGee, K., Persson, C., Rudd, C.E., and Fallman, M. (1999). YopH dephosphorylates Cas and Fyn-binding protein in macrophages. *Microb. Pathog.* **27**, 231–242.

Heesemann, J., Hantke, K., Vocke, T., Saken, E., Rakin, A., Stojiljkovic, I., and Berner, R. (1993). Virulence of *Yersinia enterocolitica* is closely associated with siderophore production, expression of an iron-repressible outer membrane polypeptide of 65,000 Da and pesticin sensitivity. *Mol. Microbiol.* **8**, 397–408.

Hinnebusch, B.J., Fischer, E.R., and Schwan, T.G. (1998). Evaluation of the role of the *Yersinia pestis* plasminogen activator and other plasmid-encoded factors in temperature-dependent blockage of the flea. *J. Infect. Dis.* **178**, 1406–1415.

Hinnebusch, B.J., Rudolph, A.E., Cherepanov, P., Dixon, J.E., Schwan, T.G., and Forsberg, A. (2002). Role of *Yersinia* murine toxin in survival of *Yersinia pestis* in the midgut of the flea vector. *Science* **296**, 733–735.

Hueck, C.J. (1998). Type III protein secretion systems in bacterial pathogens of animals and plants. *Microbiol. Mol. Biol. Rev.* **62**, 379–433.

Imler, J.L. and Hoffmann, J.A. (2001). Toll receptors in innate immunity. *Trends Cell Biol.* **11**, 304–311.

Iriarte, M. and Cornelis, G.R. (1998). YopT, a new *Yersinia* Yop effector protein, affects the cytoskeleton of host cells. *Mol. Microbiol.* **29**, 915–929.

Isberg, R.R., Hamburger, Z., and Dersch, P. (2000). Signaling and invasin-promoted uptake via integrin receptors. *Microbes Infect* **2**, 793–801.

Juris, S.J., Rudolph, A.E., Huddler, D., Orth, K., and Dixon, J.E. (2000). A distinctive role for the *Yersinia* protein kinase: actin binding, kinase activation, and cytoskeleton disruption. *Proc. Natl. Acad. Sci. USA* **97**, 9431–9436.

Juris, S.J., Shao, F., and Dixon, J.E. (2002). *Yersinia* effectors target mammalian signalling pathways. *Cell. Microbiol.* **4**, 201–211.

Kobe, B. and Kajava, A.V. (2001). The leucine-rich repeat as a protein recognition motif. *Curr. Opin. Struct. Biol.* **11**, 725–732.

Lambert de Rouvroit, C., Sluiters, C., and Cornelis, G.R. (1992). Role of the transcriptional activator, VirF, and temperature in the expression of the pYV plasmid genes of *Yersinia enterocolitica*. *Mol. Microbiol.* **6**, 395–409.

Lee, V.T. and Schneewind, O. (2002). Yop fusions to tightly folded protein domains and their effects on *Yersinia enterocolitica* type III secretion. *J. Bacteriol.* **184**, 3740–3745.

Mangel, W.F., McGrath, W.J., Toledo, D.L., and Anderson, C.W. (1993). Viral DNA and a viral peptide can act as cofactors of adenovirus virion proteinase activity. *Nature* **361**, 274–275.

May, R.C. and Machesky, L.M. (2001). Phagocytosis and the actin cytoskeleton. *J. Cell Sci.* **114**, 1061–1077.

Michiels, T., Wattiau, P., Brasseur, R., Ruysschaert, J.M., and Cornelis, G. (1990). Secretion of Yop proteins by Yersiniae. *Infect. Immun.* **58**, 2840–2849.

Miller, V.L. (2002). Connections between transcriptional regulation and type III secretion? *Curr. Opin. Microbiol.* **5**, 211–215.

Mills, S.D., Boland, A., Sory, M.P., van der Smissen, P., Kerbourch, C., Finlay, B.B., and Cornelis, G.R. (1997). *Yersinia enterocolitica* induces apoptosis in macrophages by a process requiring functional type III secretion and translocation mechanisms and involving YopP, presumably acting as an effector protein. *Proc. Natl. Acad. Sci. USA* **94**, 12,638–12,643.

Monack, D.M., Mecsas, J., Ghori, N., and Falkow, S. (1997). *Yersinia* signals macrophages to undergo apoptosis and YopJ is necessary for this cell death. *Proc. Natl. Acad. Sci. USA* **94**, 10,385–10,390.

Orth, K. (2002). Function of the *Yersinia* effector YopJ. *Curr. Opin. Microbiol.* **5**, 38–43.

Orth, K., Palmer, L.E., Bao, Z.Q., Stewart, S., Rudolph, A.E., Bliska, J.B., and Dixon, J.E. (1999). Inhibition of the mitogen-activated protein kinase kinase superfamily by a *Yersinia* effector. *Science* **285**, 1920–1923.

Orth, K., Xu, Z., Mudgett, M.B., Bao, Z.Q., Palmer, L.E., Bliska, J.B., Mangel, W.F., Staskawicz, B., and Dixon, J.E. (2000). Disruption of signaling by *Yersinia* effector YopJ, a ubiquitin-like protein protease. *Science* **290**, 1594–1597.

Palmer, L.E., Hobbie, S., Galan, J.E., and Bliska, J.B. (1998). YopJ of *Yersinia pseudotuberculosis* is required for the inhibition of macrophage TNF-alpha production and downregulation of the MAP kinases p38 and JNK. *Mol. Microbiol.* **27**, 953–965.

Palmer, L.E., Pancetti, A.R., Greenberg, S., and Bliska, J.B. (1999). YopJ of *Yersinia* spp. is sufficient to cause downregulation of multiple mitogen-activated protein kinases in eukaryotic cells. *Infect. Immun.* **67**, 708–716.

Parkhill, J., Wren, B.W., Thomson, N.R., Titball, R.W., Holden, M.T., Prentice, M.B., Sebaihia, M., James, K.D., Churcher, C., Mungall, K.L., Baker, S., Basnam, D., Bentley, S.D., Brooks, K., Cerendo-Tarraga, A.M., Chillingworth, T., Cronin, A., Davies, R.M., Davis, P., Dougan, G., Feltwell, T., Hemlin, N., Holroyd, S., Jagels, K., Karlyshev, A.V., Leather, S., Moule, S., Oyston, P.C., Quail, M., Rutherford, K., Simmons, M., Skelton, J., Stevens, K., Whitehead, S., Barrell, B.G. (2001). Genome sequence of *Yersinia pestis*, the causative agent of plague. *Nature* **413**, 523–527.

Pepe, J.C. and Miller, V.L. (1993). *Yersinia enterolitica* invasin: a primary role in the initiation of infection. *Proc. Natl. Acad. Sci. USA* **90**, 6473–6477.

Perry, R.D. and Fetherston, J.D. (1997). *Yersinia pestis* – etiologic agent of plague. *Clin. Microbiol. Rev.* **10**, 35–66.

Persson, C., Carballeira, N., Wolf-Watz, H., and Fallman, M. (1997). The PTPase YopH inhibits uptake of *Yersinia*, tyrosine phosphorylation of p130Cas and FAK, and the associated accumulation of these proteins in peripheral focal adhesions. *EMBO J.* **16**, 2307–2318.

Pettersson, J., Nordfelth, R., Dubinina, E., Bergman, T., Gustafsson, M., Magnusson, K.E., and Wolf-Watz, H. (1996). Modulation of virulence factor expression by pathogen target cell contact. *Science* **273**, 1231–1233.

Ridley, A.J. (2001). Rho proteins, PI 3-kinases, and monocyte/macrophage motility. *FEBS Lett.* **498**, 168–171.

Rohde, J.R., Luan, X.S., Rohde, H., Fox, J.M., and Minnich, S.A. (1999). The *Yersinia enterocolitica* pYV virulence plasmid contains multiple intrinsic DNA bends which melt at 37 degrees C. *J. Bacteriol.* **181**, 4198–4204.

Rosqvist, R., Forsberg, A., Rimpilainen, M., Bergman, T., and Wolf-Watz, H. (1990). The cytotoxic protein YopE of *Yersinia* obstructs the primary host defence. *Mol. Microbiol.* **4**, 657–667.

Ruckdeschel, K., Harb, S., Roggenkamp, A., Hornef, M., Zumbihl, R., Kohler, S., Heesemann, J., and Rouot, B. (1998). *Yersinia enterocolitica* impairs activation of transcription factor NF-kappa B: involvement in the induction of programmed cell death and in the suppression of the macrophage tumor necrosis factor alpha production. *J. Exp. Med.* **187**, 1069–1079.

Ruckdeschel, K., Machold, J., Roggenkamp, A., Schubert, S., Pierre, J., Zumbihl, R., Liautard, J.P., Heesemann, J., and Rouot, B. (1997). *Yersinia enterocolitica* promotes deactivation of macrophage mitogen-activated protein kinases extracellular signal-regulated kinase-1/2, p38, and c-Jun NH2-terminal kinase. Correlation with its inhibitory effect on tumor necrosis factor-alpha production. *J. Biol. Chem.* **272**, 15,920–15,927.

Ruckdeschel, K., Mannel, O., Richter, K., Jacobi, C., Trülzsch, K., Rouot, B., and Heesemann, J. (2001a). *Yersinia* outer protein P of *Yersinia enterocolitica* simultaneously blocks the nuclear factor-κ B pathhway and exploits lipopolysaccharide signaling to trigger apoptosis in macrophages. *J. Immunol.* **166**, 1823–1831.

Ruckdeschel, K., Mannel, O., and Schrottner, P. (2002). Divergence of apoptosis-inducing and preventing signals in bacteria-faced macrophages through myeloid differentiation factor 88 and IL-1 receptor-associated kinase members. *J. Immunol.* **168**, 4601–4611.

Ruckdeschel, K., Richter, K., Mannel, O., and Heesemann, J. (2001b). Arginine-143 of *Yersinia enterocolitica* YopP crucially determines isotype-related NF-kappa B suppression and apoptosis induction in macrophages. *Infect. Immun.* **69**, 7652–7662.

Ruckdeschel, K., Roggenkamp, A., Schubert, S., and Heesemann, J. (1996). Differential contribution of *Yersinia enterocolitica* virulence factors to evasion of microbicidal action of neutrophils. *Infect. Immun.* **64**, 724–733.

Sansonetti, P. (2002). Host-pathogen interactions: the seduction of molecular cross talk. *Gut* **50** (Suppl 3), III2–III8.

Sansonetti, P.J. (2001). Microbes and microbial toxins: paradigms for microbial-mucosal interactions III. Shigellosis: from symptoms to molecular pathogenesis. *Am. J. Physiol. Gastrointest. Liver Physiol.* **280**, G319–G323.

GEERTRUI DENECKER AND GUY CORNELIS

Sauvonnet, N., Lambermont, I., van der Bruggen, P., and Cornelis, G. (2002a). YopH prevents monocyte chemoattractant protein 1 expression in macrophages and T-cell proliferation through inactivation of the phosphatidylinositol 3-kinase pathway. *Mol. Microbiol.*

Sauvonnet, N., Pradet-Balade, B., Garcia-Sanz, J.A., and Cornelis, G.R. (2002b). Regulation of mRNA expression in macrophages following *Yersinia enterocolitica* infection: role of different Yop effectors. *J. Biol. Chem.* **2**, 2.

Scheid, M.P. and Woodgett, J.R. (2001). PKB/AKT: functional insights from genetic models. *Nat. Rev. Mol. Cell. Biol.* **2**, 760–768.

Schesser, K., Spiik, A.K., Dukuzumuremyi, J.M., Neurath, M.F., Pettersson, S., and Wolf-Watz, H. (1998). The yopJ locus is required for *Yersinia*-mediated inhibition of NF-kappa B activation and cytokine expression: YopJ contains a eukaryotic SH2-like domain that is essential for its repressive activity. *Mol. Microbiol.* **28**, 1067–1079.

Schulte, R., Kerneis, S., Klinke, S., Bartels, H., Preger, S., Kraehenbuhl, J.P., Pringault, E., and Autenrieth, I.B. (2000). Translocation of *Yersinia entrocolitica* across reconstituted intestinal epithelial monolayers is triggered by *Yersinia* invasin binding to beta 1 integrins apically expressed on M-like cells. *Cell. Microbiol.* **2**, 173–185.

Shao, F., Merritt, P.M., Bao, Z., Innes, R.W., and Dixon, J.E. (2002). A *Yersinia* effector and a *Pseudomonas* avirulence protein define a family of cysteine proteases functioning in bacterial pathogenesis. *Cell* **109**, 575–588.

Sing, A., Roggenkamp, A., Geiger, A.M., and Heesemann, J. (2002). *Yersinia enterocolitica* evasion of the host innate immune response by V antigen-induced IL-10 production of macrophages is abrogated in IL-10-deficient mice. *J. Immunol.* **168**, 1315–1321.

Skrzypek, E., Cowan, C., and Straley, S.C. (1998). Targeting of the *Yersinia pestis* YopM protein into HeLa cells and intracellular trafficking to the nucleus. *Mol. Microbiol.* **30**, 1051–1065.

Sorg, I., Goehring, U.M., Aktories, K., and Schmidt, G. (2001). Recombinant *Yersinia* YopT leads to uncoupling of RhoA-effector interaction. *Infect. Immun.* **69**, 7535–7543.

Sory, M.P., Boland, A., Lambermont, I., and Cornelis, G.R. (1995). Identification of the YopE and YopH domains required for secretion and internalization into the cytosol of macrophages, using the *cyaA* gene fusion approach. *Proc. Natl. Acad. Sci. USA* **92**, 11,998–12,002.

Stebbins, C.E. and Galan, J.E. (2001). Maintenance of an unfolded polypeptide by a cognate chaperone in bacterial type III secretion. *Nature* **414**, 77–81.

Underhill, D.M. and Ozinsky, A. (2002a). Phagocytosis of microbes: complexity in action. *Annu. Rev. Immunol.* **20**, 825–852.

Underhill, D.M. and Ozinsky, A. (2002b). Toll-like receptors: key mediators of microbe detection. *Curr. Opin. Immunol.* **14**, 103–110.

Viboud, G.I. and Bliska, J.B. (2001). A bacterial type III secretion system inhibits actin polymerization to prevent pore formation in host cell membranes. *EMBO J.* **20**, 5373–5382.

Von Pawel-Rammingen, U., Telepnev, M.V., Schmidt, G., Aktories, K., Wolf-Watz, H., and Rosqvist, R. (2000). GAP activity of the *Yersinia* YopE cytotoxin specifically targets the Rho pathway: a mechanism for disruption of actin microfilament structure. *Mol. Microbiol.* **36**, 737–748.

Wattiau, P., Woestyn, S., and Cornelis, G.R. (1996). Customized secretion chaperones in pathogenic bacteria. *Mol. Microbiol.* **20**, 255–262.

Wulff-Strobel, C.R., Williams, A.W., and Straley, S.C. (2002). LcrQ and SycH function together at the Ysc type III secretion system in *Yersinia pestis* to impose a hierarchy of secretion. *Mol. Microbiol.* **43**, 411–423.

Yao, T., Mecsas, J., Healy, J.I., Falkow, S., and Chien, Y. (1999). Suppression of T and B lymphocyte activation by a *Yersinia pseudotuberculosis* virulence factor, yopH. *J. Exp. Med.* **190**, 1343–1350.

Yeh, E.T., Gong, L., and Kamitani, T. (2000). Ubiquitin-like proteins: new wines in new bottles. *Gene* **248**, 1–14.

Zumbihl, R., Aepfelbacher, M., Andor, A., Jacobi, C.A., Ruckdeschel, K., Rouot, B., and Heesemann, J. (1999). The cytotoxin YopT of *Yersinia enterocolitica* induces modification and cellular redistribution of the small GTP-binding protein RhoA. *J. Biol. Chem.* **274**, 29,289–29,293.

CHAPTER 4

Stealth warfare: The interactions of EPEC and EHEC with host cells

Emma Allen-Vercoe and Rebekah DeVinney

Although the Gram-negative bacterium *Escherichia coli* is normally considered to be a harmless commensal of the gastrointestinal flora, there are some exceptions to this rule. In the past few decades it has become increasingly evident that there are serotypes of *E. coli* that may cause disease in susceptible hosts. Disease states range from the invasive infections of the urinary tract caused by uropathogenic *E. coli* (UPEC) to the more typical diarrheal disease caused by several groups of *E. coli* serotypes, including enteropathogenic *E. coli* (EPEC) and enterohemorrhagic *E. coli* (EHEC). There is an increasing realization that bacteria–host interactions with pathogenic *E. coli* are far more complex and intricate than originally imagined. Studying the ways that bacteria such as EHEC and EPEC are able to subvert host cell functions to their own ends can be thought of as a "window" through which we are able to view the inner workings of the eukaryotic cell. In this chapter we examine some of the mechanisms by which EHEC and EPEC are able to coerce their host; we also examine the sequelae of these interactions.

EPEC usually refers to *E. coli* serotypes O55:[H6], O86:H34, O111:[H2], O114:H2, O119:[H6], O127:H6, O142:H6, and O142:H34, where square brackets indicate the occurrence of nonmotile strains (Nataro and Kaper, 1998). Infection with EPEC typically causes a chronic watery diarrhea, accompanied by a low-grade fever and nausea. These symptoms can lead to rapid dehydration of the patient, and, because of this, EPEC is a leading cause of infant mortality, particularly in developing countries where rehydration therapy may be difficult. In the Western world, isolated incidences of EPEC infection are often associated with daycares and nurseries (Nataro and Kaper, 1998), where young children closely associate with one another and facilitate the spread of infection.

EMMA ALLEN-VERCOE AND REBEKAH DEVINNEY

EHEC usually refers to serotype O157:H7 and less commonly to serotype O111:H$^-$. An EHEC infection is often heralded by the onset of watery diarrhea, which progresses rapidly to severe bloody diarrhea (hemorrhagic colitis) in many patients, regardless of age (Nataro and Kaper, 1998). In the very young and very old, as well as in immunocompromised patients, the disease can be complicated by the onset of hemolytic-uremic syndrome (HUS), which is characterized by hemolytic anemia, thrombocytopenia, and renal failure. HUS is caused by the secretion by EHEC of shiga-like toxin (SLT), a potent cytotoxin with a predilection for human kidney cells. A description of the mechanisms of action of SLT and the many effects on the host cell is beyond the scope of this chapter but is reviewed by O'Loughlin and Robins-Browne, 2001. The onset of HUS, even with rapid treatment, can prove fatal to a patient. Outbreaks of EHEC infection are becoming increasingly high profile in North America and Europe. The low infectious dose required to cause infection may allow the organism to spread rapidly. A recent outbreak of EHEC O157:H7 in Walkerton, Ontario infected close to 2000 people and led to seven deaths (Kondro, 2000).

Neither EPEC nor EHEC are generally regarded as invasive pathogens; however, some limited evidence has suggested that EPEC in particular may, in fact, be able to invade host cells under the right conditions (Czerucka et al., 2000). Additionally, both EPEC and EHEC use a highly specialized mechanism that allows the bacteria to adhere tightly to the epithelial cell surface and subvert normal host cell functions by direct interaction. In this chapter, the mechanisms and consequences of both EPEC and EHEC infections of the host are discussed, and the subtle differences between EPEC and EHEC interaction with host cells, which are beginning to emerge as research progresses, are highlighted.

ADHERENCE MECHANISMS REQUIRED FOR INITIAL ATTACHMENT OF EPEC AND EHEC TO HOST CELLS

EPEC appear to have a predilection for the human ileum (Nataro and Kaper, 1998), and the first stage of adherence of EPEC to host cells is thought to involve bundle-forming pili (BFP; see Giron et al., 1991). BFP are type-4 pili encoded by a cluster of 14 genes found on a 92-kb plasmid known as the EPEC adherence factor plasmid (EAF; see Tobe et al., 1999). EPEC-expressing BFP form dense microcolonies on the surface of tissue culture monolayers in a pattern known as localized adherence (LA). Whether BFP are involved more in interbacterial adherence or adherence to the host cell is currently under debate. However, there is no doubt that BFP expression represents a significant virulence factor for EPEC. Mutant EPEC unable to produce BFP are significantly reduced in their ability to adhere to tissue culture monolayers (Frankel et al., 1998). Furthermore, human volunteers given an inoculum of mutant EPEC defective for the expression of BFP were far less likely to develop diarrhea than were volunteers given the wild-type strain (Bieber et al., 1998). Together, these factors indicate that the initial adherence of EPEC to host cells is a critical step in the subsequent development of disease.

The flagella of Gram-negative bacteria are being increasingly recognized as effectors of bacterial cell adhesion to host cells, such as in the binding of *Salmonella* to chick gut explants (Allen-Vercoe and Woodward, 1999). Recently, the flagella of EPEC have also been shown to be important in the adherence of the bacteria to epithelial cells *in vitro*, and experimental evidence suggested that a eukaryotic signal can induce flagellar expression and thus enhance adhesion (Giron et al., 2002). Additionally, it was found that EPEC serotypes previously classified as nonmotile could, in fact, elaborate flagella under the correct conditions (Giron et al., 2002).

In addition to its role in pedestal formation (discussed in what follows), the EPEC and EHEC outer membrane protein intimin may play a role in adherence host cells. The family of intimins from attaching/effacing (A/E) lesion-forming bacteria has been divided into at least five distinct types that differ from each other at the amino acid level, particularly across the putative binding domain located at the C-terminus (Frankel et al., 1995). EPEC and EHEC encode different intimin types, depending on their serotype. Intimin-α and intimin-β are usually associated with EPEC strains, whereas EHEC O157:H7 specifically expresses intimin-γ. Interestingly, intimin may be responsible for the tissue tropism exhibited by different pathogenic *E. coli*. Infection with an EHEC strain expressing EPEC intimin-α shows a pattern of adherence to human gut explants that resembles that of EPEC (Fitzhenry et al., 2002).

The eukaryotic receptor(s) for intimin has yet to be identified. However, two targets have been suggested: β_1-integrins (Frankel et al., 1996) and nucleolin (Sinclair and O'Brien, 2002). The expression of intimin by EPEC and EHEC is an important virulence determinant. In human volunteers, an EPEC intimin$^-$ mutant was shown to be significantly attenuated compared with its wild-type parent strain (Donnenberg et al., 1993). In addition, antibodies to both EPEC and EHEC intimins are often found at a high titer in individuals who have been infected with these pathogens (Loureiro et al., 1998; Jenkins et al., 2000).

The preferred site for colonization of EHEC is the colon, and the initial adherence of EHEC to host cells is thought to be quite different from that of EPEC. EHEC does not possess the EAF plasmid and thus does not elabo-rate BFP. Hence, the pattern of EHEC adherence to epithelial cells in tissue culture is more diffuse than that of EPEC, and no microcolonies are formed. In the absence of BFP, the Tir (translocated intimin receptor)-independent adhesion mediated by intimin and described herein is likely to become more important to EHEC for colonization of host cells than it is for EPEC. Indeed, an intimin⁻ mutant of EHEC O157:H7 failed to colonize human intestinal explants (Fitzhenry et al., 2002).

More recently, further EHEC adhesins have been proposed. Iha (IrgA homologue adhesin) is an outer membrane protein (OMP) with externally directed domains. Its corresponding gene, *iha*, confers an adhesive phenotype when transferred to a normally nonadherent *E. coli* K12 laboratory strain, although its role in EHEC adhesion remains to be elucidated (Tarr et al., 2000). The chromosomal gene *efa1* (EHEC factor for adherence) has also been recently implicated in the initial adherence of non-O157 EHEC to host cells, and by virtue of its low G + C content may represent part of a pathogenicity islet acquired by horizontal transmission. The product of the *efa1* gene, Efa1, has been shown to play a role in influencing the colonization of the bovine gut *in vivo* by non-O157 EHEC strains (Stevens et al., 2002). Although first characterized in EHEC, an Efa 1 homologue, LifA, has also been found in EPEC (Nicholls et al., 2000), where it confers upon the bacteria the ability to modulate host mucosal immunity in the gut (Klapproth et al., 1996; Malstrom and James, 1998; also discussed later in this chapter).

INTIMATE ATTACHMENT OF EHEC AND EPEC TO HOST CELLS, AND THE TYPE III SECRETORY APPARATUS

A hallmark of disease caused by EHEC and EPEC is the formation of A/E lesions (Fig. 4.1). These lesions are defined by the intimate attachment of the bacteria to the host cell surface and the localized destruction (efface-ment) of the brush border microvilli (Moon et al., 1983). The bacteria induce an accumulation of actin beneath their site of attachment, and this results in the formation of a pseudopod, or pedestal, upon which the bacteria are located (Fig. 4.2). A/E lesion-forming bacteria such as EHEC and EPEC uti-lize a Gram-negative specific mechanism, called a type III secretion system (TTSS), in order to inject effector molecules directly into the host cell (Hueck, 1998).

For EPEC, a region of the chromosome known as the locus for entero-cyte effacement (LEE) is sufficient for A/E lesion formation. The LEE is a 36-kb region containing 41 open reading frames (ORFs) organized into five

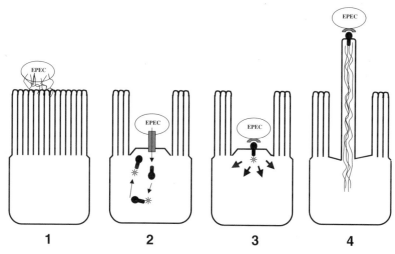

Figure 4.1. Simplified schematic of pedestal formation by EPEC. (1) EPEC uses BFP to adhere to the host cell microvilli, leading to effacement of the brush border (2), whereupon the bacterium uses its TTSS to inject Tir (black bulb shape), into the host cell, where it is phosphorylated (asterik) (3) Tir presents its intimin-binding domain to intimin expressed on the bacterial surface. The binding of Tir to intimin leads to a cascade of signaling events (4), which act to stimulate actin polymerization. A pedestal is formed on the host cell membrane, on which the bacterium resides.

polycistronic operons, termed *LEE1–LEE5* (Elliott et al., 1998; Mellies et al., 1999). *LEE1–LEE3* encode genes encoding the TTSS apparatus. *LEE4* contains the genes required for pedestal formation. These include *eae*, which codes for intimin; *tir*, which codes for Tir; and a gene that codes for the molecular chaperone for Tir, *cesT* (Abe et al., 1999). LEE5 contains genes for the type III-secreted proteins EspA, EspB, and EspD, which are necessary for pedestal formation. EspB and EspD are thought to form pore-like structures in the host cell membrane (Warawa et al., 1999; Kresse et al., 1999; Wachter et al., 1999) that interact with the filamentous EspA structures on the bacterial cell to allow the close association of the type III secretory apparatus with the host cell membrane, and the subsequent insertion of Tir (Knutton et al., 1998). Both EHEC and EPEC EspB have recently been found to interact directly with the host protein α-catenin, a cytoskeleton-associated molecule, in a Tir-independent way, and the formation of A/E lesions has been found to be dependent on this interaction (Kodama et al., 2002).

Recently, the supermolecular structure of the assembled EPEC TTSS apparatus has been elucidated by electron microscopy (Sekiya et al., 2001). It

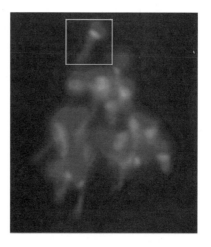

Figure 4.2. A cluster of pedestal structures induced by EPEC on the surface of a HeLa cell. Indirect immunofluorescence microscopy was used to demonstrate the formation of pedestal structures by EPEC. Long stalks of actin (green) are capped at the tip by Tir (red), upon which a single bacterium (blue) sits. The boxed region has been enlarged to clearly show a single pedestal. See color section.

was reported that the EspA protein, which has no homology to any known protein in other bacterial TTSSs other than that for EHEC, may be a major component of a sheath-like structure that is assembled by the bacteria to act as a physical bridge between bacteria and host cells. This, in turn, indicates that the assembly of EHEC and EPEC type III secretons may be carried out somewhat differently compared with other pathogens.

The Tir protein, once introduced into the host cell membrane by means of the TTSS, serves as the receptor for intimin and allows the intimate association of bacteria with host. Whereas EPEC adherence by means of BFP permits the bacteria to exist within 100–300 nm of the host cell surface, the distance between bacterial cell membrane to host cell membrane during Tir–intimin attachment can be as little as 10 nm (Nisan et al., 1998). EPEC Tir was originally thought to be a host cell protein, and its identification as a bacterial protein, injected by the bacteria in order to effect intimate adhesion by intimin, represents the first discovered example of host cell subversion of this type (Kenny et al., 1997). Work done in animal infection studies with the natural rabbit pathogen, REPEC O103:H2, has demonstrated that intimin, Tir, EspA, and EspB are essential for diarrheal disease (Abe et al., 1998; Marches et al., 2000).

Whereas some of the components of the TTSS are highly conserved between EPEC and EHEC, *eae, espA, espB, espD*, and *tir* are much less highly conserved, although the mechanism of Tir translocation is thought to be identical for both pathogens (Frankel et al., 1998). However, although it is possible to induce nonpathogenic laboratory strains of *E. coli* to form A/E lesions by introducing into them plasmids carrying the LEE region from EPEC (McDaniel and Kaper, 1997), the reciprocal experiment with the EHEC LEE region does not support either A/E lesion formation or Esp protein secretion, suggesting that, for EHEC, further factors are involved that are outside of the LEE (Elliott et al., 1999; DeVinney et al., 2001). These unidentified factors required by EHEC are a current focus for research into EHEC pathogenicity.

The first gene of *LEE1* encodes the Ler regulator, an H-NS-like protein that activates all other genes in the LEE for both EPEC and EHEC, and is required for pedestal formation (Friedberg et al., 1999; Elliott et al., 2000). The PerABC proteins encoded by the pEAF plasmid of EPEC positively regulate BFP expression (Donnenberg et al., 1992) and may also regulate the transcription of LEE genes through *ler* (Mellies et al., 1999).

The regulation of the LEE appears to be quite different for EPEC and EHEC. Both EHEC and EPEC utilize two quorum-sensing systems, termed 1 and 2, central to which are proteins termed autoinducer (AI)-1 and AI-2, respectively. During transition from the late exponential to the stationary growth phase, both EPEC and EHEC LEEs are activated by AI-2 (Sperandio et al., 1999). However, during the stationary phase, EHEC AI-1 appears to be downregulated by the expression of LEE-encoded genes through the action of a quorum-sensing regulator called SdiA (Kanamaru et al., 2000). In contrast, the AI-1 of EPEC does not appear to be affected in this manner (Abe et al., 2002). The differences in EPEC and EHEC LEE gene expression can be illustrated by the finding that, *in vitro*, EHEC Tir synthesis and secretion requires specific growth conditions that do not apply to EPEC Tir expression (DeVinney et al., 1999).

HOST CELL RESPONSES TO EHEC AND EPEC INFECTION

Pedestal formation

The formation of actin pedestals beneath adherent A/E bacteria is a host response so unusual that the presence of actin pedestals in infected tissue culture monolayers can be used as a diagnostic test for EPEC and EHEC (Knutton et al., 1989). Pedestals may grow to $10\,\mu$m above the host cell surface, at a rate of around $1\,\mu$m/min, and appear to be able to shorten and lengthen while remaining attached to the host cell (Rosenshine et al., 1996; Sanger et al., 1996). It has been found that pedestals themselves are able to propel their attached bacteria along the host cell surface at an estimated rate of 70 nm/s (Sanger et al., 1996). The exact purpose of an actin pedestal is puzzling. It may serve as a mechanism for the bacteria to avoid the host diarrheal response, or as an avoidance tactic for internalization into the host cell. However, in tissue culture, cells infected with A/E bacteria eventually round up and detach, suggesting that host cells may not be able to tolerate prolonged contact with these pathogens (Baldwin et al., 1993).

A chief effector of pedestal formation, Tir, is a 72- to 78-kDA protein that is inserted into the host cell membrane by means of the TTSS apparatus. EPEC and EHEC strains that are unable to secrete Tir do not form pedestals (DeVinney et al., 1999). Tir contains two predicted transmembrane domains that are required for the stable insertion of the protein into the host cell plasma membrane (Gauthier et al., 2000). The amino- and carboxy-termini of Tir are intracellular to the host cell, and intimin binds to a hydrophobic extracellular domain called the intimin-binding domain (IBD; see DeVinney et al., 1999; Kenny, 1999; Luo et al., 2000).

The process of pedestal formation is best understood for EPEC (Fig. 4.3). EPEC Tir is tyrosine phosphorylated at position 474 (Y474) after insertion into the host cell membrane. Whereas Tir phosphorylation is required for pedestal formation (Kenny, 1999; Goosney et al., 2000), Tir translocation is thought to be completely independent of host cell modifications. In studies of EPEC attachment to red blood cells (RBCs), tyrosine phosphorylation of Tir and pedestal formation was not seen, although Tir was inserted correctly into the RBC membrane and was able to bind intimin (Shaw et al., 2002). The Tir chaperone, CesT, functions to direct Tir to the translocation apparatus and may help to maintain Tir in a secretion-competent state (Abe et al., 1999).

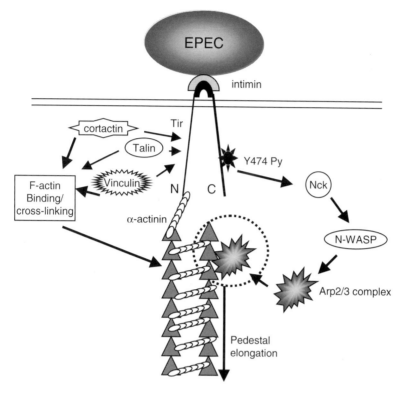

Figure 4.3. Simplified schematic of EPEC-induced signaling events leading to pedestal formation. The binding of intimin to translocated Tir leads to tyrosine phosphorylation of the protein at position 474. This phosphorylation is required for the direct binding of the cellular adaptor Nck, which signals N-WASP recruitment to Tir, and the subsequent recruitment of the Arp2/3 complex to allow pedestal formation. The cytoskeletal proteins α-actinin, villin, cortactin, and vinculin all bind directly to Tir and are thought to play a role in pedestal formation through their F-actin binding sites.

Several host cell structural proteins have been implicated in pedestal formation on the basis of observations made by use of immunofluorescence microscopy. Pedestals are predominantly formed from filamentous actin, with microfilament-associated proteins such as α-actinin, talin, ezrin, and villin also being found along the length of the pedestal (Goosney et al., 2001). The actin-stabilizing proteins tropomyosin and nonmuscle myosin II can be seen at the base of a pedestal (Sanger et al., 1996). The Tir protein serves directly as a conduit between host cell and bacteria, and much research has been focused on the mechanisms that enable this interaction, because this enigmatic protein offers us a unique window into the dynamics of host cell cytoskeletal rearrangement.

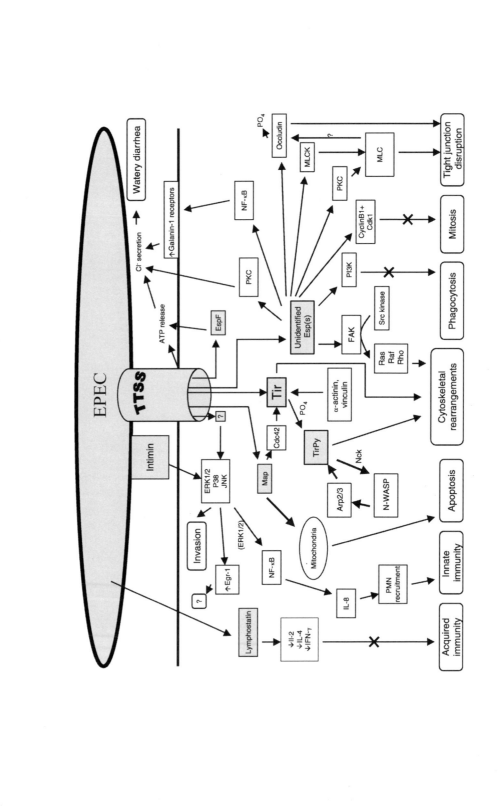

Tyrosine phosphorylation of Tir plays an essential role in pedestal formation by facilitating the direct binding of the cellular adaptor protein Nck (Gruenheid et al., 2001; Campellone et al., 2002). Nck binding leads to the recruitment of N-WASP, and the Arp2/3 complex, which in turn allows pedestal elongation (Kalman et al., 1999). Cultured cells defective for the expression of N-WASP do not support the formation of EPEC actin pedestals, highlighting the central role played by this protein (Lommel et al., 2001). The critical role played by N-WASP suggests an involvement of Rho family GTPases. Surprisingly, these do not appear to play a role in pedestal formation (Ben-Ami et al., 1998; Ebel et al., 1998).

Tir has also been demonstrated to bind directly to the focal adhesion (FA) proteins α-actinin, talin, and vinculin. FAs are anchoring structures that allow the close association of epithelial cells and fibroblasts to the extracellular matrix (ECM). Tir binds directly to α-actinin by means of its amino terminus (Goosney et al., 2000; Huang et al., 2002), in a manner that is tyrosine phosphorylation independent (Goosney et al., 2000). Tir can also bind to talin, and this association has been shown to be essential for correct F-actin focusing beneath adherent cells, possibly through the nucleation of nascent actin filaments (Cantarelli et al., 2001). Vinculin has also been found to bind to Tir directly (Freeman et al., 2000). The association of FA proteins with the pedestal structure supports the notion that pedestals are composites of FAs and microvilli (Freeman et al., 2000; Goosney et al., 2001). Recent research has shown that EPEC can disrupt FAs, by specifically dephosphorylating the FA kinase, FAK. FAK dephosphorylation requires a functional TTSS and is enhanced, but not absolutely dependent on, pedestal formation (Shifrin et al., 2002). Cortactin, an F-actin-binding protein, has been shown to accumulate underneath adherent EPEC in its tyrosine-dephosphorylated form (Cantarelli et al., 2002). In this form, cortactin is able to efficiently bind other cytoskeletal proteins and to cross-link F-actin (Huang et al., 1997), and it may thus play an important role in pedestal formation.

Several signaling molecules are thought to be involved in pedestal formation (Fig. 4.4). Phospholipase C (PLC), phosphtidyinositol 3-kinase (PI3K), and 5-lipoxygenase have all been shown to be important for the accumulation of α-actinin beneath adherent EPEC, because inhibitors of these molecules

Figure 4.4. (*facing page*). EPEC-induced signaling events in the host cell. The interaction of EPEC effector molecules (gray boxes) with host proteins (white boxes) leads to the exploitation of various signaling pathways, and an array of outcomes for the host (rounded boxes). For details of individual pathways, see text. (For clarity, some less understood pathways have been omitted.)

also inhibited α-actinin accumulation in a concentration-dependent manner (Johnson-Henry et al., 2001). Recent evidence also suggests that the F-actin-binding protein, annexin 2, is recruited underneath the sites of EPEC attachment in epithelial monolayers (Zobiack et al., 2002) and may act as a bridge between actin and the host cell membrane (Gerke and Moss, 1997). Recruitment of annexin 2 by EPEC was found to be Tir independent, raising the possibility that cytoskeletal rearrangements by EPEC may occur by at least two distinct pathways – one that is Tir mediated and another that is not (Zobiack et al., 2002).

Although EHEC and EPEC both induce pedestal formation, EHEC uses methods distinct from EPEC to recruit the actin-nucleating machinery. EHEC Tir is less than 60% identical to EPEC Tir, with the C-terminal domain showing the least homology (Paton et al., 1998). Indeed, the absence of the tyrosine residue that is targeted for phosphorylation in EPEC Tir correlates with the finding that EHEC is able to induce pedestal formation in the absence of Tir tyrosine phosphorylation (Ismaili et al., 1995; DeVinney et al., 2001). Although EHEC Tir is not tyrosine phosphorylated and does not bind Nck, both N-WASP and the Arp2/3 complex are still recruited to the EHEC pedestal (Goosney et al., 2001), which suggests that EHEC and EPEC modulate different pathways in order to form pedestals (DeVinney et al., 2001).

A further major difference is that EHEC, but not EPEC, can form pedestals with mutant EPEC Tir in which the Y474 has been replaced by phenylalanine (DeVinney et al., 2001; Kenny, 2001), suggesting that additional EHEC-specific bacterial factor(s) may be involved in pedestal formation by EHEC that is independent of Tir tyrosine phosphorylation (DeVinney et al., 2001). Although only a small proportion of EHEC strains, including EHEC O157, form pedestals that do not contain tyrosine-phosphorylated Tir, several non-O157 serotypes that are designated as EHEC have Tir sequences similar to that of EPEC and are tyrosine phosphorylated in a similar way to EPEC Tir (DeVinney et al., 2001). Because the O157:H7 serotype correlates closely with human EHEC infections, it is tempting to speculate that serotypes that form pedestals in a tyrosine-independent manner represent the more efficient pathogens, as they are less reliant on the signaling capacities of the host cell in order to subvert host cell processes to their own ends.

Inhibition of phagocytosis

Several A/E pathogens, including EPEC, EHEC, and REPEC, are able to inhibit uptake by specialist phagocytic cells such as M cells and macrophages, and they can even colonize the niches rich in these cells (Inman and Cantey, 1983; Kresse et al., 2001). An ability to inhibit phagocytosis would clearly be advantageous to the bacteria by delaying the host immune response.

The EPEC antiphagocytic phenotype has been demonstrated to be reliant on a functional TTSS and is independent of pedestal formation (Goosney et al., 1999). This study demonstrated that host protein tyrosine dephosphorylation could be observed in type III secretion-competent EPEC, and that mutants that had nonfunctional type III secretion systems were unable to elicit host protein modifications, suggesting that antiphagocytosis is reliant on a bacterial protein effector that is secreted through the type III mechanism. More recent research was able to show that EPEC blocks its uptake by inhibition of a PI3K-mediated pathway (Celli et al., 2001). Whether EHEC and other A/E pathogens utilize the same mechanism for antiphagocytosis is unknown (although considered likely). EPEC, however, has the distinction of being the first pathogen for which an antiphagocytic pathway of this type has been described.

Inhibition of mitosis

Certain strains of REPEC and EPEC (although not the reference strain E2348/69) are able to induce in HeLa cells an irreversible cytopathic effect, characterized by the progressive recruitment of stress fibers and FAs and associated with the prevention of cell proliferation (De Rycke et al., 1997; Nougayrede et al., 1999). Such effects are not related to any toxin secretion, are dependent on a functional TTSS, and require the secreted proteins EspA, EspB, and EspD, but not Tir or intimin (De Rycke et al., 1997; Nougayrede et al., 1999; Marches et al., 2000). A closer examination of the observed cytostatic effect revealed that the cell cycle is arrested in the G_2/M phase, and that this mitotic inhibition is not likely to be a consequence of the cytoskeletal arrangements themselves (Nougayrede et al., 2001).

The host cell protein, Cdk1, together with the protein cyclin B1, forms a complex that governs the transition from G_2 to mitosis. In normal cells, Cdk1 is present at a constant level throughout the cell cycle, and its dephosphorylation is required for entry into mitosis (Norbury and Nurse, 1992). Strains of REPEC and EPEC that exert a cytostatic effect appear to affect Cdk1 dephosphorylation by an as yet unknown mechanism, arresting the cell cycle in the G_2 phase (Nougayrede et al., 2001). Whether mitotic inhibition plays a role in EPEC pathogenesis, and by which exact method, can only be guessed at pending further research into this effect. However, it has been suggested that a cell-cycle arrest of the stem cells that supply cells to the intestinal villi may delay the shedding of the epithelia, thus allowing the bacteria to remain attached to its host for a prolonged period (Nougayrede et al., 2001).

Disruption of tight junctions

Although as yet incompletely understood, the pathogenesis of EPEC-induced diarrhea may involve not only the formation of A/E lesions but also disruption of epithelial barrier function. Tight junctions are specialized inter-cellular structures that form "gaskets" around epithelial cells that prevent the leakage of fluid through the intercellular gaps, and the resultant diffusion of membrane proteins. Several studies have demonstrated that EPEC infection of host cells in tissue culture consistently leads to a significant decrease in transepithelial resistance (TER), a measure of tight junction integrity, in a time-dependent manner (Spitz et al., 1995). This host response has also been shown to be dependent on a functional TTSS (Canil et al., 1993; Philpott et al., 1996), although attachment of bacteria mediated by Tir–intimin interactions was not in itself sufficient to trigger host cell membrane depolarization (Stein et al., 1996).

EPEC-induced phosphorylation of myosin light chain (MLC) seems to play a central role in the pathway that leads to eventual loss of cell membrane integrity (Manjarrez-Hernandez et al., 1996). Both EPEC and EHEC infections *in vitro* have been shown to activate protein kinase C (PKC), which is likely to lead to MLC phosphorylation (Crane and Oh, 1997; Philpott et al., 1998). Activation of MLC-kinase (MLCK) leads to further phosphorylation of MLC (Manjarrez-Hernandez et al., 1996) and contraction of the cytoskeletal rings underlying tight junctions, and thus it leads to a loss of tight junction integrity.

The host cell protein occludin is thought to play an important role in maintaining the integrity of tight junctions (McCarthy et al., 1996). EPEC infection has been shown to result in dephosphorylation of occludin and its resulting dissociation from tight junctions into the intracellular space, which would very likely contribute to tight junction permeability (Simonovic et al., 2000). Whether this occludin dephosphorylation is linked to the pathway involving the phosphorylation of MLC, or whether the two pathways are separate, is unknown. Occludin is one of many proteins identified in the makeup of tight junctions, and whether EPEC targets any other tight junction proteins has not yet been determined. Interestingly, EHEC has been shown to alter the distribution of another important tight junction-associated protein, ZO-1, again through the action of PKC (Philpott et al., 1998), which directly targets ZO-1 for phosphorylation (Stuart and Nigam, 1995).

Recently it was found that EPEC mutants deficient in their ability to express EspF, a protein that is translocated into the host cell cytoplasm by the TTSS, were deficient in their abilities to disrupt electrical resistance across polarized epithelial cells (McNamara et al., 2001). EspF appears to rely on a chaperone protein, CesF, for full translocation efficiency (Elliott et al., 2002). Once translocated into the host cell, EspF is not found distributed homogenously; instead it appears to be sequestered, perhaps by the host cell protein(s) with which it interacts (McNamara et al., 2001). Whether EspF does indeed modulate host functions through interaction with a host cell protein remains to be elucidated.

Upregulation of chloride secretion and adenosine triphosphate (ATP) release

One of the hallmark symptoms of EPEC infection is the onset of severe, watery diarrhea. PKC activation in the intestine triggers a rapid secretion of ions (particularly chloride ions, Cl^-) and fluid into the lumen (Beubler and Schirgi-Degen, 1993). Thus it was not surprising to find that EPEC is able to activate PKC (Crane and Oh, 1997), which is likely to result in the upregulation of Cl^- secretion and the subsequent development of watery diarrhea.

A further mechanism whereby watery diarrhea might be induced involves the galanin-1 receptors found on the surface of epithelial cells lining the human GI tract. These receptors, on activation, cause Cl^- secretion. Expression of the galanin-1 receptor is transcriptionally regulated by nuclear factor (NF)-κB, which is in turn activated during infection with both EHEC and EPEC (Savkovic et al., 1997; Dahan et al., 2002). Indeed, it was demonstrated that infection with both EHEC and EPEC increased the number of galanin-1 receptors by means of the activation of NF-κB, and thus elevated Cl^- secretion (Hecht et al., 1999), a factor that contributes to the etiology of infectious watery diarrhea.

In addition to NF-κB–PKC activation, a further hypothesis for the pathogenesis of diarrhea promoted by EPEC infection has emerged. It was recently demonstrated that polarized cells could be induced to release a large concentration of ATP into the apical culture medium upon infection with EPEC (Crane et al., 2002). This release of ATP is TTSS dependent, and, in particular, the EspF protein appears to play an important role (Crane et al., 2002). ATP release in response to EPEC infection may be a direct result of leakage from the ATP-rich cytosol through pores formed by the bacterial TTSS. The pore created by EPEC in the host cell membrane is between 30 and 50 Å (Ide et al., 2001), which would allow for easy passage of the ATP molecule. ATP release may also be effected by the modulation by EPEC of the cystic fibrosis transmembrane regulator (CFTR; see Crane et al., 2002), which acts as a cellular "valve" to allow for the normal release of ATP from the host cell cytosol during cell stretching or swelling or as a response to cAMP stimulation (Jiang et al., 1998).

Once extracellular, ATP is rapidly broken down into less phosphorylated nucleotides and adenosine. Adenine nucleotides released from EPEC-infected cells could spread to neighboring cells and may trigger a fluid secretory response. This could explain the apparent paradox that EPEC preferentially adheres to the villi of the gut epithelia, yet it is the crypt cells of the gut that have the capacity to generate watery diarrhea through ion secretion (Crane et al., 2002). The exact mechanism of the stimulation of ATP release by EPEC, and whether the same mechanism is utilized by EHEC, remains to be elucidated.

The secretion of Cl^- and the subsequent development of watery diarrhea may be considered an innate host defense mechanism that has evolved to rid the host of enterobacterial pathogens such as EHEC and EPEC. For the pathogen itself, this defense mechanism has the advantage that it may allow for increased spread of the bacteria from host to host.

The Ca^{2+} controversy

Early research into the effects on host cells of A/E lesion-forming EPEC demonstrated that near to the sites of bacterial attachment there seemed to be a localized elevation of Ca^{2+} within the host cell (Baldwin et al., 1991), coupled with the phosphorylation of (among other proteins) MLC (Manjarrez-Hernandez et al., 1991). Subsequent to this, increased levels of inositol triphosphate (IP) were detected in EPEC-infected cells (Dytoc et al., 1994; Foubister et al., 1994), and thus it seemed plausible that the PLC pathway, which generates IP and in turn a release of Ca^{2+} from intracellular stores, was being stimulated by EPEC infection. Additionally, it was found that the buffering of intracellular free Ca^{2+} by use of the chelating agent BAPTA could prevent A/E lesion formation by EPEC (Baldwin et al., 1991; Dytoc et al., 1994).

Recent work using more sensitive methods has been able to measure both temporal and spatial measurements of Ca^{2+} in live cells infected with both EPEC and EHEC (Bain et al., 1998). In contrast to the earlier studies just described, no increase in Ca^{2+} levels could be measured, and the addition of BAPTA to host cells was not seen to prevent the formation of A/E lesions (Bain et al., 1998). It was suggested that the alterations in cell Ca^{2+}

levels seen in previous studies was related to the cytotoxic effects of EPEC (Bain et al., 1998). Clearly, more research has to be done in this area to determine whether or not EPEC and EHEC can modulate host intracellular Ca^{2+} levels.

Activation of mitogen-activated protein (MAP) kinase cascades

The MAP kinases group includes extracellular signal-related protein kinases 1 and 2 (ERK1 and ERK2), p38 MAP kinase, and c-Jun N-terminal kinase (JNK). As well as the classical mitogenic response, MAP kinases have also been implicated in the regulation of stress responses, cytokine and chemokine responses, and cytoskeletal reorganization (Garrington and Johnson, 1999). Indeed, EHEC is able to induce the expression of the chemokine interleukin-8 (IL-8) in host cells via MAP kinase pathways (Dahan et al., 2002; discussed later).

MAP kinases play a role in the host cell response to infection with invasive bacteria such as *Salmonella* (Hobbie et al., 1997). Although invasion is not thought to be a central process required for EPEC pathogenicity, EPEC was found to be able to invade T84 epithelial cells in culture, and this invasion relied on the activation of ERK1/2 MAP kinase (Czerucka et al., 2000). Additionally, it was found that the intimate attachment of EPEC to host cells was able to activate ERK1/2, p38, and JNK signaling pathways in a manner that was dependent on both a functional TTSS and the expression of intimin (Czerucka et al., 2001). The MAP kinase-activated pathways proceed through a cascade of phosphorylation events leading to the transcriptional activation of certain genes. Recently EPEC was found to upregulate the transcription of the gene coding for the early growth response factor-1 (Egr-1) through the ERK1/2 pathway, the p38 pathway, or both (de Grado et al., 2001). Egr-1 is a protein that is activated in most cell types in response to stress, and it was found that the mouse *egr-1* gene was upregulated during *in vivo* infection with the EPEC-related pathogen, *Citrobacter rodentium* (de Grado et al., 2001), demonstrating that this effect is not isolated to cultured cells. The functional consequences of upregulation of Egr-1 production by infected cells remain to be determined and may prove to be a key part of the pathogenic strategy of these bacteria.

Modulation of apoptosis

The induction of apoptosis plays an important role in the regulation of the immune response to bacterial infections, although the process of apoptosis may be hijacked by the pathogen to its own advantage (Muller and Rudel,

2001). Several enteric pathogens are known to trigger apoptosis in host cells both *in vitro* and *in vivo* (Muller and Rudel, 2001).

For EPEC, the induction of apoptosis in host cells would seem to be counterintuitive because the bacteria rely on their adhesion to the host epithelia in order to effect A/E lesion formation. Several studies have found that EPEC may actually act to slow down the apoptotic process during infection. For example, although EPEC-infected cells eventually succumbed to cell death that had hallmark features of apoptosis, such as early expression of phosphatidylserine (PS) on host cell surfaces and internucleosomal cleavage of DNA, the time taken for this effect to become apparent was much greater than for apoptosis induced by invasive bacteria (Crane et al., 1999). More recently, a comprehensive *in vivo* study in rabbits infected with REPEC O103 demonstrated that although apoptosis physiologically occurs in the rabbit ileum, particularly at the tips of the absorptive villi, infection with REPEC did not promote the rate of apoptosis but may actually have diminished it (Heczko et al., 2001). Thus the modulation of apoptosis during infection by EPEC and EHEC could be an important bacterial strategy used both to increase the likelihood of bacterial attachment and to control the rate of (and downstream effects of) host cell death.

The stimulation of tyrosine kinases, PKC, and the transcription factor NF-κB all suppress cell death, and EPEC is able to effectively activate these factors (Crane and Oh, 1997; Savkovic et al., 1997). Interestingly, an EPEC secreted protein, EspF, has been shown to effect cell death in a manner that has features compatible with pure apoptosis (Crane et al., 2001). Conversely, EPEC BFP may play a role in the induction of apoptosis, because nonpathogenic *E. coli* strains expressing BFP genes induced significant levels of apoptosis in host epithelial cells (Abul-Milh et al., 2001).

The role that EHEC plays in the modulation of apoptosis has not been so extensively studied, although from the limited data available it appears that EHEC is also able to effect apoptosis in a manner that is dependent on bacterial attachment and occurs in a similar timeframe to apoptotic cell death induced by EPEC (Barnett Foster et al., 2000). The best-characterized toxin of EHEC, SLT, is also known to induce apoptosis (Jones et al., 2000), although the rate of this toxin-mediated apoptosis is much slower than that triggered by EPEC and EHEC attachment (Barnett Foster et al., 2000). Interestingly, this study also pointed to a rationale for EPEC and EHEC adhesion, because one of the physiological effects of apoptotic cell death is the upregulation of levels of phosphatidylethanolamine (PE) in the cell membrane, a molecule used by both EPEC and EHEC for bacterial attachment (Barnett Foster et al., 1999).

Disruption of mitochondrial membrane potential

In agreement with the observation of host cell apoptotis in response to EPEC infection, it has recently emerged that the gene *orf19*, upstream of *tir*, codes for a protein that seems to interfere with the mitochondrial ability to maintain its membrane potential, a proapoptotic trigger (Kenny and Jepson, 2000). The precise mechanism of action of the Orf19 protein (since renamed Map, for mitochondrial-associated protein) and the consequences of its interactions with mitochondria have yet to be elucidated, although the delivery of Map into the host cell cytoplasm by means of the TTSS and its subsequent targeting directly to the mitochondria represents the first discovered example of bacteria–host cell interactions of this type (Kenny and Jepson, 2000). Whether EHEC is able to target mitochondria in a similar manner remains to be elucidated, but this seems likely if one considers the possession by EHEC of a Map homologue (Perna et al., 2001).

Although it may seem counterintuitive for EPEC to possess mechanisms that act both to prevent and induce apoptosis, in fact it demonstrates the extraordinary ability of this pathogen to control a central host cell function. The downstream effects of modulation of apoptosis are diverse and may be advantageous to the bacterium at different stages of infection. It is possible that bacterial genes that affect apoptosis are regulated in response to diverse environmental stimuli encountered during the disease process.

Modulation of Cdc42 activity

Early during infection, EPEC is able to trigger transient filopodia-like cytoskeletal rearrangements distinct from those stimulated by Tir–intimin pedestal formation (Kenny et al., 2002). Interestingly, it was found that if Map expression was increased by expression of *orf19* from a plasmid, then the formation of filopodia in infected cells was stimulated, although this stimulation was independent of any observable effect on the mitochondria (Kenny et al., 2002). This second role for Map was demonstrated to be dependent on the activation of the host GTPase, Cdc42 (Kenny et al., 2002), a molecule that is not required for pedestal formation triggered by Tir and intimin (Kalman et al., 1999), although it plays a role in cytoskeletal modification in uninfected cells.

Surprisingly, the transient nature of filopodia formation was found to be a result of further interactions of Cdc42 with Tir, which serve to deactivate the host GTPase. In this respect, Tir may also have a dual function, behaving as a GTPase-activating protein (GAP) and a signal-transducing molecule, as well as a receptor for intimin. To modulate Cdc42 activity, Tir seems to require its interaction with intimin, although this function of the Tir–intimin complex is separate from its ability to induce pedestal formation (Kenny et al., 2002).

This fascinating interplay between two bacterial proteins that seem to act on host cell functions in an opposing manner appears to be counterintuitive at first. It is thought that the Cdc42-dependent signaling mediated by Map is inhibitory to pedestal formation (Kenny et al., 2002), and because pedestal formation is important to virulence, a bacterial mechanism to downregulate the Cdc42-mediated pathway is necessary. Why Map induces Cdc42 in the first place is a mystery that remains to be solved, although Map is undoubtedly important to the infection process (because, if it were not, its function would likely have been abolished by natural selection).

Modification of host immune responses

Innate immunity

Infection of host cells with microbial pathogens, including EPEC and EHEC, leads to a proinflammatory response by means of the triggering of various cellular signals that activate the NF-κB pathway and stimulate expression of IL-8 (Kresse et al., 2001; Savkovic et al., 2001). EPEC has been found to exploit the ERK1/2 pathway and to induce inflammation via these routes (Czerucka et al., 2000, 2001; Savkovic et al., 2001). EHEC infection *in vitro* results in phosphorylation of ERK1/2 as well as two further groups of the MAPK family, p38 and JNK, which activate NF-κB as well as the transcription factor, AP-1 (Dahan et al., 2002). These transcription factors in turn regulate the expression of IL-8. It is thought that the stimulation of these pathways is likely to be effected by the secretion of as yet unknown bacterial factors by means of the EPEC–EHEC TTSS (Dahan et al., 2002).

When histological specimens from the intestines of animals infected with EPEC are examined, there is evidence for a dramatic infiltration by inflammatory cells, in particular, polymorphonuclear leukocytes (PMNs) (Moon et al., 1983). The exact advantage to the pathogen of PMN attraction to the site of infection remains to be elucidated, although it is known that recruited PMNs release prostaglandins, which in turn increase the activity of adenylate cyclase in intestinal cells. As a result, cAMP levels increase, leading to a release of Cl$^-$ by the cells. As already discussed, release of Cl$^-$ contributes to the development of watery diarrhea, a hallmark of both EPEC and EHEC infection and a likely mechanism by which the bacteria facilitate their spread from host to host.

One further advantage of PMN infiltration, to SLT-producing EHEC in particular, has recently been described. The genes encoding SLT in EHEC are carried as part of lysogenic bacteriophages, and it is thought that the induction of these phages during infection contributes to EHEC pathogenesis by increasing the transcription and copy number of the phage genes (Plunkett et al., 1999). It was recently found that phage induction during EHEC infection may be due, at least in part, to the hydrogen peroxide released by neutrophils as a component of their antibacterial arsenal (Wagner et al., 2001). Indeed, SLT has been shown to inhibit neutrophil apoptosis (Liu et al., 1999), which may prolong H_2O_2 release and promote production of SLT.

Adaptive immunity

The severity of EPEC and EHEC infections tends to correlate with the age of the patient, with infants and young children being most at risk for the development of disease. It has been found that infants are more likely to develop diarrhea during their first colonization with EPEC than they are during subsequent exposures to the pathogen (Cravioto et al., 1990), although whether this is also true of EHEC infections of infants is unknown. In adult volunteers infected with EPEC, there was no specific effect of prior EPEC infection on the incidence of diarrhea, although disease severity was reduced in individuals who were reinfected with a homologous EPEC strain (Donnenberg et al., 1998). Effects such as these may be the result of acquired (adaptive) immunity in the host.

Several studies have been undertaken in an effort to identify the antibody responses generated to specific EPEC and EHEC antigens during the acute and later stages of disease. Given the key roles that EHEC and EPEC factors, such as Esps, Tir, intimin, and (for EPEC) BFP, play in the virulence of these pathogens, studies have usually focused on the immune responses generated to these bacterial factors in both natural and volunteer infection studies.

Human volunteers experimentally infected with EPEC developed a strong IgG response against intimin (Nataro and Kaper, 1998), although in subsequent studies of natural infections of both EHEC and EPEC, an antibody response to intimin has varied in strength from strong to undetectable (Li et al., 2000). Immune responses to EspA and EspB were similarly variable in patients naturally infected with EPEC (Martinez et al., 1999). In sera from patients infected with EHEC O157:H7, there was little reactivity to EspA and EspB during the acute phase of illness, although this titer was increased in the later stages of disease. Tir was shown to induce a significant antibody response in EHEC infections (Li et al., 2000), and this response was found to persist after infection (Li et al., 2000). BFP was found to elicit an IgG response, but not an IgA response, in children infected with BFP$^+$ EPEC strains (Martinez et al., 1999).

Interestingly, a study of secretory IgA purified from breast milk from mothers who were living in areas where EPEC was endemic showed that these antibodies prevented the adherence of EPEC to cultured epithelial cells (Cravioto et al., 1991). The antibodies reacted to many EPEC proteins, several of them unknown (Manjarrez-Hernandez et al., 2000). This illustrates, in part, the complexity of the human immune response to EPEC (which is likely to be reflected by that to EHEC) and indicates that there is still much to be learned before a suitable molecular target is identified for vaccine production.

Lysates of both EPEC and EHEC are known to inhibit lymphokine production by lymphoid cells from multiple sites (Malstrom and James, 1998). The nature of this inhibition was recently partially characterized when it was found that a large gene present in EPEC encoded a toxin that specifically acted to inhibit lymphocyte proliferation, and IL-2, IL-4, and gamma interferon (IFN-γ) production (Klapproth et al., 2000). The toxin was named lymphostatin and is one of the largest bacterial toxins known. Homologues of *lifA* (encoding lymphostatin) are found in most strains of EPEC, some strains of EHEC (but not O157 strains) where it is designated Efa1, and *C. rodentium* (Klapproth et al., 2000), correlating with an A/E ability of the pathogen.

The mechanism of action of lymphostatin is yet to be determined because it is difficult to purify (Klapproth et al., 2000). However, its effects are limited to lymphokine expression and appear to cause no changes in the epithelial cell cytoskeleton (Klapproth et al., 2000). A potential consequence of lymphostatin expression may be the suppression of an adaptive immune response to the bacteria, thus prolonging the infection and increasing the likelihood that the pathogen will be passed on to new hosts.

CONCLUSIONS

By now, it should be clear that the pathogenesis of EPEC and EHEC infection is not a simple process. EPEC and EHEC have evolved myriad strategies to subvert host cell functions that, when combined, give rise to the etiology of the diseases that they cause. The host is undoubtedly not passive during infection with these bacteria, but the outcome of infection rests on the delicate balance of host cell processes, some of which are competing in a molecular tug of war controlled by the host on one side and the bacteria on the other. With the advent of increasing bacterial resistance to antibiotics, we can no longer rely on antimicrobial compounds for the effective treatment of bacterial infections. To design better therapeutics, we will need to target the mechanisms the bacteria use to control their hosts.

Although we have been forced to find alternative methods to treat infections with bacteria such as EPEC and EHEC, this coercion has not been without benefit. The study of bacteria–host cell interactions has given new dimension to the field of cell biology, and as we learn more about the ways in which bacteria interact with us, we will undoubtedly discover levels of complexity within the host cell far beyond those we have studied to date. Pathogens themselves have become the cell biologist's working tools and are providing fascinating insights into the workings of the eukaryotic cell.

ACKNOWLEDGMENTS

As a result of space limitations, we could not reference original sources for all data cited in this review; original sources can be found in the reviews cited. Research in the DeVinney Lab is supported by Canadian Institutes for Health Research, and the Alberta Heritage Foundation for Medical Research (AHFMR). E. Allen-Vercoe is an AHFMR postdoctoral fellow, and R. DeVinney is an AHFMR scholar.

REFERENCES

Abe, A., de Grado, M., Pfuetzner, R.A., Sanchez-Sanmartin, C., Devinney, R., Puente, J.L., Strynadka, N.C., and Finlay, B.B. (1999). Enteropathogenic *Escherichia coli* translocated intimin receptor, Tir, requires a specific chaperone for stable secretion. *Mol. Microbiol.* **33**, 1162–1175.

Abe, A., Heczko, U., Hegele, R.G., and Finlay, B.B. (1998). Two enteropathogenic *Escherichia coli* type III secreted proteins, EspA and EspB, are virulence factors. *J. Exp. Med.* **188**, 1907–1916.

Abe, H., Tatsuno, I., Tobe, T., Okutani, A., and Sasakawa, C. (2002). Bicarbonate ion stimulates the expression of locus of enterocyte effacement-encoded genes in enterohemorrhagic *Escherichia coli* O157:H7. *Infect. Immun.* **70**, 3500–3509.

Abul–Milh, M., Wu, Y., Lau, B., Lingwood, C.A., and Foster, D.B. (2001). Induction of epithelial cell death including apoptosis by enteropathogenic *Escherichia coli* expressing bundle-forming pili. *Infect. Immun.* **69**, 7356–7364.

Allen-Vercoe, E. and Woodward, M.J. (1999). The role of flagella, but not fimbriae, in the adherence of *Salmonella enterica* serotype Enteritidis to chick gut explant. *J. Med. Microbiol.* **48**, 771–780.

Bain, C., Keller, R., Collington, G.K., Trabulsi, L.R., and Knutton, S. (1998). Increased levels of intracellular calcium are not required for the formation of attaching and effacing lesions by enteropathogenic and enterohemorrhagic *Escherichia coli. Infect. Immun.* **66**, 3900–3908.

Baldwin, T.J., Lee-Delaunay, M.B., Knutton, S., and Williams, P.H. (1993). Calcium-calmodulin dependence of actin accretion and lethality in cultured HEp-2 cells infected with enteropathogenic *Escherichia coli. Infect. Immun.* **61**, 760–763.

Baldwin, T.J., Ward, W., Aitken, A., Knutton, S., and Williams, P.H. (1991). Elevation of intracellular free calcium levels in HEp-2 cells infected with enteropathogenic *Escherichia coli. Infect. Immun.* **59**, 1599–1604.

Barnett Foster, D., Abul-Milh, M., Huesca, M., and Lingwood, C.A. (2000). Enterohemorrhagic *Escherichia coli* induces apoptosis which augments bacterial binding and phosphatidylethanolamine exposure on the plasma membrane outer leaflet. *Infect. Immun.* **68**, 3108–3115.

Barnett Foster, D., Philpott, D., Abul-Milh, M., Huesca, M., Sherman, P.M., and Lingwood, C.A. (1999). Phosphatidylethanolamine recognition promotes enteropathogenic *E. coli* and enterohemorrhagic *E. coli* host cell attachment. *Microb. Pathog.* **27**, 289–301.

Ben-Ami, G., Ozeri, V., Hanski, E., Hofmann, F., Aktories, K., Hahn, K.M., Bokoch, G.M., and Rosenshine, I. (1998). Agents that inhibit Rho, Rac, and Cdc42 do not block formation of actin pedestals in HeLa cells infected with enteropathogenic *Escherichia coli. Infect. Immun.* **66**, 1755–1758.

Beubler, E. and Schirgi-Degen, A. (1993). Stimulation of enterocyte protein kinase C by laxatives in-vitro. *J. Pharm. Pharmacol.* **45**, 59–62.

Bieber, D., Ramer, S.W., Wu, C.Y., Murray, W.J., Tobe, T., Fernandez, R., and Schoolnik, G.K. (1998). Type IV pili, transient bacterial aggregates, and virulence of enteropathogenic *Escherichia coli. Science* **280**, 2114–2118.

Campellone, K.G., Giese, A., Tipper, D.J., and Leong, J.M. (2002). A tyrosine-phosphorylated 12-amino-acid sequence of enteropathogenic *Escherichia coli* Tir binds the host adaptor protein Nck and is required for Nck localization to actin pedestals. *Mol. Microbiol.* **43**, 1227–1241.

Canil, C., Rosenshine, I., Ruschkowski, S., Donnenberg, M.S., Kaper, J.B., and Finlay, B.B. (1993). Enteropathogenic *Escherichia coli* decreases the transepithelial electrical resistance of polarized epithelial monolayers. *Infect. Immun.* **61**, 2755–2762.

Cantarelli, V.V., Takahashi, A., Yanagihara, I., Akeda, Y., Imura, K., Kodama, T., Kono, G., Sato, Y., and Honda, T. (2001). Talin, a host cell protein, interacts directly with the translocated intimin receptor, Tir, of enteropathogenic

Escherichia coli, and is essential for pedestal formation. *Cell. Microbiol.* **3**, 745–751.

Cantarelli, V.V., Takahashi, A., Yanagihara, I., Akeda, Y., Imura, K., Kodama, T., Kono, G., Sato, Y., Iida, T., and Honda, T. (2002). Cortactin is necessary for F-actin accumulation in pedestal structures induced by enteropathogenic *Escherichia coli* infection. *Infect. Immun.* **70**, 2206–2209.

Celli, J., Olivier, M., and Finlay, B.B. (2001). Enteropathogenic *Escherichia coli* mediates antiphagocytosis through the inhibition of PI 3-kinase-dependent pathways. *EMBO J.* **20**, 1245–1258.

Crane, J.K., Majumdar, S., and Pickhardt, D.F. III (1999). Host cell death due to enteropathogenic *Escherichia coli* has features of apoptosis. *Infect. Immun.* **67**, 2575–2584.

Crane, J.K., McNamara, B.P., and Donnenberg, M.S. (2001). Role of EspF in host cell death induced by enteropathogenic *Escherichia coli*. *Cell. Microbiol.* **3**, 197–211.

Crane, J.K. and Oh, J.S. (1997). Activation of host cell protein kinase C by enteropathogenic *Escherichia coli*. *Infect. Immun.* **65**, 3277–3285.

Crane, J.K., Olson, R.A., Jones, H.M., and Duffey, M.E. (2002). Release of ATP during host cell killing by enteropathogenic *E. coli* and its role as a secretory mediator. *Am. J. Physiol. Gastrointest. Liver Physiol.* **283**, G74–G86.

Cravioto, A., Reyes, R.E., Trujillo, F., Uribe, F., Navarro, A., De La Roca, J.M., Hernandez, J.M., Perez, G., and Vazquez, V. (1990). Risk of diarrhea during the first year of life associated with initial and subsequent colonization by specific enteropathogens. *Am. J. Epidemiol.* **131**, 886–904.

Cravioto, A., Tello, A., Villafan, H., Ruiz, J., del Vedovo, S., and Neeser, J.R. (1991). Inhibition of localized adhesion of enteropathogenic *Escherichia coli* to HEp-2 cells by immunoglobulin and oligosaccharide fractions of human colostrum and breast milk. *J. Infect. Dis.* **163**, 1247–1255.

Czerucka, D., Dahan, S., Mograbi, B., Rossi, B., and Rampal, P. (2001). Implication of mitogen-activated protein kinases in T84 cell responses to enteropathogenic *Escherichia coli* infection. *Infect. Immun.* **69**, 1298–1305.

Czerucka, D., Dahan, S., Mograbi, B., Rossi, B., and Rampal, P. (2000). *Saccharomyces boulardii* preserves the barrier function and modulates the signal transduction pathway induced in enteropathogenic *Escherichia coli*-infected T84 cells. *Infect. Immun.* **68**, 5998–6004.

Dahan, S., Busuttil, V., Imbert, V., Peyron, J.F., Rampal, P., and Czerucka, D. (2002). Enterohemorrhagic *Escherichia coli* infection induces interleukin-8 production via activation of mitogen-activated protein kinases and the transcription factors NF-kappa B and AP-1 in T84 cells. *Infect. Immun.* **70**, 2304–2310.

de Grado, M., Rosenberger, C.M., Gauthier, A., Vallance, B.A., and Finlay, B.B. (2001). Enteropathogenic *Escherichia coli* infection induces expression of the early growth response factor by activating mitogen-activated protein kinase cascades in epithelial cells. *Infect. Immun.* **69**, 6217–6224.

De Rycke, J., Comtet, E., Chalareng, C., Boury, M., Tasca, C., and Milon, A. (1997). Enteropathogenic *Escherichia coli* O103 from rabbit elicits actin stress fibers and focal adhesions in HeLa epithelial cells, cytopathic effects that are linked to an analog of the locus of enterocyte effacement. *Infect. Immun.* **65**, 2555–2563.

DeVinney, R., Puente, J.L., Gauthier, A., Goosney, D., and Finlay, B.B. (2001). Enterohaemorrhagic and enteropathogenic *Escherichia coli* use a different Tir-based mechanism for pedestal formation. *Mol. Microbiol.* **41**, 1445–1458.

DeVinney, R., Stein, M., Reinscheid, D., Abe, A., Ruschkowski, S., and Finlay, B.B. (1999). Enterohemorrhagic *Escherichia coli* O157:H7 produces Tir, which is translocated to the host cell membrane but is not tyrosine phosphorylated. *Infect. Immun.* **67**, 2389–2398.

Donnenberg, M.S., Giron, J.A., Nataro, J.P., and Kaper, J.B. (1992). A plasmid-encoded type IV fimbrial gene of enteropathogenic *Escherichia coli* associated with localized adherence. *Mol. Microbiol.* **6**, 3427–3437.

Donnenberg, M.S., Tacket, C.O., Losonsky, G., Frankel, G., Nataro, J.P., Dougan, G., and Levine, M.M. (1998). Effect of prior experimental human enteropathogenic *Escherichia coli* infection on illness following homologous and heterologous rechallenge. *Infect. Immun.* **66**, 52–58.

Donnenberg, M.S., Tzipori, S., McKee, M.L., O'Brien, A.D., Alroy, J., and Kaper, J.B. (1993). The role of the *eae* gene of enterohemorrhagic *Escherichia coli* in intimate attachment in vitro and in a porcine model. *J. Clin. Invest.* **92**, 1418–1424.

Dytoc, M., Fedorko, L., and Sherman, P.M. (1994). Signal transduction in human epithelial cells infected with attaching and effacing *Escherichia coli* in vitro. *Gastroenterology* **106**, 1150–1161.

Ebel, F., von Eichel-Streiber, C., Rohde, M., and Chakraborty, T. (1998). Small GTP-binding proteins of the Rho- and Ras-subfamilies are not involved in the actin rearrangements induced by attaching and effacing *Escherichia coli*. *FEMS Microbiol. Lett.* **163**, 107–112.

Elliott, S.J., O'Connell, C.B., Koutsouris, A., Brinkley, C., Donnenberg, M.S., Hecht, G., and Kaper, J.B. (2002). A gene from the locus of enterocyte effacement that is required for enteropathogenic *Escherichia coli* to increase tight-junction permeability encodes a chaperone for EspF. *Infect. Immun.* **70**, 2271–2277.

Elliott, S.J., Sperandio, V., Giron, J.A., Shin, S., Mellies, J.L., Wainwright, L., Hutcheson, S.W., McDaniel, T.K., and Kaper, J.B. (2000). The locus of enterocyte effacement (LEE)-encoded regulator controls expression of both LEE- and non-LEE-encoded virulence factors in enteropathogenic and enterohemorrhagic *Escherichia coli*. *Infect. Immun.* **68**, 6115–6126.

Elliott, S.J., Wainwright, L.A., McDaniel, T.K., Jarvis, K.G., Deng, Y.K., Lai, L.C., McNamara, B.P., Donnenberg, M.S., and Kaper, J.B. (1998). The complete sequence of the locus of enterocyte effacement (LEE) from enteropathogenic *Escherichia coli* E2348/69. *Mol. Microbiol.* **28**, 1–4.

Elliott, S.J., Yu, J., and Kaper, J.B. (1999). The cloned locus of enterocyte effacement from enterohemorrhagic *Escherichia coli* O157:H7 is unable to confer the attaching and effacing phenotype upon *E. coli* K-12. *Infect. Immun.* **67**, 4260–4263.

Fitzhenry, R.J., Pickard, D.J., Hartland, E.L., Reece, S., Dougan, G., Phillips, A.D., and Frankel, G. (2002). Intimin type influences the site of human intestinal mucosal colonisation by enterohaemorrhagic *Escherichia coli* O157:H7. *Gut* **50**, 180–185.

Foubister, V., Rosenshine, I., and Finlay, B.B. (1994). A diarrheal pathogen, enteropathogenic *Escherichia coli* (EPEC), triggers a flux of inositol phosphates in infected epithelial cells. *J. Exp. Med.* **179**, 993–998.

Frankel, G., Candy, D.C., Fabiani, E., Adu-Bobie, J., Gil, S., Novakova, M., Phillips, A.D., and Dougan, G. (1995). Molecular characterization of a carboxy-terminal eukaryotic-cell-binding domain of intimin from enteropathogenic *Escherichia coli*. *Infect. Immun.* **63**, 4323–4328.

Frankel, G., Philips, A.D., Novakova, M., Batchelor, M., Hicks, S., and Dougan, G. (1998). Generation of *Escherichia coli* intimin derivatives with differing biological activities using site-directed mutagenesis of the intimin C-terminus domain. *Mol. Microbiol.* **29**, 559–570.

Frankel, G., Phillips, A.D., Novakova, M., Field, H., Candy, D.C., Schauer, D.B., Douce, G., and Dougan, G. (1996). Intimin from enteropathogenic *Escherichia coli* restores murine virulence to a *Citrobacter rodentium eaeA* mutant: induction of an immunoglobulin A response to intimin and EspB. *Infect. Immun.* **64**, 5315–5325.

Frankel, G., Phillips, A.D., Rosenshine, I., Dougan, G., Kaper, J.B., and Knutton, S. (1998). Enteropathogenic and enterohaemorrhagic *Escherichia coli*: more subversive elements. *Mol. Microbiol.* **30**, 911–921.

Freeman, N.L., Zurawski, D.V., Chowrashi, P., Ayoob, J.C., Huang, L., Mittal, B., Sanger, J.M., and Sanger, J.W. (2000). Interaction of the enteropathogenic *Escherichia coli* protein, translocated intimin receptor (Tir), with focal adhesion proteins. *Cell Motil. Cytoskeleton* **47**, 307–318.

Friedberg, D., Umanski, T., Fang, Y., and Rosenshine, I. (1999). Hierarchy in the expression of the locus of enterocyte effacement genes of enteropathogenic *Escherichia coli*. *Mol. Microbiol.* **34**, 941–952.

Garrington, T.P. and Johnson, G.L. (1999). Organization and regulation of mitogen-activated protein kinase signaling pathways. *Curr. Opin. Cell. Biol.* **11**, 211–218.

Gauthier, A., de Grado, M., and Finlay, B.B. (2000). Mechanical fractionation reveals structural requirements for enteropathogenic *Escherichia coli* Tir insertion into host membranes. *Infect. Immun.* **68**, 4344–4348.

Gerke, V. and Moss, S.E. (1997). Annexins and membrane dynamics. *Biochim. Biophys. Acta* **1357**, 129–154.

Giron, J.A., Ho, A.S., and Schoolnik, G.K. (1991). An inducible bundle-forming pilus of enteropathogenic *Escherichia coli*. *Science* **254**, 710–713.

Giron, J.A., Torres, A.G., Freer, E., and Kaper, J.B. (2002). The flagella of enteropathogenic *Escherichia coli* mediate adherence to epithelial cells. *Mol. Microbiol.* **44**, 361–379.

Goosney, D.L., Celli, J., Kenny, B., and Finlay, B.B. (1999). Enteropathogenic *Escherichia coli* inhibits phagocytosis. *Infect. Immun.* **67**, 490–495.

Goosney, D.L., DeVinney, R., and Finlay, B.B. (2001). Recruitment of cytoskeletal and signaling proteins to enteropathogenic and enterohemorrhagic *Escherichia coli* pedestals. *Infect. Immun.* **69**, 3315–3322.

Goosney, D.L., DeVinney, R., Pfuetzner, R.A., Frey, E.A., Strynadka, N.C., and Finlay, B.B. (2000). Enteropathogenic *E. coli* translocated intimin receptor, Tir, interacts directly with alpha-actinin. *Curr. Biol.* **10**, 735–738.

Gruenheid, S., DeVinney, R., Bladt, F., Goosney, D., Gelkop, S., Gish, G.D., Pawson, T., and Finlay, B.B. (2001). Enteropathogenic *E. coli* Tir binds Nck to initiate actin pedestal formation in host cells. *Nat. Cell Biol.* **3**, 856–859.

Hecht, G., Marrero, J.A., Danilkovich, A., Matkowskyj, K.A., Savkovic, S.D., Koutsouris, A., and Benya, R.V. (1999). Pathogenic *Escherichia coli* increase Cl-secretion from intestinal epithelia by upregulating galanin-1 receptor expression. *J. Clin. Invest.* **104**, 253–262.

Heczko, U., Carthy, C.M., O'Brien, B.A., and Finlay, B.B. (2001). Decreased apoptosis in the ileum and ileal Peyer's patches: a feature after infection with rabbit enteropathogenic *Escherichia coli* O103. *Infect. Immun.* **69**, 4580–4589.

Hobbie, S., Chen, L.M., Davis, R.J., and Galan, J.E. (1997). Involvement of mitogen-activated protein kinase pathways in the nuclear responses and cytokine production induced by *Salmonella typhimurium* in cultured intestinal epithelial cells. *J. Immunol.* **159**, 5550–5559.

Huang, C., Ni, Y., Wang, T., Gao, Y., Haudenschild, C.C., and Zhan, X. (1997). Down-regulation of the filamentous actin cross-linking activity of cortactin by Src-mediated tyrosine phosphorylation. *J. Biol. Chem.* **272**, 13,911–13,915.

Huang, L., Mittal, B., Sanger, J.W., and Sanger, J.M. (2002). Host focal adhesion protein domains that bind to the translocated intimin receptor (Tir) of enteropathogenic *Escherichia coli* (EPEC). *Cell. Motil. Cytoskeleton* **52**, 255–265.

Hueck, C.J. (1998). Type III protein secretion systems in bacterial pathogens of animals and plants. *Microbiol. Mol. Biol. Rev.* **62**, 379–433.

Ide, T., Laarmann, S., Greune, L., Schillers, H., Oberleithner, H., and Schmidt, M.A. (2001). Characterization of translocation pores inserted into plasma membranes by type III-secreted Esp proteins of enteropathogenic *Escherichia coli*. *Cell. Microbiol.* **3**, 669–679.

Inman, L.R. and Cantey, J.R. (1983). Specific adherence of *Escherichia coli* (strain RDEC-1) to membranous (M) cells of the Peyer's patch in *Escherichia coli* diarrhea in the rabbit. *J. Clin. Invest.* **65**, 1–8.

Ismaili, A., Philpott, D.J., Dytoc, M.T., and Sherman, P.M. (1995). Signal transduction responses following adhesion of verocytotoxin-producing *Escherichia coli*. *Infect. Immun.* **63**, 3316–3326.

Jenkins, C., Chart, H., Smith, H.R., Hartland, E.L., Batchelor, M., Delahay, R.M., Dougan, G., and Frankel, G. (2000). Antibody response of patients infected with verocytotoxin-producing *Escherichia coli* to protein antigens encoded on the LEE locus. *J. Med. Microbiol.* **49**, 97–101.

Jiang, Q., Mak, D., Devidas, S., Schwiebert, E.M., Bragin, A., Zhang, Y., Skach, W.R., Guggino, W.B., Foskett, J.K., and Engelhardt, J.F. (1998). Cystic fibrosis transmembrane conductance regulator-associated ATP release is controlled by a chloride sensor. *J. Cell Biol.* **143**, 645–657.

Johnson-Henry, K., Wallace, J.L., Basappa, N.S., Soni, R., Wu, G.K., and Sherman, P.M. (2001). Inhibition of attaching and effacing lesion formation following enteropathogenic *Escherichia coli* and Shiga toxin-producing *E. coli* infection. *Infect. Immun.* **69**, 7152–7158.

Jones, N.L., Islur, A., Haq, R., Mascarenhas, M., Karmali, M., Purdue, M.H., Z.B.W., and Sherman, P. (2000). *Escherichia coli* Shiga toxins induce apoptosis in epithelial cells that is regulated by the Bcl-2 family. *Am. J. Physiol. Gastrointest. Liver Physiol.* **278**, G811–G819.

Kalman, D., Weiner, O.D., Goosney, D.L., Sedat, J.W., Finlay, B.B., Abo, A., and Bishop, J.M. (1999). Enteropathogenic *E. coli* acts through WASP and Arp2/3 complex to form actin pedestals. *Nat. Cell Biol.* **1**, 389–391.

Kanamaru, K., Tatsuno, I., Tobe, T., and Sasakawa, C. (2000). Regulation of virulence factors of enterohemorrhagic *Escherichia coli* O157:H7 by self-produced extracellular factors. *Biosci. Biotechnol. Biochem.* **64**, 2508–2511.

Kenny, B. (2001). The enterohaemorrhagic *Escherichia coli* (serotype O157:H7) Tir molecule is not functionally interchangeable for its enteropathogenic *E. coli* (serotype O127:H6) homologue. *Cell. Microbiol.* **3**, 499–510.

Kenny, B. (1999). Phosphorylation of tyrosine 474 of the enteropathogenic *Escherichia coli* (EPEC) Tir receptor molecule is essential for actin nucleating activity and is preceded by additional host modifications. *Mol. Microbiol.* **31**, 1229–1241.

Kenny, B., DeVinney, R., Stein, M., Reinscheid, D.J., Frey, E.A., and Finlay, B.B. (1997). Enteropathogenic *E. coli* (EPEC) transfers its receptor for intimate adherence into mammalian cells. *Cell* **91**, 511–520.

Kenny, B., Ellis, S., Leard, A.D., Warawa, J., Mellor, H., and Jepson, M.A. (2002). Co-ordinate regulation of distinct host cell signalling pathways by multifunctional enteropathogenic *Escherichia coli* effector molecules. *Mol. Microbiol.* **44**, 1095–1107.

Kenny, B. and Jepson, M. (2000). Targeting of an enteropathogenic *Escherichia coli* (EPEC) effector protein to host mitochondria. *Cell. Microbiol.* **2**, 579–590.

Klapproth, J.M., Donnenberg, M.S., Abraham, J.M., and James, S.P. (1996). Products of enteropathogenic *E. coli* inhibit lymphokine production by gastrointestinal lymphocytes. *Am. J. Physiol.* **271**, G841–G848.

Klapproth, J.M., Scaletsky, I.C., McNamara, B.P., Lai, L.C., Malstrom, C., James, S.P., and Donnenberg, M.S. (2000). A large toxin from pathogenic *Escherichia coli* strains that inhibits lymphocyte activation. *Infect. Immun.* **68**, 2148–2155.

Knutton, S., Baldwin, T., Williams, P.H., and McNeish, A.S. (1989). Actin accumulation at sites of bacterial adhesion to tissue culture cells: basis of a new diagnostic test for enteropathogenic and enterohemorrhagic *Escherichia coli*. *Infect. Immun.* **57**, 1290–1298.

Knutton, S., Rosenshine, I., Pallen, M.J., Nisan, I., Neves, B.C., Bain, C., Wolff, C., Dougan, G., and Frankel, G. (1998). A novel EspA-associated surface organelle of enteropathogenic *Escherichia coli* involved in protein translocation into epithelial cells. *EMBO J.* **17**, 2166–2176.

Kodama, T., Akeda, Y., Kono, G., Takahashi, A., Imura, K., Iida, T., and Honda, T. (2002). The EspB protein of enterohaemorrhagic *Escherichia coli* interacts directly with alpha-catenin. *Cell. Microbiol.* **4**, 213–222.

Kondro, W. (2000). *E. coli* outbreak deaths spark judicial inquiry in Canada. *Lancet* **355**, 2058.

Kresse, A.U., Guzman, C.A., and Ebel, F. (2001). Modulation of host cell signalling by enteropathogenic and Shiga toxin-producing *Escherichia coli*. *Int. J. Med. Microbiol.* **291**, 277–285.

Kresse, A.U., Rohde, M., and Guzman, C.A. (1999). The EspD protein of enterohemorrhagic *Escherichia coli* is required for the formation of bacterial surface appendages and is incorporated in the cytoplasmic membranes of target cells. *Infect. Immun.* **67**, 4834–4842.

Li, Y., Frey, E., Mackenzie, A.M., and Finlay, B.B. (2000). Human response to *Escherichia coli* O157:H7 infection: antibodies to secreted virulence factors. *Infect. Immun.* **68**, 5090–5095.

Liu, J., Akahoshi, T., Sasahana, T., Kitasato, H., Namai, R., Sasaki, T., Inoue, M., and Kondo, H. (1999). Inhibition of neutrophil apoptosis by verotoxin 2 derived from *Escherichia coli* O157:H7. *Infect. Immun.* **67**, 6203–6205.

Lommel, S., Benesch, S., Rottner, K., Franz, T., Wehland, J., and Kuhn, R. (2001). Actin pedestal formation by enteropathogenic *Escherichia coli* and intracellular motility of *Shigella flexneri* are abolished in N-WASP-defective cells. *EMBO Rep.* **2**, 850–857.

Loureiro, I., Frankel, G., Adu-Bobie, J., Dougan, G., Trabulsi, L.R., and Carneiro-Sampaio, M.M. (1998). Human colostrum contains IgA antibodies reactive to enteropathogenic *Escherichia coli* virulence-associated proteins: intimin, BfpA, EspA, and EspB. *J. Pediatr. Gastroenterol. Nutr.* **27**, 166–171.

Luo, Y., Frey, E.A., Pfuetzner, R.A., Creagh, A.L., Knoechel, D.G., Haynes, C.A., Finlay, B.B., and Strynadka, N.C. (2000). Crystal structure of enteropathogenic *Escherichia coli* intimin-receptor complex. *Nature* **405**, 1073–1077.

Malstrom, C. and James, S. (1998). Inhibition of murine splenic and mucosal lymphocyte function by enteric bacterial products. *Infect. Immun.* **66**, 3120–3127.

Manjarrez-Hernandez, H.A., Amess, B., Sellers, L., Baldwin, T.J., Knutton, S., Williams, P.H., and Aitken, A. (1991). Purification of a 20 kDa phosphoprotein from epithelial cells and identification as a myosin light chain. Phosphorylation induced by enteropathogenic *Escherichia coli* and phorbol ester. *FEBS Lett.* **292**, 121–127.

Manjarrez-Hernandez, H.A., Baldwin, T.J., Williams, P.H., Haigh, R., Knutton, S., and Aitken, A. (1996). Phosphorylation of myosin light chain at distinct sites and its association with the cytoskeleton during enteropathogenic *Escherichia coli* infection. *Infect. Immun.* **64**, 2368–2370.

Manjarrez-Hernandez, H.A., Gavilanes-Parra, S., Chavez-Berrocal, E., Navarro-Ocana, A., and Cravioto, A. (2000). Antigen detection in enteropathogenic *Escherichia coli* using secretory immunoglobulin A antibodies isolated from human breast milk. *Infect. Immun.* **68**, 5030–5036.

Marches, O., Nougayrede, J.P., Boullier, S., Mainil, J., Charlier, G., Raymond, I., Pohl, P., Boury, M., DeRycke, J., Milon, A., and Oswald, E. (2000). Role of *tir* and intimin in the virulence of rabbit enteropathogenic *Escherichia coli* serotype O103:H2. *Infect. Immun.* **68**, 2171–2182.

Martinez, M.B., Taddei, C.R., Ruiz-Tagle, A., Trabulsi, L.R., and Giron, J.A. (1999). Antibody response of children with enteropathogenic *Escherichia coli* infection to the bundle-forming pilus and locus of enterocyte effacement-encoded virulence determinants. *J. Infect. Dis.* **179**, 269–274.

McCarthy, K.M., Skare, I.B., Stankewich, M.C., Furuse, M., Tsukita, S., Rogers, R.A., Lynch, R.D., and Schneeberger, E.E. (1996). Occludin is a functional component of the tight junction. *J. Cell Sci.* **109**, 2287–2298.

McDaniel, T.K. and Kaper, J.B. (1997). A cloned pathogenicity island from enteropathogenic *Escherichia coli* confers the attaching and effacing phenotype on *E. coli* K-12. *Mol. Microbiol.* **23**, 399–407.

McNamara, B.P., Koutsouris, A., O'Connell, C.B., Nougayrede, J.P., Donnenberg, M.S., and Hecht, G. (2001). Translocated EspF protein from enteropathogenic *Escherichia coli* disrupts host intestinal barrier function. *J. Clin. Invest.* **107**, 621–629.

Mellies, J.L., Elliott, S.J., Sperandio, V., Donnenberg, M.S., and Kaper, J.B. (1999). The Per regulon of enteropathogenic *Escherichia coli*: identification of a regulatory cascade and a novel transcriptional activator, the locus of enterocyte effacement (LEE)-encoded regulator (Ler). *Mol. Microbiol.* **33**, 296–306.

Moon, H.W., Whipp, S.C., Argenzio, R.A., Levine, M.M., and Giannella, R.A. (1983). Attaching and effacing activities of rabbit and human enteropathogenic *Escherichia coli* in pig and rabbit intestines. *Infect. Immun.* **41**, 1340–1351.

Muller, A. and Rudel, T. (2001). Modification of host cell apoptosis by viral and bacterial pathogens. *Int. J. Med. Microbiol.* **291**, 197–207.

Nataro, J.P. and Kaper, J.B. (1998). Diarrheagenic *Escherichia coli*. *Clin. Microbiol. Rev.* **11**, 142–201.

Nicholls, L., Grant, T.H., and Robins-Browne, R.M. (2000). Identification of a novel genetic locus that is required for *in vitro* adhesion of a clinical isolate of enterohaemorrhagic *Escherichia coli* to epithelial cells. *Mol. Microbiol.* **35**, 275–288.

Nisan, I., Wolff, C., Hanski, E., and Rosenshine, I. (1998). Interaction of enteropathogenic *Escherichia coli* with host epithelial cells. *Folia Microbiol.* **43**, 247–252.

Norbury, C. and Nurse, P. (1992). Animal cell cycles and their control. *Annu. Rev. Biochem.* **61**, 441–470.

Nougayrede, J.P., Boury, M., Tasca, C., Marches, O., Milon, A., Oswald, E., and De Rycke, J. (2001). Type III secretion-dependent cell cycle block caused in HeLa cells by enteropathogenic *Escherichia coli* O103. *Infect. Immun.* **69**, 6785–6795.

Nougayrede, J.P., Marches, O., Boury, M., Mainil, J., Charlier, G., Pohl, P., De Rycke, J., Milon, A., and Oswald, E. (1999). The long-term cytoskeletal rearrangement induced by rabbit enteropathogenic *Escherichia coli* is Esp dependent but intimin independent. *Mol. Microbiol.* **31**, 19–30.

EMMA ALLEN-VERCOE AND REBEKAH DEVINNEY

O'Longhlin, E.V. and Robins-Browne, R.M. (2001). Effect of Shiga toxin and Shiga-like toxins on eucaryotic cells. *Microbes Infect.* **3**, 493–507.

Paton, A.W., Manning, P.A., Woodrow, M.C., and Paton, J.C. (1998). Translocated intimin receptors (Tir) of Shiga-toxigenic *Escherichia coli* isolates belonging to serogroups O26, O111, and O157 react with sera from patients with hemolytic-uremic syndrome and exhibit marked sequence heterogeneity. *Infect. Immun.* **66**, 5580–5586.

Perna, N.T., Plunkett, G. III, Burland, V., Mau, B., Glasner, J.D., Rose, D.J., Mayhew, G.F., Evans, P.S., Gregor, J., Kirkpatrick, H.A., Posfai, G., Hackett, J., Klink, S., Boutin, A., Shao, Y., Miller, L., Grotbeck, E.J., Davis, N.W., Lim, A., Dimalanta, E.T., Potamousis, K.D., Apodaca, J., Anantharaman, T.S., Lin, J., Yen, G., Schwartz, D.C., Welch, R.A., and Blattner, F.R. (2001). Genome sequence of enterohaemorrhagic *Escherichia coli* O157:H7. *Nature* **409**, 529–533.

Philpott, D.J., McKay, D.M., Mak, W., Perdue, M.H., and Sherman, P.M. (1998). Signal transduction pathways involved in enterohemorrhagic *Escherichia coli*-induced alterations in T84 epithelial permeability. *Infect. Immun.* **66**, 1680–1687.

Philpott, D.J., McKay, D.M., Sherman, P.M., and Perdue, M.H. (1996). Infection of T84 cells with enteropathogenic *Escherichia coli* alters barrier and transport functions. *Am. J. Physiol.* **270**, G634–G645.

Plunkett, G. III, Rose, D.J., Durfee, T.J., and Blattner, F.R. (1999). Sequence of Shiga toxin 2 phage 933W from *Escherichia coli* O157:H7: Shiga toxin as a phage late-gene product. *J. Bacteriol.* **181**, 1767–1778.

Rosenshine, I., Donnenberg, M.S., Kaper, J.B., and Finlay, B.B. (1992). Signal transduction between enteropathogenic *Escherichia coli* (EPEC) and epithelial cells: EPEC induces tyrosine phosphorylation of host cell proteins to initiate cytoskeletal rearrangement and bacterial uptake. *EMBO J.* **11**, 3551–3560.

Rosenshine, I., Ruschkowski, S., Stein, M., Reinscheid, D.J., Mills, S.D., and Finlay, B.B. (1996). A pathogenic bacterium triggers epithelial signals to form a functional bacterial receptor that mediates actin pseudopod formation. *EMBO J.* **15**, 2613–2624.

Sanger, J.M., Chang, R., Ashton, F., Kaper, J.B., and Sanger, J.W. (1996). Novel form of actin-based motility transports bacteria on the surfaces of infected cells. *Cell. Motil. Cytoskeleton* **34**, 279–287.

Savkovic, S.D., Koutsouris, A., and Hecht, G. (1997). Activation of NF-kappa B in intestinal epithelial cells by enteropathogenic *Escherichia coli*. *Am. J. Physiol.* **273**, C1160–C1167.

Savkovic, S.D., Ramaswamy, A., Koutsouris, A., and Hecht, G. (2001). EPEC-activated ERK1/2 participate in inflammatory response but not tight junction barrier disruption. *Am. J. Physiol. Gastrointest. Liver Physiol.* **281**, G890–G898.

Sekiya, K., Ohishi, M., Ogino, T., Tamano, K., Sasakawa, C., and Abe, A. (2001). Supermolecular structure of the enteropathogenic *Escherichia coli* type III secretion system and its direct interaction with the EspA-sheath-like structure. *Proc. Natl. Acad. Sci. USA* **98**, 11,638–11,643.

Shaw, R.K., Daniell, S., Frankel, G., and Knutton, S. (2002). Enteropathogenic *Escherichia coli* translocate Tir and form an intimin-Tir intimate attachment to red blood cell membranes. *Microbiology* **148**, 1355–1365.

Shifrin, Y., Kirschner, J., Geiger, B., and Rosenshine, I. (2002). Enteropathogenic *Escherichia coli* induces modification of the focal adhesions of infected host cells. *Cell. Microbiol.* **4**, 235–243.

Simonovic, I., Rosenberg, J., Koutsouris, A., and Hecht, G. (2000). Enteropathogenic *Escherichia coli* dephosphorylates and dissociates occludin from intestinal epithelial tight junctions. *Cell. Microbiol.* **2**, 305–315.

Sinclair, J.F. and O'Brien, A.D. (2002). Cell surface-localized nucleolin is a eukaryotic receptor for the adhesin intimin-gamma of enterohemorrhagic *Escherichia coli* O157:H7. *J. Biol. Chem.* **277**, 2876–2885.

Sperandio, V., Mellies, J.L., Nguyen, W., Shin, S., and Kaper, J.B. (1999). Quorum sensing controls expression of the type III secretion gene transcription and protein secretion in enterohemorrhagic and enteropathogenic *Escherichia coli*. *Proc. Natl. Acad. Sci. USA* **96**, 15,196–15,201.

Spitz, J., Yuhan, R., Koutsouris, A., Blatt, C., Alverdy, J., and Hecht, G. (1995). Enteropathogenic *Escherichia coli* adherence to intestinal epithelial monolayers diminishes barrier function. *Am. J. Physiol.* **268**, G374–G379.

Stein, M.A., Mathers, D.A., Yan, H., Baimbridge, K.G., and Finlay, B.B. (1996). Enteropathogenic *Escherichia coli* markedly decreases the resting membrane potential of Caco-2 and HeLa human epithelial cells. *Infect. Immun.* **64**, 4820–4825.

Stevens, M.P., van Diemen, P.M., Frankel, G., Phillips, A.D., and Wallis, T.S. (2002). Efa1 influences colonization of the bovine intestine by Shiga toxin-producing *Escherichia coli* serotypes O5 and O111. *Infect. Immun.* **70**, 5158–5166.

Stuart, R.O. and Nigam, S.K. (1995). Regulated assembly of tight junctions by protein kinase C. *Proc. Natl. Acad. Sci. USA* **92**, 6072–6076.

Tarr, P.I., Bilge, S.S., Vary, J.C. Jr., Jelacic, S., Habeeb, R.L., Ward, T.R., Baylor, M.R., and Besser, T.E. (2000). Iha: a novel *Escherichia coli* O157:H7 adherence-conferring molecule encoded on a recently acquired chromosomal island of conserved structure. *Infect. Immun.* **68**, 1400–1407.

Tobe, T., Hayashi, T., Han, C.G., Schoolnik, G.K., Ohtsubo, E., and Sasakawa, C. (1999). Complete DNA sequence and structural analysis of the enteropathogenic *Escherichia coli* adherence factor plasmid. *Infect. Immun.* **67**, 5455–5462.

EMMA ALLEN-VERCOE AND REBEKAH DEVINNEY

Wachter, C., Beinke, C., Mattes, M., and Schmidt, M.A. (1999). Insertion of EspD into epithelial target cell membranes by infecting enteropathogenic *Escherichia coli*. *Mol. Microbiol.* **31**, 1695–1707.

Wagner, P.L., Acheson, D.W., and Waldor, M.K. (2001). Human neutrophils and their products induce Shiga toxin production by enterohemorrhagic *Escherichia coli*. *Infect. Immun.* **69**, 1934–1937.

Warawa, J., Finlay, B.B., and Kenny, B. (1999). Type III secretion-dependent hemolytic activity of enteropathogenic *Escherichia coli*. *Infect. Immun.* **67**, 5538–5540.

Zobiack, N., Rescher, U., Laarmann, S., Michgehl, S., Schmidt, M.A., and Gerke, V. (2002). Cell-surface attachment of pedestal-forming enteropathogenic *E. coli* induces a clustering of raft components and a recruitment of annexin 2. *J. Cell Sci.* **115**, 91–98.

STEALTH WARFARE: THE INTERACTIONS OF EPEC AND EHEC WITH HOST CELLS

CHAPTER 5

Molecular ecology and cell biology of
Legionella pneumophila

Maëlle Molmeret, Dina M. Bitar, and Yousef Abu Kwaik

Legionella pneumophila, a Gram-negative bacillus that is ubiquitous in aquatic environments, is responsible for Legionnaires' disease. It is a facultative intracellular pathogen that can replicate within eukaryotic host cells such as protozoan and macrophages. In water, *L. pneumophila* grows within protozoan hosts. There are at least 13 species of amoebae and 2 species of ciliated protozoa that support intracellular replication of *L. pneumophila* (Fields, 1996). Among the most predominant amoebae in water sources are *hartmannellae* and *acanthamoebae*, which have also been isolated from water sources associated with Legionnaires' disease outbreaks (Fields, 1996). Interaction between *L. pneumophila* and protozoa is considered to be central to the pathogenesis and ecology of *L. pneumophila* (Rowbotham, 1986; Harb et al., 2000). In humans, *L. pneumophila* reaches the lungs after inhalation of contaminated aerosol droplets (Fields, 1996; Fliermans, 1996; also see Fig. 5.1). The main sources of contaminated water droplets are hot water and air-conditioning systems, but the bacteria have been isolated from fountains, spas, pools, dental and hospital units, and other man-made water systems (Fliermans, 1996; also see Fig. 5.1). No person-to-person transmission has been described. Once in the lungs, *L. pneumophila* are ingested in alveolar macrophages, the major site of bacterial replication. This results in an acute and severe pneumonia. In addition to Legionnaires' disease, *L. pneumophila* also causes Pontiac fever, which is a self-limiting flu-like illness that is not well understood but is not lethal. Approximately one half of the 48 species of *Legionella* have been associated with human disease. *L. pneumophila* is responsible for 90% of cases of Legionnaires' disease. However, all the *Legionella* species under appropriate conditions may be capable of intracellular growth and infliction of human disease. Infections that are due to less common species of legionellae

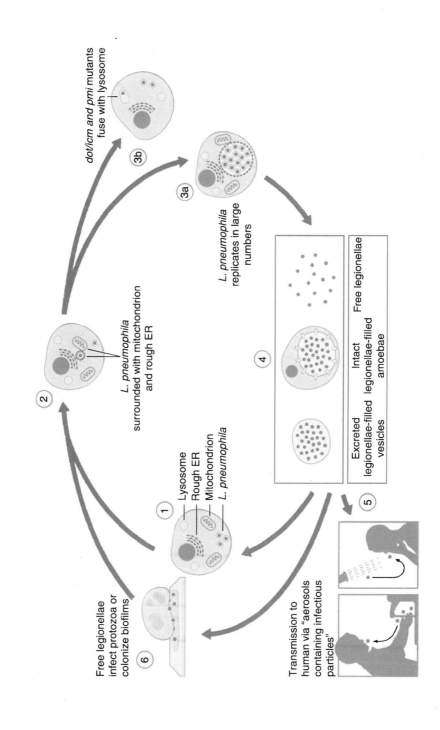

dot/icm and *pmi* mutants fuse with lysosome

3b

3a

L. pneumophila replicates in large numbers

2

L. pneumophila surrounded with mitochondrion and rough ER

1

Lysosome
Rough ER
Mitochondrion
L. pneumophila

4

Excreted legionellae-filled vesicles

Intact legionellae-filled amoebae

Free legionellae

5

Transmission to human via "aerosols containing infectious particles"

6

Free legionellae infect protozoa or colonize biofilms

are not frequently diagnosed and reported, and they are less studied than *L. pneumophila* (Fields et al., 2002).

The unique intracellular fate of *L. pneumophila* is one of the interesting aspects of this organism. Unlike phagosomes containing inert particles or avirulent bacteria, the *L. pneumophila*-containing vacuoles avoid the "normal" endocytic pathway, recruiting rough endoplasmic reticulum (RER) and mitochondria, to reside in a specialized vacuole allowing intracellular replication (Horwitz and Silverstein, 1980; Horwitz, 1983a, 1983b, 1984; Horwitz and Maxfield, 1984; also see Fig. 5.2). The formation of this specialized vacuole is directed by the type IV secretion system encoded by the *dot/icm* genes. The *dot* (defect in organelle trafficking)/*icm* (intracellular multiplication) loci consist of 23 genes located in two chromosomal loci of *L. pneumophila*. These genes have been identified independently by two different laboratories (Segal et al., 1998; Segal and Shuman, 1998; Vogel et al., 1998). An analysis of the predicted amino acid sequences of the *dot/icm* genes has revealed several characteristics that indicate a role in conjugal transfer of DNA, which has been confirmed by conjugation studies (Segal and Shuman, 1998; Vogel et al., 1998). This apparatus has also been shown to be involved in proper maturation of the *L. pneumophila*-containing phagosome in mammalian and protozoan cells, directing the biogenesis of the specialized vacuole in which *Legionella* replicate (Swanson and Isberg, 1995b; Segal and Shuman, 1999; Molmeret et al., 2002a). The *dot/icm* genes are also required for macropinocytosis in A/J mice macrophages (Watarai et al., 2001b), upregulation of phagocytosis in human-derived macrophages (Hilbi et al., 2001), induction of apoptosis in macrophages (Gao and Abu Kwaik, 1999a, 1999b; Zink et al., 2002), and

---

Figure 5.1. (*facing page*). The environmental life of *L. pneumophila* within protozoa.
(1) *L. pneumophila* from biofilms with other bacteria, or in suspension, infecting protozoa.
(2) Following entry, *L. pneumophila* resides in a membrane-bound vacuole that recruits host cell organelles, such as the mitochondria and the rough endoplasmic reticulum, and does not fuse with lysosomes. Mutants such as *dot/icm* mutants fuse to lysosomes (3b); *L. pneumophila* replicates within specialized vacuoles and reaches large numbers (3a).
(4) The infectious particle is not known but may include excreted *Legionella*-filled vesicles, intact *Legionella*-filled amoebae, or free legionellae that have lysed their host cell.
(5) Transmission to humans occurs by mechanical means, such as faucets and showerheads. Infection in humans occurs by inhalation of the infectious particle and establishment of infection in the lungs. (6) Legionellae that have escaped their host cell may survive in suspension for long periods of time, reinfect other protozoa, or recolonize biofilms. (This figure was taken from *ASM News* 66 (10): 609–616, 2000.)

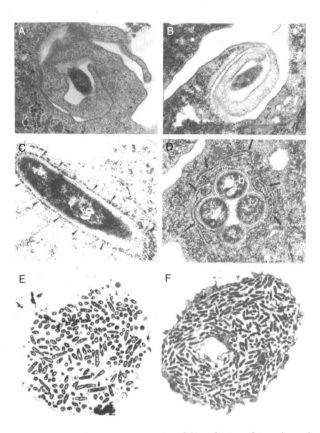

Figure 5.2. Transmission electron micrographs of the infection of *Acanthamoeba polyphaga* (top panel) and U937 macrophages (bottom panel) by *L. pneumophila*. Coiling phagocytosis (A and B); formation of the RER-surrounded phagosome (C and D); and late stages of the infection (E and F). Note that in E and F there is no intact phagosomal membrane. (This figure was adapted from *ASM News* 66 (10): 609–616, 2000.)

pore-formation-mediated cytotoxicity in both protozoan and mammalian cells (Alli et al., 2000; Gao and Abu Kwaik, 2000; Molmeret et al., 2002a, 2002b).

ECOLOGY OF LEGIONELLAE

After isolation of *L. pneumophila* from the air-conditioning system during the first outbreak in Philadelphia, Pennsylvania (1976), the bacteria have been isolated from numerous sources in the environment. Legionellae species have been repeatedly shown to be ubiquitous, particularly in aquatic environments (Fields, 1996).

MAËLLE MOLMERET, DINA M. BITAR, AND YOUSEF ABU KWAIK

In the environment, legionellae species cannot multiply extracellularly, and they have been shown to be parasites of protozoa (Harb et al., 2000). In 1980, Rowbotham was the first to describe the ability of *L. pneumophila* to multiply intracellularly within protozoa (Rowbotham, 1980). The 13 species of amoebae and 2 species of ciliated protozoa that allow intracellular bacterial replication have been shown to be potential environmental hosts for legionellae species (Abu Kwaik et al., 1998). This rather sophisticated host–parasite interaction indicates a tremendous adaptation of legionellae to parasitize protozoa. This host–parasite interaction is also central to the pathogenesis and ecology of these bacteria.

There are now at least 46 species of legionellae [Benin, et al., 2002]. In addition, 12 phylogenetic groups of bacteria belonging to 5 species have been designated as *Legionella*-like amoebal pathogens (LLAP; Adeleke et al., 1996). The LLAPs are genetically related to legionellae and many of them have been associated with Legionnaires' disease (Birtles et al., 1996; Marrie et al., 2001). In contrast to legionellae species, the LLAPs cannot be cultured *in vitro* on artificial media. The LLAPs are isolated by coculture with protozoa (Fields, 1996). LLAPs have been isolated from sputum samples derived from patients with Legionnaires' disease on the basis of the ability of these bacteria to multiply in protozoa, because they cannot be grown on artificial media (Adeleke et al., 1996). The recent developments in using the polymerase chain reaction for bacterial identification in environmental samples will facilitate better identification of legionellae species and LLAPs.

Many strategies have been used to eradicate legionellae from sources of infection in water and plumbing systems that have been associated with disease outbreaks. These strategies include chemical biocides such as chlorine, overheating of the water, and UV irradiation (Biurrun et al., 1999; Kool et al., 1999; Muraca et al., 1987). Such interventions have been successful for short periods of time after which the bacteria are again found in these sources (Yamamoto et al., 1991; Biurrun et al., 1999). Thus, eradication of *L. pneumophila* from the environmental sources of infection requires continuous treatment of the water with agents such as monochloramine or copper-silver ions in addition to maintenance of the water temperature above ~55°C [Kool et al., 1999; Kusnetsov et al., 2001]. It is clear that the sophisticated association of legionellae with protozoa is a major factor in continuous presence of the bacteria in the environment. Compared with *L. pneumophila* grown *in vitro*, amoebae-grown bacteria have been shown to be highly resistant to chemical disinfectants and to treatment with biocides (Barker et al., 1992). Amoebae-grown *L. pneumophila* have been shown to manifest a dramatic increase in their resistance to harsh environmental conditions, such as fluctuation in temperature, osmolarity, pH, and exposure to oxidizing agents (Abu Kwaik et al., 1997). Protozoa have been shown to release vesicles containing *L. pneumophila* that are highly resistant to biocides (Berk et al., 1998).

The ability of *L. pneumophila* to survive within amoebic cysts further contributes to the resistance of *L. pneumophila* to physical and biochemical agents used in bacterial eradication (Barker et al., 1992, 1995). It is most likely that eradication of the bacteria from the environment should start by preventing protozoan infection, an integral part of the infectious cycle of *L. pneumophila*. Further characterization of the mechanisms of bacterial invasion into protozoa may allow the design of strategies to block the protozoan receptor from attachment to legionellae and thus prevent bacterial entry. Extracellular *L. pneumophila* will be more susceptible to environmental conditions and will not be protected from biocides and disinfectants. Furthermore, blockage of bacterial entry into amoebae will render bacteria less infective and virulent to mammalian cells. Alternatively, treatment of water sources contaminated with *L. pneumophila* with "safe" agents that block certain essential bacterial metabolic pathways, such as the peptidoglycan biosynthesis pathway, may prove to be useful (Harb et al., 1998).

It has been proposed that the infectious particle for Legionnaires' disease is amoebae infected with the bacteria (Rowbotham, 1980; also see Fig. 5.1). Although this has not yet been proven, there are many lines of evidence to suggest that protozoa play major roles in the transmission of *L. pneumophila*. First, many protozoan hosts have been identified that allow intracellular bacterial replication, which is the only means of bacterial amplification in the environment (Fields, 1996; Abu Kwaik et al., 1998; Harb et al., 2000). Second, in outbreaks of Legionnaires' disease, amoebae and bacteria have been isolated from the same source of infection and the isolated amoebae support intracellular replication of the bacteria (Fields et al., 1990). Third, as discussed, following intracellular replication within protozoa, *L. pneumophila* exhibit a dramatic increase in resistance to harsh conditions, such as high temperature, acidity, and high osmolarity, which may facilitate bacterial survival in the environment (Abu Kwaik et al., 1997). Fourth, intracellular *L. pneumophila* within protozoa are more resistant to chemical disinfection and biocides compared with bacteria grown *in vitro* bacteria (Barker et al., 1992, 1993, 1995). Fifth, protozoa have been shown to release vesicles of respirable size that contain numerous *L. pneumophila*. The vesicles are resistant to freeze-thawing and sonication, and the bacteria within the vesicles are highly resistant to biocides (Berk et al., 1998). Sixth, following their release from the protozoan host, the bacteria exhibit a dramatic increase in their infectivity for mammalian cells *in vitro* (Cirillo et al., 1994).

In addition, it has been demonstrated that intracellular bacteria within *Hartmanella vermiformis* are dramatically more infectious and are highly lethal in mice (Brieland et al., 1997). Seventh, the number of bacteria isolated from the source of infection of Legionnaires' disease is usually very low or undetectable, and thus enhanced infectivity of intracellular bacteria within protozoa may compensate for the low infectious dose (O'Brein and Bhopal, 1993). Eighth, viable but nonculturable *L. pneumophila* can be resuscitated by coculture with protozoa (Steinert et al., 1997). This observation may suggest that failure to isolate the bacteria from environmental sources of infection may be due to this "dormant" phase of the bacteria that cannot be recovered on artificial media. Ninth, there has been no documented case of bacterial transmission between individuals. The only source of transmission is environmental droplets generated from many human-made devices, such as shower heads, water fountains, whirlpools, and cooling towers of air-conditioning systems (Fields, 1996).

ADHERENCE AND ENTRY MECHANISMS

Initial interactions between *L. pneumophila* and its primitive protozoan hosts

Bacterial attachment to *H. vermiformis* is mediated by adherence to a protozoan receptor that has been characterized as a galactose/N-acetyl-galactosamine (Gal/GalNAc) lectin with similarity to the β_2 integrin-like Gal/GalNAc lectin of the pathogenic protozoan *Entamoebae histolytica* (Mann et al., 1991; Venkataraman et al., 1997; Harb et al., 1998). Integrins are heterodimeric protein tyrosine kinase receptors that undergo tyrosine phosphorylation upon engagement to ligands, which subsequently results in recruitment and rearrangements of the cytoskeleton. Interestingly, attachment of *L. pneumophila* to the Gal/GalNAc of *H. vermiformis* triggers signal transduction events in *H. vermiformis* that are manifested in dramatic tyrosine dephosphorylation of the lectin receptor and other proteins (Venkataraman et al., 1997). Moreover, in addition to these manipulations of the signal transduction of *H. vermiformis* by *L. pneumophila*, bacterial invasion is also associated with specific induction of gene expression in the protozoa, and inhibition of this gene expression blocks entry of the bacteria (Abu Kwaik et al., 1994). Following this initial host–parasite interaction, uptake of *L. pneumophila* by protozoan cells occurs by both conventional and coiling phagocytosis (in which the bacterium is surrounded by a multilayer coil-like structure; Abu Kwaik, 1996; Bozue and Johnson, 1996; also see Fig. 5.2).

Invasion of *H. vermiformis* by *L. pneumophila* requires host protein synthesis, because eukaryotic protein synthesis inhibitors (cycloheximide) block the entry process (Abu Kwaik et al., 1994). The uptake of *L. pneumophila* into *H. vermiformis* mainly occurs through cup-shaped invaginations (or zipper phagocytosis) on the surface of the amoeba, although some coiling phagocytosis also occurs (Venkataraman et al., 1998). Such invaginations are known to be microfilament dependent (Venkataraman et al., 1998). However, the entry of the bacteria into *H. vermiformis* is not inhibited by microfilament inhibitors such as cytochalasin D (King et al., 1991; Harb et al., 1998). Methylamine, which is an inhibitor of receptor-mediated endocytosis, inhibits the entry of *L. pneumophila* into *H. vermiformis* (King et al., 1991). Apparently, infection of *Acanthamoeba polyphaga* by *L. pneumophila* occurs through a different mechanism. It is not inhibited by galactose or N-acetylgalactosamine (Harb et al., 1998). The 170-kDa Gal/GalNAc-inhibitable lectin is only mildly dephosphorylated in *A. polyphaga* upon attachment of *L. pneumophila* (Harb et al., 1998). Furthermore, host protein synthesis by *A. polyphaga* is not required for invasion by *L. pneumophila* (Harb et al., 1998). The uptake of the bacteria is not inhibited by cytoskeleton-disrupting agents. The role of this form of phagocytosis in the intracellular fate of *L. pneumophila* is not fully understood, because human macrophages are able to phagocytose heat- or formalin-killed *L. pneumophila* by coiling phagocytosis (Horwitz, 1984).

ATTACHMENT AND ENTRY TO MAMMALIAN CELLS

Invasion and intracellular replication of *L. pneumophila* within pulmonary cells in the alveoli is the hallmark of Legionnaires' disease (Abu Kwaik, 1998b). These alveolar cells include macrophages, and type I and II epithelial cells. Attachment of *L. pneumophila* into macrophages has been shown to be mediated, at least in part, through the attachment of complement-coated bacteria to the complement receptor (Payne and Horwitz, 1987), although non-complement-mediated uptake also occurs (Elliott and Winn, 1986; Rodgers and Gibson, 1993). The host cell receptor involved in non-complement-mediated uptake in macrophages and epithelial cells is not known.

Uptake of *L. pneumophila* by monocytes and macrophages has been shown to occur through conventional and coiling phagocytosis (Horwitz, 1984; Weinbaum et al., 1984; Elliott and Winn, 1986; Rechnitzer and Blom, 1989; Dowling et al., 1992; also see Fig. 5.2). Because heat-killed and formalin-killed *L. pneumophila* are also taken up by coiling phagocytosis (Horwitz, 1984) but are targeted to the lysosomes (Horwitz and Maxfield, 1984), this mode of uptake may not play a role in subsequent pathogenicity of the bacteria. Many clinical isolates of *L. pneumophila* have been shown to be taken up exclusively by conventional phagocytosis (Elliott and Winn, 1986; Rechnitzer and Blom, 1989). In addition, other species of legionellae, such as *L. micdadei*, which is the second most common species of legionellae that causes Legionnaires' disease, is taken up exclusively by conventional phagocytosis (Rechnitzer and Blom, 1989). The bacterial ligand that mediates the coiling mode of phagocytosis is not known. Moreover, the phagocytic receptor that binds the bacteria seems to play some role in determining the fate of the intracellular bacteria because opsonization with antibodies reduces intracellular growth (Horwitz and Silverstein, 1981a; Nash et al., 1984; Payne and Horwitz, 1987).

Studies have focused on the genetic aspects of the uptake of *L. pneumophila* in its host cells. *L. pneumophila* mutants impaired in different loci, such as *rtxA* and *enhC*, display significantly reduced entry into host cells, compared with wild-type bacteria (Cirillo et al., 2000). Recently, it has been shown that the enhanced phagocytosis of *L. pneumophila* by mammalian cells is *dot/icm* dependent (Hilbi et al., 2001). Interestingly, the *dot/icm* genes delay uptake and induce macropinocytosis in A/J mice macrophages (Watarai et al., 2001b). Macropinosomes containing *L. pneumophila* are induced transiently and shrink rapidly (5–15 min; Watarai et al., 2001b), and this mode of uptake is linked to the *lgn1* locus on chromosome 13 of mice (Watarai et al., 2001b). With the exception of A/J mice, most of the inbred mouse strains are not permissive to *L. pneumophila* infection; neither are macrophages isolated from these mice (Yamamoto et al., 1992; Beckers et al., 1995). The difference between these two mice strains is located on chromosome 13 and is linked to a single locus, *lgn1* (Dietrich et al., 1995; Beckers et al., 1997). In macrophages of nonpermissive strains of mice, the macropinocytic uptake of *L. pneumophila* is reduced (Watarai et al., 2001b). The *lgn1* allele makes the bacteria behave as if they are lacking the *dot/icm* system (Watarai et al., 2001b). Thus, the *lgn1* allele is required for *dot/icm*-dependent macropinocytosis and delayed uptake by mice macrophages (Watarai et al., 2001b).

INTRACELLULAR TRAFFICKING

Intracellular survival and replication within host cells

During the first few minutes after entry into amoebae, the bacterium is enclosed in a phagosome surrounded by mitochondria and host cell vesicles (Abu Kwaik, 1996; also see Fig. 5.2). The bacterial phagosome is blocked from fusion to the lysosomes (Bozue and Johnson, 1996). In addition, the phagosome is surrounded by a multilayer membrane derived from the RER of amoebae (Abu Kwaik, 1996; also see Fig. 5.2). Following formation of this phagosome within protozoan cells, bacterial replication is initiated. The 4-h period prior to initiation of intracellular replication may be the time required to recruit these host cell organelles that may be required for replication. Alternatively, the 4-h period may be a lag phase of metabolic and environmental adjustment of the bacteria to a new niche.

Similar to the protozoan infection, within 5 min following entry of the bacteria into macrophages and monocytes, the *L. pneumophila* phagosome is surrounded by host cell organelles such as mitochondria, vesicles, and the RER (Horwitz, 1983b; Swanson and Isberg, 1995a; Tilney et al., 2001; also see Fig. 5.2). Also similar to the trafficking of *L. pneumophila* within protozoa, the phagosome within mammalian macrophages does not fuse to lysosomes (Horwitz, 1983a, 1984; Horwitz and Maxfield, 1984; also see Fig. 5.2). The role of the RER in the intracellular infection is not known, but the RER is not required as a source of protein for the bacteria (Abu Kwaik, 1998a). Interestingly, examination of the intracellular infection of macrophages, alveolar epithelial cells, and protozoa by another *Legionella* species, *L. micdadei*, showed that, within all of these host cells, the bacteria were localized to RER-free phagosomes (Gao et al., 1999). Whether other *Legionella* species replicate within RER-free phagosomes is still to be determined.

Macrophages, peripheral blood monocytes, and alveolar epithelial cells support intracellular replication of *L. pneumophila* (Nash et al., 1984; Gao et al., 1998b). Although alveolar epithelial cells, which constitute more than 95% of the alveolar surface (Gao et al., 1998b), have been shown to allow intracellular replication of *L. pneumophila*, their role in the pathogenesis has been largely overlooked.

Role of the *dot/icm* genes in evasion of the endocytic pathway

The Dot/Icm type IV secretion system is the main virulence system of *L. pneumophila*. Because the Dot/Icm secretion system is ancestrally related to type IV secretion systems that mediate conjugal DNA transfer between bacteria (Christie and Vogel, 2000), *L. pneumophila* may utilize this transporter to transfer macromolecules into the host cell to evade endocytic fusion (Roy and Tilney, 2002). The *dot/icm* loci may be involved in the insertion of a pore in the host cellular membrane to transfer the effector proteins (Kirby and Isberg, 1998; Kirby et al., 1998). The effector molecules involved in intracellular trafficking and evasion of the lysosomal fusion within mammalian cells are limited to the phagosome harboring the bacterium, which does not alter the biology of endocytic fusion in the rest of the cell (Coers et al., 1999). With few exceptions, the function of individual Dot/Icm proteins is unknown.

The DotA protein was the first to be described (Berger et al., 1994). It is a polytopic inner membrane protein with eight hydrophobic transmembrane domains (Roy and Isberg, 1997). *dotA* mutants are defective in all virulence activities that require the Dot/Icm complex (Berger and Isberg, 1993; Berger et al., 1994; Swanson and Isberg, 1995a, 1995b; Kirby et al., 1998; Coers et al., 2000). These data are supported by the fact that the DotA sequence possesses significant similarities with that of TraY (Segal et al., 1999; Komano et al., 2000), a component of the type IV transporter required for the conjugal transfer of plasmids ColIbP9 and R64. However, Nagai and Roy have shown that the DotA protein is also secreted through the Dot/Icm transporter into the culture supernatant during growth of *L. pneumophila* in liquid broth by means of a functional type IV secretion system (Nagai and Roy, 2001). Electron micrographs also show that DotA is part of the oligomer that could be the membrane channel (Roy and Isberg, 1997; Nagai and Roy, 2001). Mutants defective in DotA protein expression are unable to form pores in host cell membranes (Coers et al., 2000).

As DotH/IcmK and DotO/IcmB in growing *L. pneumophila* cultures are mainly associated with the membranous fraction, and as *dot/icm* products may be required during direct contact with host cells, the location on the surface of *L. pneumophila* of DotH and DotO proteins has been examined (Watarai et al., 2001a). These proteins are surface exposed and associated with a fibrous structure on *L. pneumophila* after exposure to bone-marrow-derived macrophages. In contrast, during broth culture, this fibrous structure seems to be absent (Watarai et al., 2001a). However, with the use of *dotA*, *dotB*, *dotH*, and *dotO* mutants, it has been shown that the exposure of DotO/DotH on the bacterial surface is not dependent on other Dot/Icm proteins, including the DotH protein, for the DotO exposure and vice versa (Watarai et al., 2001a). This result shows that the surface exposure of DotH and DotO after contact with macrophages is not dependent on an intact Dot/Icm secretion system (Watarai et al., 2001a). The surface exposure of these two proteins does not involve bacterial contact with the target cell because bacteria incubated in medium that have been conditioned by bone-marrow-derived macrophages for 24 h yield almost identical results (Watarai et al., 2001a).

In addition, DotH and DotO surface exposure on *L. pneumophila* requires intracellular growth of the bacteria in macrophages and is observed late in the infection process, mostly when there are more than 30 bacteria per phagosome (Watarai et al., 2001a). In fact, surface-exposed DotH and DotO disappear after uptake but reappear following intracellular growth. The exposure of these two proteins increases *L. pneumophila* uptake into cells (Watarai et al., 2001a) and may be necessary to promote bacterial escape from the phagosome and to facilitate the initiation of a new infection in macrophages (Watarai et al., 2001a). These data may also suggest that macrophage components are able to induce changes in the *L. pneumophila* envelope, and DotH/DotO export may occur as a response to the target macrophages just before uptake to allow efficient initiation of intracellular growth.

Dot/Icm proteins such as IcmR, IcmQ, IcmX, or IcmW do not present sequence homology to other protein components of the type IV secretion system (Segal and Shuman, 1998; Vogel et al., 1998; Christie and Vogel, 2000). IcmX, a periplasmic protein, is required for pore-forming activity (Matthews and Roy, 2000). The *icmX* gene is also required for biogenesis of the replicative phagosomes containing *L. pneumophila* (Matthews and Roy, 2000). A truncated IcmX product is secreted into culture supernatants by wild-type *L. pneumophila* growing in liquid media, but the transport of the protein into eukaryotic host cells has not been detected (Matthews and Roy, 2000).

icmS and *icmW* mutants are not defective in pore-forming activities, but their phagosomes fuse to lysosomes (Coers et al., 2000). IcmS and IcmW are small proteins required for the trafficking of the *Legionella*-containing phagosome, and they may function as chaperones to help the secretion of proteins through the type IV secretion system (Coers et al., 2000; Nagai and Roy, 2001). It has also been shown that IcmS and IcmW interact, suggesting that they may be components of a protein complex that is required for modulating phagosome biogenesis (Coers et al., 2000). Interestingly, phagosomes harboring *icmS* and *icmW* mutants still recruit host vesicles, including the RER. It is still to be confirmed whether recruitment of host vesicles does not prevent the phagosome from lysosomal fusion (Roy and Tilney, 2002).

The IcmR protein may also be a chaperone (Coers et al., 2000). An *icmR* mutant has undetectable pore-forming activity and can partially evade the endocytic pathway, but it eventually fuses with the lysosomes (Coers et al., 2000). These data suggest that IcmR may be a cofactor for another protein effector involved in evasion of lysosomal fusion (Roy and Tilney, 2002). The *icmR* mutant that evades endocytic fusion does not recruit host vesicles, and the phagosome is not surrounded by the RER (Roy and Tilney, 2002). Thus, effector molecules that recruit host cell vesicles may be different from the ones involved in evasion of the endocytic pathway (Roy and Tilney, 2002).

Similar to *dotA* and *icmX* mutants, an *icmQ* mutant is defective in all virulence activities (Coers et al., 2000). Furthermore, the *icmQ* gene, like the *icmR* and *icmS* genes, encodes soluble protein. The IcmR and IcmQ proteins interact as protein chaperone–substrate. The presence of IcmR, which has chaperone characteristics (it is acidic, small, and predicted to have a hydrophobic alpha-helix in its C-terminal domain) affects the physical state of IcmQ directly (Dumenil and Isberg, 2001). IcmR prevents IcmQ from participating in the formation of high-molecular-weight complexes by dissociating IcmQ homopolymers (Dumenil and Isberg, 2001).

It has been shown that most of the *dot/icm* genes required for intracellular growth within human cells are also required for intracellular growth in the protozoan host *A. castellanii* (Segal and Shuman, 1999). In addition, enhanced phagocytosis by *A. castellanii* of wild-type *L. pneumophila* has also been demonstrated to be dependent on *dot/icm* genes (Hilbi et al., 2001). Although some loci have been shown to be essential only for the intracellular growth of *L. pneumophila* in macrophages (Gao et al., 1998a), numerous loci have been identified as essential for survival and intracellular replication of *L. pneumophila* in *A. polyphaga* or *H. vermiformis* and in macrophages (Cianciotto and Fields, 1992; Gao et al., 1997; Stone et al., 1999).

Another host, *Dictyostelium discoideum*, a unicellular organism that lives in soil, has been shown to support the intracellular multiplication of *L. pneumophila* (Solomon and Isberg, 2000; Solomon et al., 2000). In the amoebal form, the cells are highly motile and are very active in phagocytosis. This model is interesting because genetic tools are available for analysis of host–pathogen interactions, such as the existence of plasmids replicating in this haploid organism as well as extensive sequence DNA information (Solomon and Isberg, 2000; Solomon et al., 2000). The intracellular fate of *L. pneumophila* is very similar in infected *D. discoideum* to that in macrophages, including the recruitment of RER and evasion of lysosomal fusion (Solomon et al., 2000). The growth of *L. pneumophila* in *D. discoideum* depends on *dot/icm* gene functions (Solomon et al., 2000). The analysis of growth of wild-type and three isogenic *dot/icm* mutant strains of *L. pneumophila* in *D. discoideum* indicates that intracellular growth requires the products of multiple loci, because *dotH*, *dotI*, and *dotO* mutants all failed to grow and lost viability over the course of 4 days. (Solomon et al., 2000). The similarity between the infection by *L. pneumophila* of different protozoa supports the idea that the ability of *L. pneumophila* to parasitize macrophages and hence to cause human disease is a consequence of its prior adaptation to intracellular growth within protozoa.

RECRUITMENT OF RER

In 1982, Katz and Hashemi showed that the *L. pneumophila*-containing phagosome resembles the ER membrane. By means of electron microscopy, replicating *L. pneumophila* within macrophages were visualized located within organelles morphologically identical with the RER (Katz and Hashemi, 1982; also see Fig. 5.2). Later, several studies demonstrated that upon internalization of *L. pneumophila* by the host cell, the *Legionella*-containing vacuole recruits organelles such as vesicles, mitochondria, and ER (Fig. 5.2; also see Katz and Hashemi, 1982; Horwitz, 1983b; Bozue and Johnson, 1996).

With the use of fluorescent markers specific for the ER, it has been shown that the *L. pneumophila*-containing vacuoles may resemble nascent autophagosomes (Swanson and Isberg, 1995b). Autophagy is a physiologically important cellular process for the degradation of unwanted organelles and cellular components, during which the autophagosome is formed from RER and fuses to lysosomes (Dorn et al., 2002). The hypothesis that *L. pneumophila* exploits the autophagy machinery in host cells and establishes an intracellular niche favorable for replication (Swanson and Isberg, 1995b) has been challenged recently (see the paragraphs that follow; also Tilney et al., 2001; Roy and Tilney, 2002).

Recent studies suggest that fusion (Roy and Tilney, 2002), or exchange of lipid bilayer with ER vesicles on the *L. pneumophila*-containing phagosome (Tilney et al., 2001) occurs, allowing the phagosomal membrane to become as thin as the ER membrane, with similar characteristics (Tilney et al., 2001; Roy and Tilney, 2002). It has been shown that, within 5 min of uptake, host vesicles come into contact with wild-type *Legionella*-containing phagosomes and flatten along the surface of the phagosome, and this process is completed within 15 min (Tilney et al., 2001). This does not occur in *dot/icm* mutant-containing phagosomes (Tilney et al., 2001). This is consistent with earlier studies that have shown that, after 4 h of infection, there are only few vesicles associated with the phagosomal membrane, but there are ribosomes studding the phagosomal membrane (Horwitz, 1983b). Interestingly, the thickness of the phagosome membrane becomes similar to the ER membrane within 15 min (Tilney et al., 2001). Thus, within 15 min of infection, the phagosomal membrane resembles that of the ER. The ribosomes at 6 h stud the phagosomal membrane, and *L. pneumophila* is thought to be located within the RER (Tilney et al., 2001). However, these studies rely completely on the thickness of the membranes of the ER and the phagosome membranes to provide evidence that *L. pneumophila* is located within the RER (Tilney et al., 2001). Immunocytochemical studies should be performed to confirm these observations. It is likely that the recruitment of the ER may be involved in the biogenesis of the phagosome that is dependent on the type IV secretion system, because the *dot/icm* mutants are unable to recruit RER and their phagosomes fuse to the lysosomes (Swanson and Isberg, 1995b).

Interestingly, a recent study showed that recruitment of RER to maturing phagosomes may be part of regular phagoytosis (Gagnon et al., 2002). Electron micrographs of latex beads and pathogens such as *Salmonella* within macrophages have shown that ER membranes fuse with plasmalemma, underneath the phagocytic cup, and successive waves of ER are recruited to the phagosomes of latex beads and bacteria. In neutrophils, the bacteria are quickly killed and the ER is not involved in the turnover of membrane for phagocytosis (Gagnon et al., 2002). However, because the *dot/icm* system is essential for RER recruitment (Swanson and Isberg, 1995b), it is likely that *L. pneumophila* utilizes a specific pathogen-regulated process to recruit vesicles and that this process is distinct from regular phagocytosis.

THE ARF PROTEIN

The protein ADP ribosylation factor-1 (ARF1), a highly conserved small GTP-binding protein that acts as a key regulator of vesicle traffic from ER to Golgi, is found on phagosomes that contain wild-type *L. pneumophila* but not *dot/icm* mutants (Nagai et al., 2002). These data suggest that a protein injected through the type IV secretion system may be required for ARF1 recruitment. Searching the *L. pneumophila* genome for proteins that have homology to ARF-specific guanine nucleotide exchange factors (GEFs), Nagai et al. have identified a protein, RalF, that has a sec7-homology domain, known to be sufficient to stimulate the exchange of GDP for GTP (Nagai et al., 2002). It has been shown that RalF is injected through the phagosomal membrane by a process that requires the *dot/icm* system (Nagai et al., 2002). However, phagosomes containing *ralF* mutants that do not recruit ARF1 evade fusion to lysosomes, and the bacteria replicate intracellularly within macrophages and amoebae (Nagai et al., 2002). Thus, RalF is not essential for transport of *L. pneumophila* to the ER (Nagai et al., 2002; Roy and Tilney, 2002).

Pore-forming activity

Kirby et al. were the first to describe the pore-forming ability of *L. pneumophila* in host cell membranes (Kirby et al., 1998). This ability was documented by lysis of macrophages and red blood cells, which is dependent on the Dot/Icm secretion system (Kirby et al., 1998). Because *dot/icm* mutants are defective in evasion of lysosomal fusion, it has been proposed that the pore-forming activity is required for export of effector molecules necessary for evasion of the lysosomal fusion (Kirby et al., 1998). Upon initial contact with the host cell, *L. pneumophila* may insert a pore into the plasma membrane to deliver bacterial effector molecules into the host cell (Kirby and Isberg, 1998; Kirby et al., 1998). This concept is supported by the fact that many *dot/icm* mutants are defective in both intracellular replication and pore-formation-mediated cytotoxicity (Kirby and Isberg, 1998; Kirby et al., 1998). Moreover, some *dot/icm* mutants are defective in trafficking and intracellular replication but not in pore-forming activity (Zuckman et al., 1999). Thus, the pore-forming activity is not sufficient for phagosomal trafficking (Zuckman et al., 1999).

We have identified five spontaneous mutants that are unable to egress from the macrophages but are able to grow as well as the wild-type strains within these cells (Alli et al., 2000). These mutants, designated *rib* (release of intracellular bacteria), are defective in the pore-forming toxin activity as shown by the contact-dependent hemolysis assay and by permeability to propidium iodide upon infection of macrophages and epithelial cells (Alli et al., 2000). The *rib* mutants are also defective in acute cytotoxic lethality to A/J mice and fail to cause alveolar inflammation (Alli et al., 2000; Molmeret et al., 2002a), thus indicating a key role for the pore-forming toxin in the pulmonary pathology of the bacterium. We have further documented that the Rib toxin is not required for intracellular trafficking and replication (Alli et al., 2000; Gao and Abu Kwaik, 2000; Molmeret et al., 2002b).

In addition to defects in evasion of lysosomal fusion, *dot/icm* mutants are also defective in induction of apoptosis (Zink et al., 2002), enhancement of phagocytosis by human-derived macrophages (Hilbi et al., 2001), and induction of macropinocytic delayed uptake by A/J mice macrophages (Watarai et al., 2001b). In contrast, *rib* mutants are defective in pore-forming toxin but are not defective in evasion of lysosomal fusion, intracellular replication, or induction of apoptosis (Alli et al., 2000; Molmeret et al., 2002b; Zink et al., 2002). These observations may suggest that there are at least two pores inserted into host membranes through the type IV secretion system at different stages of the infection (Molmeret and Abu Kwaik, 2002). The first pore, is the "invasion and trafficking pore," which is inserted upon initial contact with the host cell to deliver effectors into the host cell cytoplasm and allow the establishment of the intracellular infection. The *dot/icm* genes necessary for the assembly of the secretion apparatus are constitutively expressed and are required for this step (Hales and Shuman, 1999). The second pore is the "egress pore," which is turned off during exponential intracellular replication but is triggered upon cessation of replication and is essential for lysis of the host cell (Figs. 5.3 and 5.4; also see Alli et al., 2000). This is consistent with the fact that, upon transition into the postexponential growth *in vitro*, *L. pneumophila* becomes cytotoxic (Byrne and Swanson, 1998). Thus, the *rib*

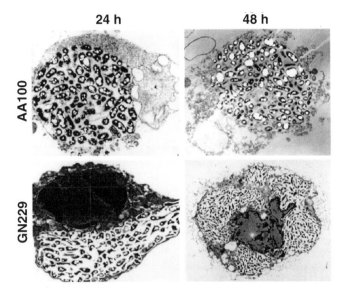

24 h   **48 h**

AA100

GN229

Figure 5.3. The *rib* mutants' defect in cytolysis of the host cell is due to a defect in necrosis-mediated killing. Representative transmission electron micrographs of infected U937 macrophages at 24 h and 48 h postinfection by the wild-type strain AA100 and the GN229 mutant. The original magnifications were 7,000× and 5,000× for the 24-hp and 48-h infections, respectively. (This figure was adapted from Alli et al., *Infect Immun.* *68* (11): 6431–6441, 2000.)

mutants retain the "invasion and trafficking pore" but are defective in the "cytolysin/egress pore," because they replicate within but fail to egress from the host cells (Figs. 5.3 and 5.4).

The phenotype of the *rib* mutants is attributable to a point deletion in the *icmT* gene that is predicted to result in a truncated protein of 54 amino acids instead of the 86 amino acids of the native protein (Molmeret et al., 2002a, 2002b). In contrast, an *icmT* null mutant is defective in both intracellular replication and pore formation (Molmeret et al., 2002a, 2002b). It has been shown that *icmT* expression is induced at the stationary phase compared with the exponential phase (Gal-Mor et al., 2002). We have shown that IcmT is bifunctional and that the carboxy terminus is essential for the pore-forming "cytolysin/egress pore" activity and the amino terminus is essential to export effectors involved in various pathogenic traits (Molmeret et al., 2002a, 2002b).

We hypothesized that, upon transition to the postexponential phase of growth, the Rib toxin activity is triggered, resulting in insertions of pore first in the phagosomal membrane, leading to its disruption and bacterial egress

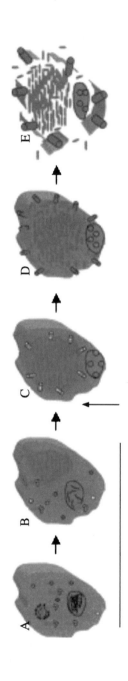

1. Caspase-3-mediated apoptosis
   DNA fragmentation
2. Expression of the pore-forming
   activity is turned off

Expression of the pore-forming activity is
triggered

Figure 5.4. A model of growth-phase-dependent cytolysis of mammalian cells by *L. pneumophila* upon termination of intracellular bacterial replication to egress from the spent host cell. During early stages of formation of the mitochondria and rough endoplasmic reticulum-surrounded phagosome (A) and during exponential intracellular replication (B), expression of the pore-forming activity is turned off, but caspase-3-mediated apoptosis is triggered. Upon transition to the postexponential phase of growth, expression of the pore-forming activity is triggered, which results in insertions of pores in the phagosomal membrane first (C), leading to its disruption (D). This is followed by insertions of pores in the plasma membrane (E), leading to osmotic lysis of the cell and release of the intracellular bacteria. (This figure was adapted from Alli et al., *Infct. Immun.* 68 (11): 6431–6440, 2000.) See color section.

into the cytoplasm (Figs. 5.3 and 5.4). To test this hypothesis, we examined late stages of the intracellular infection of macrophages and amoebae by electron microscopy. Our data showed that disruption of the phagosomal membrane was detectable 12 h postinfection in both *A. polyphaga* and macrophages (unpublished data). After 12 and 18 h postinfection in both host cells, vesicles and organelles from the host cytoplasm entered into the *Legionella*-containing phagosome. Between 18 h and 24 h, most of the bacteria present in this disrupted phagosomal membrane are surrounded by cytoplasmic organelles, and not by distinct phagosomal membrane (unpublished data). These data show that the phagosomal membrane is disrupted first; there is no simultaneous lysis of both the phagosomal and the plasma membranes. Whether this disruption is the result of a mechanic process or is linked to the pore-forming activity of the type IV secretion system (Kirby et al., 1998; Molmeret and Abu Kwaik, 2002; Molmeret et al., 2002b) is not known. Our data are consistent with a previous study in which Katz and Hashemi (1982) had observed on electron micrographs that, when present in macrophages at numbers greater than 25 per cell, *L. pneumophila* was usually dispersed within the cytoplasm of the host cell.

GENE REGULATION IN *L. PNEUMOPHILA*

Growth phase and gene regulation

The phase of growth has a dramatic effect on the phenotype of *L. pneumophila*, whether the bacteria are cultivated in host cells or in microbiological medium (Rowbotham, 1986; Byrne and Swanson, 1998). Robowtham was the first to observe in amoeba two distinct phenotypic phases of *L. pneumophila*, the replicative phase and the "active infective phase" (Rowbotham, 1986). When *L. pneumophila* is released from eukaryotic cells, it is short, thick, and highly motile, with a smooth and thick cell wall (Rowbotham, 1986). The cells have a different composition of membrane fatty acids, and they differ in LPS and outer membrane protein profiles (Barker et al., 1993). After growth in host cells, *L. pneumophila* becomes more resistant to biocides and antibiotics (Barker et al., 1992, 1995), more invasive to host cells, and more virulent in the guinea pig model (Cirillo et al., 1994, 1999; Brieland et al., 1997).

In contrast to exponentially growing bacteria, *L. pneumophila* obtained from postexponential cultures expresses traits that are correlated with virulence, including sodium sensitivity, cytotoxicity, osmotic resistance, motility, and the capacity to evade phagosome–lysosome fusion (Rowbotham, 1980, 1986; Byrne and Swanson, 1998; Hammer and Swanson, 1999). During the replication phase within host cells, *L. pneumophila* organisms are sodium resistant and aflagellated (Rowbotham, 1980, 1986; Byrne and Swanson, 1998; Hammer and Swanson, 1999). When *L. pneumophila* bacteria egress from host cells, they are flagellated and sodium sensitive (Rowbotham, 1980, 1986; Byrne and Swanson, 1998; Hammer and Swanson, 1999). It has been hypothesized that amino acid limitation induces the virulent phenotype (Rowbotham, 1980, 1986; Byrne and Swanson, 1998; Hammer and Swanson, 1999). When *L. pneumophila* enters into the postexponential growth phase or is subjected to amino acid limitation, the bacteria accumulate the stringent response signal, guanosine 3',5'-bispyrophosphate (ppGpp), through the ppGpp synthetase, RelA (Bachman and Swanson, 2001). When *relA* from *Escherichia coli* is expressed in *L. pneumophila*, ppGpp accumulates and the bacteria express virulent traits independent of nutrient supply or cell density (Bachman and Swanson, 2001). The accumulation of ppGpp increases the amount of the stationary-phase sigma factor RpoS, which triggers the expression of the stationary-phase genes (Bachman and Swanson, 2001).

A *rpoS* mutant replicates within monocyte HL 60 and THP-1 cells but is attenuated in *A. castellanii* (Hales and Shuman, 1999). Therefore, some RpoS-regulated traits could be critical for efficient transmission or infection in this amoebae model (Hales and Shuman, 1999). Sodium sensitivity and maximal expression of flagellin also requires RpoS (Bachman and Swanson, 2001). *L. pneumophila* in the postexponenetial phase becomes cytotoxic by an RpoS-independent pathway (Bachman and Swanson, 2001). To replicate efficiently in macrophages, both an RpoS-independent and RpoS-dependent mechanism are utilized by *L. pneumophila* (Bachman and Swanson, 2001). Thus, when nutrient levels and other conditions are favorable, *L. pneumophila* replicates within host cells, and when amino acids become rare, intracellular bacteria express several traits that permit escape from the host cell, survival in the environment, and transmission to a new host.

The transmission phenotype

Because flagellin is expressed at the postexponential phase, Hammer et al. (2002) screened for *L. pneumophila* mutants deficient in flagellin expression in order to identify genes involved in the phenotypic transition at the postexponential phase. Five activators of virulence have been identified: LetA and LetS, a two-component regulator homologous to GacA/S of *Pseudomonas*, or SirA/BarA of *Salmonella*, that represses the flagellar regulon; the stationary-phase sigma factor RpoS; the flagellar sigma factor FliA, required for both motility and contact-dependent cytoxicity (Heuner et al., 2002); and a novel locus, *letE* (Hammer et al., 2002).

In the postexponential phase, mutants in *letA*, *letS*, *fliA*, and *letE* are nonmotile or show poor motility, are not cytotoxic nor sodium sensitive (except *fliA*), and are not efficient at infecting macrophages (except *letE*; Hammer et al., 2002). In contrast, intracellular growth is independent of these genes (Hammer et al., 2002). Amino acid depletion or ppGpp accumulation triggers a LetA/LetS expression and RpoS-dependent cellular differentiation (Hammer et al., 2002). The *letE* locus does not appear to produce any protein and may encode for a regulatory RNA that may act as a regulator of *letA*/*letS* expression (Hammer et al., 2002).

The *relA* mutant, producing undetectable levels of ppGpp in the cells during the stationary phase, is unable to produce pigment as it becomes flagellated. Although a previous study has shown that RpoS, which accumulates when RelA is activated, is required for intracellular growth in *A. castellanii* (Hales and Shuman, 1999). Zusman et al. (2002) have shown that *relA* gene product is dispensable for intracellular growth in HL-60-derived human macrophages and in *A. castellanii*. Moreover, it has also been shown that RelA and RpoS have minor effects on expression of some of the *dot*/*icm* genes (*icmT*, *icmR*, *icmQ*, *icmP*, *icmM*, *icmJ*, *icmF*, *icmV*, and *icmW*; see Zusman et al., 2002). Thus, the role of RpoS in intracellular infection seems to be specific to the host cell.

DNA regulatory elements of the *dot*/*icm* genes

Several studies showed that *dot*/*icm* expression is regulated during the intracellular infection of *L. pneumophila* within the host cells. The *dot*/*icm* system upregulates phagocytosis of *L. pneumophila* (Hilbi et al., 2001), and surface exposure of DotO/DotH on *L. pneumophila* is induced at the earlier and later stages of the infection (Watarai et al., 2001a).

Gal-Mor et al. (2002) examined DNA regulatory elements that may control the expression of the *dot*/*icm* genes, using a promoter fusion to *lacZ* in *L. pneumophila*. They showed that the expression levels of different *dot*/*icm* genes are distinct from one another. The *icmR*, *icmF*, *icmV*, and *icmW* genes have high levels of expression in both the exponential and postexponential phase of growth, whereas *icmR* and *icmF* have higher levels in the stationary phase than in the exponential phase. The *icmT*, *icmP*, *icmQ*, *icmM*, and *icmJ* genes show low levels of expression in both exponential and postexponential phases of growth. However, *icmT* and *icmP* have a higher level of expression at the stationary phase.

A total of 12 regulatory elements have been identified (Gal-Mor et al., 2002); 10 promoter elements of *icmT*, *icmP*, *icmQ*, and *icmM* genes have low expression levels. These promoters contain 6-bp putative consensus sequence TATACT, located from 32 to 74 bp from the ATG codons, which is essential for their expression. −10 promoter elements of *icmV, icmW,* and *icmR* genes that have high expression levels have also been identified. The *icmR* locus contains at least three regulatory elements, and regulatory elements have been also identified for *icmW, icmV,* and *icmF* that have high expression levels. A 9-bp putative consensus sequence CTATAGTAT has been observed. In addition, *icmV* and *icmW* have an overlapping regulatory region (Gal-Mor et al., 2002).

The reduced effect of an insertion in *rpoS* or *relA* on expression of *icmP* only among the 9 *icm::lacZ* fusions tested has shown that these two genes are not involved in the regulation of the *dot/icm* genes (Hales and Shuman, 1999; Zusman et al., 2002) and that there may be other factors necessary for this regulation. The −10 promoter elements found in some *dot/icm* genes have extensive homology to one another and are probably recognized by *L. pneumophila* RpoD (Gal-Mor et al., 2002). An examination of the *L. pneumophila* genome sequence has shown that *L. pneumophila* contains homologs of at least six sigma factors (RpoD, RpoH, RpoF, RpoE, RpoS, and RpoN). The promoter sequences of RpoH and RpoF are different from the −10 promoter sequences of the *dot/icm* genes (Gal-Mor et al., 2002). In addition, the promoters recognized by the factors RpoE and RpoN are also different from the promoters of the *dot/icm* genes (Gal-Mor et al., 2002). As RpoS is not involved in the expression of the *dot/icm* genes except for the moderate expression of *icmP* (Zusman et al., 2002), it has been proposed that these −10 regulatory elements of the *dot/icm* genes are recognized by the vegetative sigma factor RpoD (Gal-Mor et al., 2002).

APOPTOTIC OR NECROTIC CELL DEATH

Induction of apoptosis by *L. pneumophila* in mammalian but not protozoan host cells

Apoptosis requires a cascade of activation of a family of cysteine proteases (caspases) that specifically cleave proteins after aspartate residues (Anderson, 1997). Among them, caspase-3 plays a central role, allowing caspase-activated DNase to enter the nucleus and degrade chromosomal DNA (Enari et al., 1998). Muller et al. (1996) have shown that *L. pneumophila* induce apoptosis in HL-60 human-derived macrophages after 24–48 h, at a multiplicity of infection (MOI) of 10–100. The induction of apoptosis in mammalian cells is mediated by activation of caspase-3 that is dose dependent and is maximal at

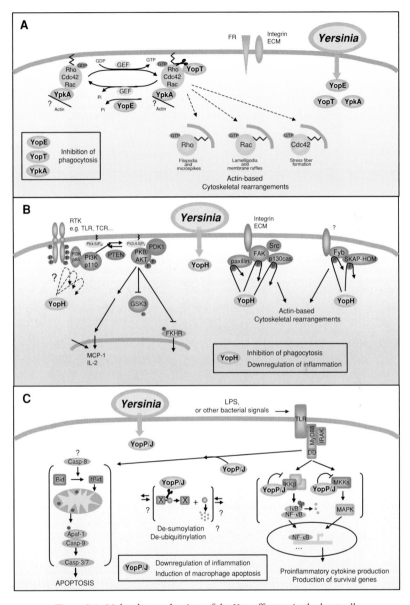

Figure 3.4. Molecular mechanism of the Yop effectors in the host cell.

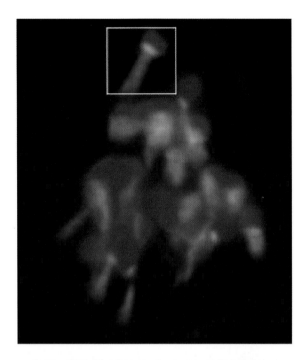

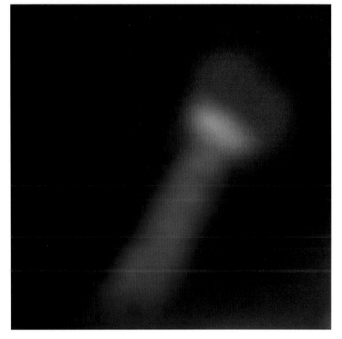

Figure 4.2. A cluster of pedestal structures induced by EPEC on the surface of a HeLa cell.

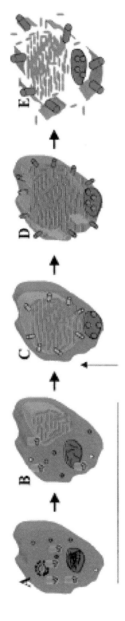

1. Caspase-3-mediated apoptosis
   DNA fragmentation
2. Expression of the pore-forming
   activity is turned off

Expression of the pore-forming activity is
triggered

Figure 5.4. A model of growth-phase-dependent cytolysis of mammalian cells by *L. pneumophila* upon termination of intracellular bacterial replication to egress from the spent host cell.

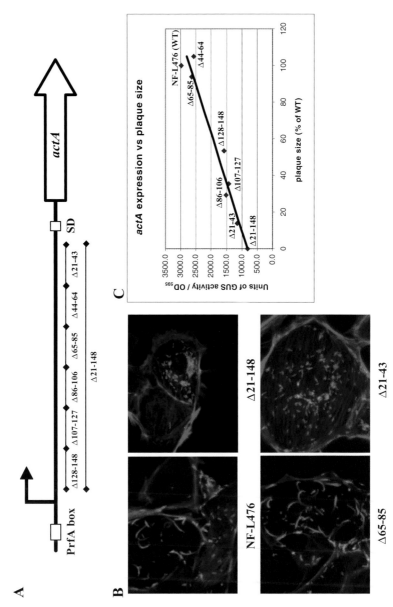

Figure 6.3. Correlation of *actA* expression levels with *L. monocytogenes* cell-to-cell spread.

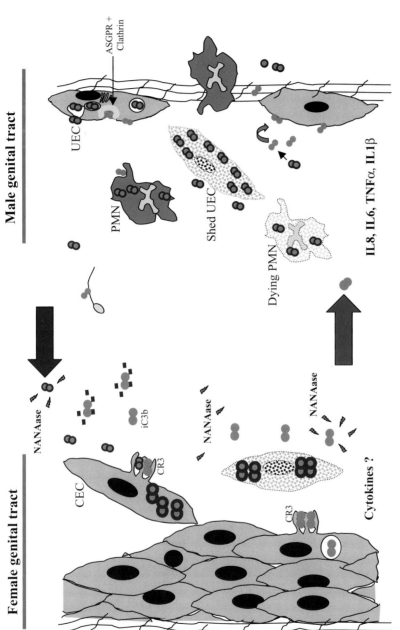

Figure 7.8. The different pathogenic interactions of the gonococcus in men and women.

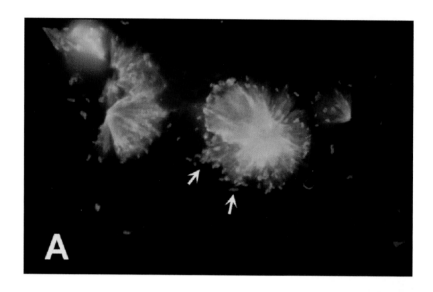

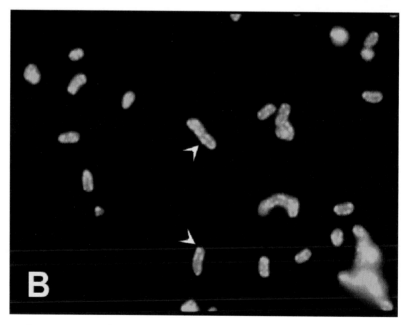

Figure 9.4. Microtubule asters and the *A. actinomycetemcomitans* kinesin-like entity.

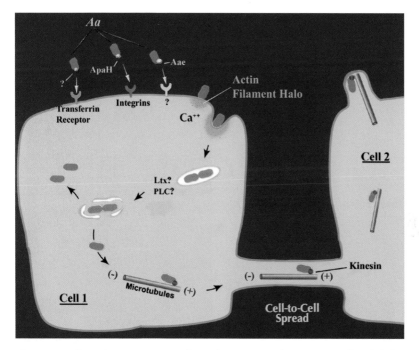

Figure 9.5. Schematic representation of *A. actinomycetemcomitans* invasion of epithelial cells.

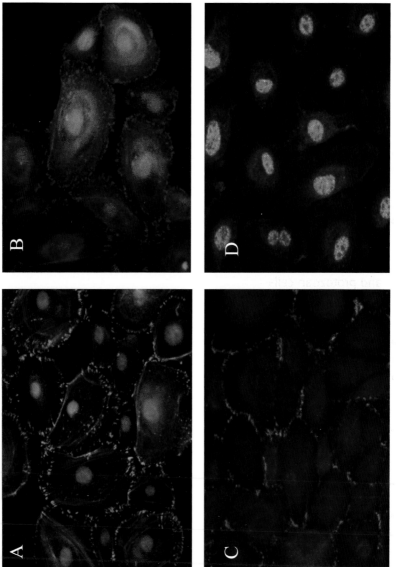

Figure 10.1. Immunofluorescence microscopy of *P. gingivalis* invasion of GEC.

3 h postinfection at MOI 50 (Gao and Abu Kwaik, 1999a, 1999b; also see Fig. 5.4). In alveolar epithelial cells and macrophages, the induction of apoptosis is dose dependent but not largely growth phase regulated and can be induced by extracellular bacteria (Gao and Abu Kwaik, 1999a, 1999b). The dot/icm mutants of L. pneumophila are defective in inducing caspase-3 activation and, thus, apoptosis (Gao and Abu Kwaik, 1999a; Zink et al., 2002). Therefore, the Dot/Icm type IV secretion system of L. pneumophila is essential for the induction of apoptosis in mammalian cells (Gao and Abu Kwaik, 1999a; Zink et al., 2002). The pore-forming toxin is not required for the induction of apoptosis, but upon entry into the postexponential growth phase it enhances the ability of the bacteria to induce apoptosis (Zink et al., 2002).

Interestingly L. pneumophila induces apoptosis in human phagocytes but not in protozoan host cells such as A. castellanii (Hagele et al., 1998; Gao and Abu Kwaik, 2000). Moreover, although A. polyphaga is capable of undergoing apoptosis, upon stimulation by actinomycin D, L. pneumophila does not induce apoptosis in A. polyphaga (Gao and Abu Kwaik, 1999b, 2000).

Induction of necrosis by *L. pneumophila* in mammalian and protozoan cells

The dot/icm-regulated Rib pore-forming toxin is essential for L. pneumophila induction of necrosis, killing, and exiting the host cells (Alli et al., 2000; Gao and Abu Kwaik, 2000; Molmeret et al., 2002a). At high MOI, mutants defective in the Rib pore-forming toxin replicate like the wild-type strains within the protozoan and mammalian cells but are trapped within these cells and cannot be released (Figs. 5.3 and 5.4; also see Alli et al., 2000; Gao and Abu Kwaik, 2000). The expression of the pore-forming activity by L. pneumophila grown in vitro and within macrophages is completely repressed during exponential growth but is temporally activated upon entry into postexponential phase (Fig. 5.4; also see Alli et al., 2000). All five rib mutants mentioned earlier that possess identical point deletions following a poly(T) stretch in the icmT gene are also defective in acute lethality in A/J mice (Alli et al., 2000; Molmeret et al., 2002b). Therefore, the Rib pore-forming activity is not required for phagosomal trafficking and intracellular multiplication of L. pneumophila within the cells, but it is essential for induction of cytolysis of the infected macrophages (Alli et al., 2000; Gao and Abu Kwaik, 2000). Thus the C-terminus of IcmT is essential for pore-formation-mediated cytolysis.

L. pneumophila kills A. polyphaga through temporal induction of necrosis, which is mediated by the Rib pore-forming activity of L. pneumophila.

Wild-type intracellular *L. pneumophila* causes necrosis-mediated cytolysis of *A. polyphaga* within 48 h after infection (Gao and Abu Kwaik, 2000). However, in *A. polyphaga*, at low MOI the *rib* mutants are defective in intracellular multiplication (Molmeret et al., 2002b). These results show that the C-terminus of IcmT is essential for pore formation and intracellular trafficking and multiplication within *A. polyphaga*, as confirmed by fusion of phagosomes harboring the *rib* mutants to lysosomes (Molmeret et al., 2002a). Experiments performed with an IcmT null mutant that is defective in intracellular trafficking, intracellular multiplication, and egress from both protozoan and mammalian cells suggest that IcmT is bifunctional (Molmeret et al., 2002a, 2002b). The C-terminus of IcmT is essential for pore-formation-mediated cytolysis in mammalian cells, and the N-terminus is required for intracellular trafficking (Molmeret et al., 2002a). It is unlikely that IcmT is an effector or a common regulator, but it is possible that this protein is a cofactor involved in the export of different substrates with roles in pore-forming toxicity and intracellular trafficking (Molmeret et al., 2002a). Effectors of this type IV secretion system remain to be identified.

The ability to lyse host cells and to egress from them is a fundamental step in the pathogenic cycle of intracellular bacteria and determines the overall consequences of the infection of an organism. Apoptosis and necrosis are the two commonly observed types of cell death. Necrosis is characterized by physical damage that causes cell death. Apoptosis is a regulated suicide program of the cell that manifests morphological and biochemical features distinct from those of necrosis (Cohen, 1993). Killing of mammalian cells by *L. pneumophila* has been proposed to occur in two phases (Gao and Abu Kwaik, 1999a, 1999b; also see Fig 5.4). In the first phase, *L. pneumophila* induces apoptosis in macrophages, monocytes, and alveolar epithelial cells during the early stages of the infection (Hagele et al., 1998; Gao and Abu Kwaik, 1999b), which is mediated through the activation of caspase-3 (Gao and Abu Kwaik, 1999a). Induction of apoptosis is largely independent of the bacterial growth phase (Gao and Abu Kwaik, 1999a). The second phase is mediated through rapid induction of necrosis by *L. pneumophila* upon entry into the postexponential phase of growth, when the bacteria become cytotoxic (Fig. 5.4). Our working model of bacterial egress can be presented in three steps (Fig. 5.4). First, upon exiting the exponential phase of intracellular growth, an "egress pore" is inserted into the phagosomal membrane, leading to its disruption. Second, the bacteria egress into cytoplasm. Third, disruption of organelles and the plasma membrane occurs, culminating in lysis of the host cells and bacterial egress.

IMMUNE RESPONSE TO *L. PNEUMOPHILA* INFECTION

When *L. pneumophila* is inhaled into the lungs, acute alveolitis and bronchiolitis can be observed (Winn and Myerowitz, 1981). The alveolar exudate typically consists of polymorphonuclear cells and macrophages, red blood cells, and cellular debris (Glavin et al., 1979). *L. pneumophila* are mainly intracellular, and the fate of the bacteria depends on the host cells. Within macrophages, most of the bacteria are intact in large vacuoles containing a large number of bacilli (Glavin et al., 1979). Within neutrophils, *L. pneumophila* most often appears degraded (Katz and Hashemi, 1982).

The role of alveolar epithelial cells in Legionnaires' disease has not been well studied. *L. pneumophila* has been shown to replicate in a phagosome within many epithelial cells *in vitro*, including type I and II alveolar epithelial cells (Gao et al., 1998b). There is no reason why *L. pneumophila* is not expected to replicate at the foci of infection within alveolar epithelial cells, particularly because these constitute most of the alveolar surface, where the foci of infection are recognized. In addition, these cells are potential sites of intracellular replication during activation of macrophages by IFN-γ, which inhibits intracellular replication of *L. pneumophila* within monocytes and macrophages (Byrd and Horwitz, 1989).

Although the bacteria bind complement component C3, they are resistant to complement-mediated killing (Horwitz and Silverstein, 1981a, 1981b; Payne and Horwitz, 1987). *L. pneumophila* is also resistant when treated with complement and specific antibodies in presence of polymorphonuclear cells (Horwitz and Silverstein, 1981a, 1981b). Different results have been obtained with monocytes in experiments performed at a high multiplicity of infection (Horwitz and Silverstein, 1981a). In the presence of both antibody and complement, phagocytosis of *L. pneumophila* is more efficient and monocytes kill approximately half of the inoculum of the bacteria (Horwitz and Silverstein, 1981a). In addition, antibodies promote fusion of the infected vacuole to lysosomes as compared with cells treated with complement alone (Horwitz and Maxfield, 1984). However, adherence studies of *L. pneumophila* to U937 macrophages, guinea pig alveolar macrophages, and MRC-5 cells in absence of serum have shown that neither complement nor antibody is required for binding (Gibson et al., 1993). Overall, *L. pneumophila* is relatively resistant to innate and humoral immune responses and the infection is most likely controlled by cell-mediated immunity.

Important roles of Th1-type cytokines, such as tumor necrosis factor alpha (TNF-α), IFN-γ, and interleukin-12 (IL-12), have been demonstrated in the murine A/J model of *L. pneumophila* infection (Brieland et al., 1994, 1995; Gebran et al., 1994b). Susa et al. (1998) have examined cytokines and the role of CD4 and CD8 cells in Legionnaires' disease. After injection of 10^6 CFU of *L. pneumophila* in A/J mice, a challenge that allows the survival of the infected animals, inflammatory cells are recruited into the lung on the second day, and, by the third day of infection, macrophages, B cells, NK cells, and large mononuclear cells are mainly present (Susa et al., 1998). T lymphocytes infiltrate subsequently (Susa et al., 1998). The *L. pneumophila* infection results in a rapid upregulation of systemic concentrations of IFN-γ, TNF-α, IL-1β, IL-4, and IL-6 (but not IL-2); then these levels decrease on the third day (Susa et al., 1998). Recruitment of T cells is necessary for the clearance of the bacteria. The depletion of CD4+ and CD8+ T cells leads to increased lethality to mice. Moreover, another study has shown that nitric oxide (NO) produced by numerous cells, such as macrophages and neutrophils, may regulate IL-6 production in *L. pneumophila*-infected macrophages (Yamamoto et al., 1996). Indeed, when macrophages were primed with IFN-γ, bacterial replication was inhibited and NO was produced in large amounts (Yamamoto et al., 1996). An examination of cytokine levels in *L. pneumophila*-infected macrophages primed with IFN-γ revealed a moderate increase of IL-6 production (Yamamoto et al., 1996).

After intratracheal injection of *L. pneumophila*, the increase of bacteria in the lungs by 48 h is accompanied by a massive accumulation of neutrophils (Tateda et al., 2001b). Neutrophils are recruited early during the infection of *L. pneumophila* in animal models and patients (Brieland et al., 1995). However, *Legionella* are partially resistant to neutrophil killing, particularly when the bacteria are opsonized (Horwitz and Silverstein, 1980, 1981a; Katz and Hashemi, 1983). Neutrophils have the ability to produce immunoregulatory cytokines–chemokines, including IL-12, which may affect Th1/Th2 host responses (Cassatella, 1995). Previous studies have shown a protective role for Th1 cytokines such as IFN-γ and IL-12 during *L. pneumophila* infection (Gebran et al., 1994b; Brieland et al., 1995). In contrast, the Th2 cytokine, IL-10, facilitates growth of *L. pneumophila* in macrophages through the IL-10-mediated suppression of Th1 cytokines (Tateda et al., 2001a). A recent study shows that the CXC chemokine receptor 2, a receptor for chemokine-mediated neutrophil accumulation, may play a role in host defense against *L. pneumophila*, because blockade of this receptor enhances mortality in the A/J mouse model (Susa et al., 1998). The neutrophils may have a protective effect through immunomodulatory actions in *L. pneumophila* infection (Tateda et al., 2001b). Early recruitment of neutrophils may contribute to T1 polarization in a murine model (Tateda et al., 2001a).

IFN-γ is recognized as an important cytokine in both innate and cell-mediated immune responses. Previous studies have shown that endogenous IFN-γ is induced in response to *Legionella* infection (Brieland et al., 1994; Susa et al., 1998). Moreover, treatment with IFN-γ is able to inhibit the intracellular replication of *L. pneumophila* (Nash et al., 1988). When the cultures are supplemented with iron-saturated transferrin, the IFN-γ effect is neutralized (Byrd and Horwitz, 1989). Administration of IFN-γ in a murine model of *L. pneumophila* induces the expression of IFN-γ and IL-12 from natural killer cells (Deng et al., 2001). Moreover, intrapulmonary adenovirus-mediated IFN-γ gene therapy, in a nonlethal murine model of *L. pneumophila* pneumonia, results in a 10-fold decrease in lung bacterial CFU at 48 h postinfection, compared with controls that did not receive the gene. Thus, even in immunocompetent hosts, expression of IFN-γ promotes *L. pneumophila* clearance, independent of cell recruitment and proinflammatory cytokine induction. Alveolar macrophages from uninfected animals treated *in vivo* with the adenovirus-mediated IFN-γ inhibit the intracellular growth of *L. pneumophila* (Deng et al., 2001). Therefore, IFN-γ-secreting cells such as T cells and NK cells may directly contribute to bacteriostasis or killing, either by downregulating transferrin receptors (Byrd and Horwitz, 1989, 2000; Gebran et al., 1994a) or by endogenous NO that may regulate IL-6 production (Yamamoto et al., 1996). TNF-α also helps to control the infection by means of endogenous NO. As a consequence of the *L. pneumophila* infection, a rapid nonspecific immune response is followed by a slower-developing specific immune response necessary for final eradication of the infection. CD4+ and CD8+ T cells play a role in both phases. During the first phase, T cells might produce IFN and IL-6, whereas in the second phase they support humoral immunity and specific T-cell-mediated immunity (Kaufmann, 1993; Susa et al., 1998).

CONCLUSION

L. pneumophila is an intracellular pathogen that utilizes protozoa in aquatic environments to replicate within and to be protected from adverse conditions. Legionnaires' disease has become a threat to humans after our industrialization and production of man-made devices that generate aerosols, which allows transmission of the bacteria to humans. Thus, the normal life cycle in protozoa stops when the bacteria encounter human host cells such as macrophages. Although *L. pneumophila* utilizes different ligands to enter the host cells and is unable to induce apoptosis in protozoan cells such as *A. polyphaga* or *A. castellanii*, the remarkable similarities in the intracellular infection of the two evolutionarily distant host cells (macrophages and protozoa) suggest that *L. pneumophila* may utilize similar molecular mechanisms to manipulate processes of these host cells (Gao et al., 1997). It has been hypothesized that *L. pneumophila* has evolved as a protozoan parasite in the environment and its adaptation to this primitive phagocytic unicellular host was sufficient to allow the bacteria to survive and replicate within the biologically similar phagocytic cells of the more evolved mammalian host (Cianciotto and Fields, 1992; Abu Kwaik, 1996).

The type IV secretion system is the key virulence factor of this organism, allowing it to invade the host cells, replicate, evade the endocytic pathway, induce apoptosis, and egress from the host cells (Gao and Abu Kwaik, 1999a, 1999b, 2000; Alli et al., 2000; Hilbi et al., 2001; Watarai et al., 2001b; Molmeret et al., 2002a, 2002b; Zink et al., 2002). Not all the functions of the Dot/Icm proteins have been understood, and no effectors other than RalF have yet been identified. However, understanding these host–parasite interactions at the molecular level will be of considerable help in controlling the replication of this bacterium, both environmentally in the protozoa and in human infection.

REFERENCES

Abu Kwaik, Y., Fields, B.S., and Engleberg, N.C. (1994). Protein expression by the protozoan *Hartmannella vermiformis* upon contact with its bacterial parasite *Legionella pneumophila*. *Infect. Immun.* **62**, 1860–1866.

Abu Kwaik, Y. (1996). The phagosome containing *Legionella pneumophila* within the protozoan *Hartmanella vermiformis* is surrounded by the rough endoplasmic reticulum. *Appl. Environ. Microbiol.* **62**, 2022–2028.

Abu Kwaik, Y., Gao, L.-Y., Harb, O.S., and Stone, B.J. (1997). Transcriptional regulation of the macrophage-induced gene (*gspA*) of *Legionella pneumophila* and phenotypic characterization of a null mutant. *Mol. Microbiol.* **24**, 629–642.

Abu Kwaik, Y. (1998a). Induced expression of the *Legionella pneumophila* gene encoding a 20-kilodalton protein during intracellular infection. *Infect. Immun.* **66**, 203–212.

Abu Kwaik, Y. (1998b). Fatal attraction of mammalian cells to *Legionella pneumophila*. *Mol. Microbiol.* **30**, 689–696.

Abu Kwaik, Y., Gao, L.-Y., Stone, B.J., Venkataraman, C., and Harb, O.S. (1998). Invasion of protozoa by *Legionella pneumophila* and its role in bacterial ecology and pathogenesis. *Appl. Environ. Microbiol.* **64**, 3127–3133.

Adeleke, A., Pruckler, J., Benson, R., Rowbotham, T., Halablab, M., and Fields, B.S. (1996). *Legionella*-like amoebal pathogens – phylogenetic status and possible role in respiratory disease. *Emerg. Infect. Dis.* **2**, 225–229.

Alli, O.A.T., Gao, L.-Y., Pedersen, L.L., Zink, S., Radulic, M., Doric, M., and Abu Kwaik, Y. (2000). Temporal pore formation-mediated egress from macrophages and alveolar epithelial cells by *Legionella pneumophila*. *Infect. Immun.* **68**, 6431–6440.

Anderson, P. (1997). Kinase cascades regulating entry into apoptosis. *Microbiol. Mol. Biol. Rev.* **61**, 33–46.

Bachman, M.A. and Swanson, M.S. (2001). RpoS co-operates with other factors to induce *Legionella pneumophila* virulence in the stationary phase. *Mol. Microbiol.* **40**, 1201–1214.

Barker, J., Brown, M.R.W., Collier, P.J., Farrell, I., and Gilbert, P. (1992). Relationships between *Legionella pneumophila* and *Acanthamoebae polyphaga*: physiological status and susceptibility to chemical inactivation. *Appl. Environ. Microbiol.* **58**, 2420–2425.

Barker, J., Lambert, P.A., and Brown, M.R.W. (1993). Influence of intra-amoebic and other growth conditions on the surface properties of *Legionella pneumophila*. *Infect. Immun.* **61**, 3503–3510.

Barker, J., Scaife, H., and Brown, M.R.W. (1995). Intraphagocytic growth induces an antibiotic-resistant phenotype of *Legionella pneumophila*. *Antimicrob. Agents Chemother.* **39**, 2684–2688.

Beckers, M.C., Yoshida, S., Morgan, K., Skamene, E., and Gros, P. (1995). Natural resistance to infection with *Legionella pneumophila*: chromosomal localization of the *Lgn1* susceptibility gene. *Mamm. Genome* **6**, 540–545.

Beckers, M.C., Ernst, E., Diez, E., Morissette, C., Gervais, F., Hunter, K., Housman, D., Yoshida, S., Skamene, E., and Gros, P. (1997). High-resolution linkage map of mouse chromosome 13 in the vicinity of the host resistance locus *Lgn1*. *Genomics* **39**, 254–263.

Benin, A.L., Benson, R.F., and Besser, R.E. (2002). Trends in Legionnaires disease, 1980–1998: declining mortality and new patterns of diagnosis. *Clin. Inf. Dis.* **35**, 1039–1046.

Berger, K.H. and Isberg, R.R. (1993). Two distinct defects in intracellular growth complemented by a single genetic locus in *Legionella pneumophila*. *Mol. Microbiol.* **7**, 7–19.

Berger, K.H., Merriam, J., and Isberg, R.R. (1994). Altered intracellular targeting properties associated with mutations in the *Legionella pneumophila dotA* gene. *Mol. Microbiol.* **14**, 809–822.

Berk, S.G., Ting, R.S., Turner, G.W., and Ashburn, R.J. (1998). Production of respirable vesicles containing live *Legionella pneumophila* cells by two *Acanthamoeba* spp. *Appl. Environ. Microbiol.* **64**, 279–286.

Birtles, R.J., Rowbotham, T.J., Raoult, D., and Harrison, T.G. (1996). Phylogenetic diversity of intra-amoebal legionellae as revealed by 16S rRNA gene sequence comparison. *Microbiology* **142**, 3525–3530.

Biurrun, A., Caballero, L., Pelaz, C., Leon, E., and Gago, A. (1999). Treatment of a *Legionella pneumophila*-colonized water distribution system using copper-silver ionization and continuous chlorination. *Infect. Control Hosp. Epidemiol.* **20**, 426–428.

Bozue, J.A., and Johnson, W. (1996). Interaction of *Legionella pneumophila* with *Acanthamoeba catellanii*: uptake by coiling phagocytosis and inhibition of phagosome-lysosome fusion. *Infect. Immun.* **64**, 668–673.

Brieland, J., Freeman, P., Kunkel, R., Chrisp, C., Hurley, M., Fantone, J., and Engleberg, N.C. (1994). Replicative *Legionella pneumophila* lung infection in intratracheally inoculated A/J mice: a murine model of human Legionnaires' disease. *Am. J. Pathol.* **145**, 1537–1546.

Brieland, J.K., Remick, D.G., Freeman, P.T., Hurley, M.C., Fantone, J.C., and Engleberg, N.C. (1995). *In vivo* regulation of replicative *Legionella pneumophila* lung infection by endogenous tumor necrosis factor alpha and nitric oxide. *Infect. Immun.* **63**, 3253–3258.

Brieland, J.K., Fantone, J.C., Remick, D.G., LeGendre, M., McClain, M., and Engleberg, N.C. (1997). The role of *Legionella pneumophila*-infected *Hartmanella vermiformis* as an infectious particle in a murine model of Legionnaires' disease. *Infect. Immun.* **65**, 4892–4896.

Byrd, T.F. and Horwitz, M.A. (1989). Interferon gamma-activated human monocytes downregulate transferrin receptors and inhibit the intracellular multiplication of *Legionella pneumophila* by limiting the availability of iron. *J. Clin. Invest.* **83**, 1457–1465.

Byrd, T.F. and Horwitz, M.A. (2000). Aberrantly low transferrin receptor expression on human monocytes is associated with nonpermissiveness for *Legionella pneumophila* growth. *J. Infect. Dis.* **181**, 1394–1400.

Byrne, B. and Swanson, M.S. (1998). Expression of *Legionella pneumophila* virulence traits in response to growth conditions. *Infect. Immun.* **66**, 3029–3034.

Cassatella, M.A. (1995). The production of cytokines by polymorphonuclear neutrophils. *Immunol. Today* **16**, 21–26.

Christie, P.J. and Vogel, J.P. (2000). Bacterial type IV secretion: conjugation systems adapted to deliver effector molecules to host cells. *Trends Microbiol.* **8**, 354–360.

Cianciotto, N.P. and Fields, B.S. (1992). *Legionella pneumophila mip* gene potentiates intracellular infection of protozoa and human macrophages. *Proc. Natl. Acad. Sci. USA* **89**, 5188–5191.

Cirillo, J.D., Tompkins, L.S., and Falkow, S. (1994). Growth of *Legionella pneumophila* in *Acanthamoeba castellanii* enhances invasion. *Infect. Immun.* **62**, 3254–3261.

Cirillo, J.D., Cirillo, S.L., Yan, L., Bermudez, L.E., Falkow, S., and Tompkins, L.S. (1999). Intracellular growth in *Acanthamoeba castellanii* affects monocyte entry mechanisms and enhances virulence of *Legionella pneumophila*. *Infect. Immun.* **67**, 4427–4434.

Cirillo, S.L., Lum, J., and Cirillo, J.D. (2000). Identification of novel loci involved in entry by *Legionella pneumophila*. *Microbiology* **146**, 1345–1359.

Coers, J., Monahan, C., and Roy, C.R. (1999). Modulation of phagosome biogenesis by *Legionella pneumophila* creates an organelle permissive for intracellular growth. *Nature Cell Biol.* **1**, 451–453.

Coers, J., Kagan, J.C., Matthews, M., Nagai, H., Zuckman, D.M., and Roy, C.R. (2000). Identification of Icm protein complexes that play distinct roles in the biogenesis of an organelle permissive for *Legionella pneumophila* intracellular growth. *Mol. Microbiol.* **38**, 719–736.

Cohen, J.J. (1993). Mechanisms of apoptosis. *Immunol. Today* **14**, 126–130.

Deng, J.C., Tateda, K., Zeng, X., and Standiford, T.J. (2001). Transient transgenic expression of gamma interferon promotes *Legionella pneumophila* clearance in immunocompetent hosts. *Infect. Immun.* **69**, 6382–6390.

Dietrich, W.F., Damron, D.M., Isberg, R.R., Lander, E.S., and Swanson, M.S. (1995). *Lgn1*, a gene that determines susceptibility to *Legionella pneumophila*, maps to mouse chromosome 13. *Genomics* **26**, 443–450.

Dorn, B.R., Dunn, W.A. Jr., and Progulske-Fox, A. (2002). Bacterial interactions with the autophagic pathway. *Cell. Microbiol.* **4**, 1–10.

Dowling, J.N., Saha, A.K., and Glew, R.H. (1992). Virulence factors of the family *Legionellaceae*. *Microbiol. Rev.* **56**, 32–60.

Dumenil, G. and Isberg, R.R. (2001). The *Legionella pneumophila* IcmR protein exhibits chaperone activity for IcmQ by preventing its participation in high-molecular-weight complexes. *Mol. Microbiol.* **40**, 1113–1127.

Elliott, J.A. and Winn, W.C. Jr. (1986). Treatment of alveolar macrophages with cytochalasin D inhibits uptake and subsequent growth of *Legionella pneumophila*. *Infect. Immun.* **51**, 31–36.

Enari, M., Sakahira, H., Yokoyama, H., Okawa, K., Iwamatsu, A., and Nagata, S. (1998). A caspase-activated DNase that degrades DNA during apoptosis, and its inhibitor ICAD. *Nature* **391**, 43–50.

Fields, B.S., Nerad, T.A., Sawyer, T.K., King, C.H., Barbaree, J.M., Martin, W.T., Morrill, W.E., and Sanden, G.N. (1990). Characterization of an axenic strain of *Hartmannella vermiformis* obtained from an investigation of nosocomial legionellosis. *J. Protozool.* **37**, 581–583.

Fields, B.S. (1996). The molecular ecology of legionellae. *Trends. Microbiol.* **4**, 286–290.

Fields, B.S., Benson, R.F., and Besser, R.E. (2002). *Legionella* and Legionnaires' disease: 25 years of investigation. *Clin. Microbiol. Rev.* **15**, 506–526.

Fliermans, C.B. (1996). Ecology of *Legionella*: from data to knowledge with a little wisdom. *Microb. Ecol.* **32**, 203–228.

Gagnon, E., Duclos, S., Rondeau, C., Chevet, E., Cameron, P.H., Steele-Mortimer, O., Paiement, J., Bergeron, J.J., and Desjardins, M. (2002). Endoplasmic reticulum-mediated phagocytosis is a mechanism of entry into macrophages. *Cell* **110**, 119–131.

Gal-Mor, O., Zusman, T., and Segal, G. (2002). Analysis of DNA regulatory elements required for expression of the *Legionella pneumophila icm* and *dot* virulence genes. *J. Bacteriol.* **184**, 3823–3833.

Gao, L.-Y., Harb, O.S., and Abu Kwaik, Y. (1997). Utilization of similar mechanisms by *Legionella pneumophila* to parasitize two evolutionarily distant hosts, mammalian and protozoan cells. *Infect. Immun.* **65**, 4738–4746.

Gao, L.-Y., Harb, O.S., and Abu Kwaik, Y. (1998a). Identification of macrophage-specific infectivity loci (*mil*) of *Legionella pneumophila* that are not required for infectivity of protozoa. *Infect. Immun.* **66**, 883–892.

Gao, L.-Y., Stone, B.J., Brieland, J.K., and Abu Kwaik, Y. (1998b). Different fates of *Legionella pneumophila pmi* and *mil* mutants within human-derived macrophages and alveolar epithelial cells. *Microb. Pathog.* **25**, 291–306.

Gao, L.-Y. and Abu Kwaik, Y. (1999a). Activation of caspase-3 in *Legionella pneumophila*-induced apoptosis in macrophages. *Infect. Immun.* **67**, 4886–4894.

Gao, L.-Y. and Abu Kwaik, Y. (1999b). Apoptosis in macrophages and alveolar epithelial cells during early stages of infection by *Legionella pneumophila* and its role in cytopathogenicity. *Infect. Immun.* **67**, 862–870.

Gao, L.-Y., Susa, M., Ticac, B., and Abu Kwaik, Y. (1999). Heterogeneity in intracellular replication and cytopathogenicity of *Legionella pneumophila* and *Legionella micdadei* in mammalian and protozoan cells. *Microb. Pathogen.* **27**, 273–287.

Gao, L.-Y. and Abu Kwaik, Y. (2000). The mechanism of killing and exiting the protozoan host *Acanthamoeba polyphaga* by *Legionella pneumophila*. *Environ. Microbiol.* **2**, 79–90.

Gebran, S.J., Newton, C., Yamamoto, Y., Widen, R., Klein, T.W., and Friedman, H. (1994a). Macrophage permissiveness for *Legionella pneumophila* growth modulated by iron. *Infect. Immun.* **62**, 564–568.

Gebran, S.J., Yamamoto, Y., Newton, C., Klein, T.W., and Friedman, H. (1994b). Inhibition of *Legionella pneumophila* growth by gamma interferon in permissive A/J mouse macrophages: role of reactive oxygen species, nitric oxide, tryptophan, and iron(III). *Infect. Immun.* **62**, 3197–3205.

Gibson, F.C. III, Tzianabos, O.A., and Rodgers, F.G. (1993). Adherence of *Legionella pneumophila* to U-937 cells, guinea-pig alveolar macrophages, and MRC-5 cells by a novel, complement-independent binding mechanism. *Can. J. Microbiol.* **39**, 718–722.

Glavin, F.L., Winn, W.C., and Graighead, J.E. (1979). Ultrastructure of lung in Legionnaires' disease. *Ann. Intern. Med.* **90**, 555–559.

Hagele, S., Hacker, J., and Brand, B.C. (1998). *Legionella pneumophila* kills human phagocytes but not protozoan host cells by inducing apoptotic cell death. *FEMS Microbiol. Lett.* **169**, 51–58.

Hales, L.M. and Shuman, H.A. (1999). The *Legionella pneumophila rpoS* gene is required for growth within *Acanthamoeba castellanii. J. Bacteriol.* **181**, 4879–4889.

Hammer, B.K. and Swanson, M.S. (1999). Co-ordination of *Legionella pneumophila* virulence with entry into stationary phase by ppGpp. *Mol. Microbiol.* **33**, 721–731.

Hammer, B.K., Tateda, E.S., and Swanson, M.S. (2002). A two-component regulator induces the transmission phenotype of stationary-phase *Legionella pneumophila. Mol. Microbiol.* **44**, 107–118.

Harb, O.S., Venkataraman, C., Haack, B.J., Gao, L.-Y., and Abu Kwaik, Y. (1998). Heterogeneity in the attachment and uptake mechanisms of the Legionnaires' disease bacterium, *Legionella pneumophila*, by protozoan hosts. *Appl. Environ. Microbiol.* **64**, 126–132.

Harb, O.S., Gao, L.-Y., and Abu Kwaik, Y. (2000). From protozoa to mammalian cells: a new paradigm in the life cycle of intracellular bacterial pathogens. *Environ. Microbiol.* **2**, 251–265.

Heuner, K., Dietrich, C., Skriwan, C., Steinert, M., and Hacker, J. (2002). Influence of the alternative sigma(28) factor on virulence and flagellum expression of *Legionella pneumophila. Infect. Immun.* **70**, 1604–1608.

Hilbi, H., Segal, G., and Shuman, H.A. (2001). *Icm/dot*-dependent upregulation of phagocytosis by *Legionella pneumophila. Mol. Microbiol.* **42**, 603–617.

Horwitz, M.A. and Silverstein, S.C. (1980). Legionnaires' disease bacterium (*Legionella pneumophila*) multiplies intracellularly in human monocytes. *J. Clin. Invest.* **66**, 441–450.

Horwitz, M.A. and Silverstein, S.C. (1981a). Interaction of the Legionnaires' disease bacterium (*Legionella pneumophila*) with human phagocytes. II. Antibody promotes binding of *L. pneumophila* to monocytes but does not inhibit intracellular multiplication. *J. Exp. Med.* **153**, 398–406.

Horwitz, M.A. and Silverstein, S.C. (1981b). Interaction of the Legionnaires' disease bacterium (*Legionella pneumophila*) with human phagocytes. I. *L. pneumophila* resists killing by polymorphonuclear leukocytes, antibody, and complement. *J. Exp. Med.* **153**, 386–397.

Horwitz, M.A. (1983a). The Legionnaires' disease bacterium (*Legionella pneumophila*) inhibits phagosome-lysosome fusion in human monocytes. *J. Exp. Med.* **158**, 2108–2126.

Horwitz, M.A. (1983b). Formation of a novel phagosome by the Legionnaires' disease bacterium (*Legionella pneumophila*) in human monocytes. *J. Exp. Med.* **158**, 1319–1331.

Horwitz, M.A. (1984). Phagocytosis of the Legionnaires' disease bacterium (*Legionella pneumophila*) occurs by a novel mechanism: engulfment within a pseudopod coil. *Cell* **36**, 27–33.

Horwitz, M.A. and Maxfield, F.R. (1984). *Legionella pneumophila* inhibits acidification of its phagosome in human monocytes. *J. Cell Biol.* **99**, 1936–1943.

Katz, S.M. and Hashemi, S. (1982). Electron microscopic examination of the inflammatory response to *Legionella pneumophila* in guinea pigs. *Lab. Invest.* **46**, 24–32.

Katz, S.M. and Hashemi, S. (1983). Electron microscopic examination of the inflammatory response of guinea pig neutrophils and macrophages to *Legionella pneumophila. Adv. Exp. Med. Biol.* **162**, 327–333.

Kaufmann, S.H.E. (1993). Immunity to intracellular bacteria. *Annu. Rev. Immunol.* **11**, 129–163.

King, C.H., Fields, B.S., Shotts, E.B. Jr., and White, E.H. (1991). Effects of cytochalasin D and methylamine on intracellular growth of *Legionella pneumophila* in amoebae and human monocyte-like cells. *Infect. Immun.* **59**, 758–763.

Kirby, J.E. and Isberg, R.R. (1998). Legionnaires' disease: the pore macrophage and the legion of terror within. *Trends Microbiol.* **6**, 256–258.

Kirby, J.E., Vogel, J.P., Andrews, H.L., and Isberg, R.R. (1998). Evidence for poreforming ability by *Legionella pneumophila. Mol. Microbiol.* **27**, 323–336.

Komano, T., Yoshida, T., Narahara, K., and Furuya, N. (2000). The transfer region of IncI1 plasmid R64: similarities between R64 *tra* and *legionella icm/dot* genes. *Mol. Microbiol.* **35**, 1348–1359.

Kool, J.L., Carpenter, J.C., and Fields, B.S. (1999). Effect of monochloramine disinfection of municipal drinking water on risk of nosocomial Legionnaires' disease. *Lancet* **353**, 272–277.

Kusnetsov, J., Iivanainen, E., Elomaa, N., Zacheus, O., and Martikainen, P.J. (2001). Copper and silver ions more effective against legionellae than against mycobacteria in a hospital warm water system. *Water Res.* **35**, 4217–4225.

Mann, B.J., Torian, B.E., Vedvick, T.S., and Petri, W.A.J. (1991). Sequence of a cysteine-rich galactose-specific lectin of *Entamoeba histolytica. Proc. Natl. Acad. Sci. USA* **88**, 3248–3252.

Marrie, T.J., Raoult, D., La Scola, B., Birtles, R.J., and de Carolis, E. (2001). *Legionella*-like and other amoebal pathogens as agents of communityacquired pneumonia. *Emerg. Infect. Dis.* **7**, 1026–1029.

Matthews, M. and Roy, C.R. (2000). Identification and subcellular localization of the *legionella pneumophila* IcmX protein: a factor essential for establishment of a replicative organelle in eukaryotic host cells. *Infect. Immun.* **68**, 3971–3982.

Molmeret, M. and Abu Kwaik, Y. (2002). How does *Legionella pneumophila* exit the host cell? *Trends Microbiol.* **10**, 258–260.

Molmeret, M., Alli, O.A., Radulic, M., Susa, M., Doric, M., and Kwaik, Y.A. (2002a). The C-terminus of IcmT is essential for pore formation and for intracellular trafficking of *Legionella pneumophila* within *Acanthamoeba polyphaga*. *Mol. Microbiol.* **43**, 1139–1150.

Molmeret, M., Alli, O.A., Zink, S., Flieger, A., Cianciotto, N.P., and Kwaik, Y.A. (2002b). *icmT* is essential for pore formation-mediated egress of *Legionella pneumophila* from mammalian and protozoan cells. *Infect. Immun.* **70**, 69–78.

Muller, A., Hacker, J., and Brand, B. (1996). Evidence for apoptosis of human macrophage-like HL-60 cells by *Legionella pneumophila* infection. *Infect. Immun.* **64**, 4900–4906.

Muraca, P., Stout, J.E., and Yu, V.L. (1987). Comparative assessment of chlorine, heat, ozone, and UV light for killing *Legionella pneumophila* within a model plumbing system. *Appl. Environ. Microbiol.* **53**, 447–453.

Nagai, H. and Roy, C.R. (2001). The DotA protein from *Legionella pneumophila* is secreted by a novel process that requires the Dot/Icm transporter. *EMBO J.* **20**, 5962–5970.

Nagai, H., Kagan, J.C., Zhu, X., Kahn, R.A., and Roy, C.R. (2002). A bacterial guanine nucleotide exchange factor activates ARF on *Legionella* phagosomes. *Science* **295**, 679–682.

Nash, T.W., Libby, D.M., and Horwitz, M.A. (1984). Interaction between the Legionnaires' disease bacterium (*Legionella pneumophila*) and human alveolar macrophages. Influence of antibody, lymphokines, and hydrocortisone. *J. Clin. Invest.* **74**, 771–782.

Nash, T.W., Libby, D.M., and Horwitz, M.A. (1988). IFN-gamma-activated human alveolar macrophages inhibit the intracellular multiplication of *Legionella pneumophila*. *J. Immunol.* **140**, 3978–3981.

O'Brein, S.J. and Bhopal, R.S. (1993). Legionnaires' disease: the infective dose paradox. *Lancet* **342**, 5–6.

Payne, N.R. and Horwitz, M.A. (1987). Phagocytosis of *Legionella pneumophila* is mediated by human monocyte complement receptors. *J. Exp. Med.* **166**, 1377–1389.

Rechnitzer, C. and Blom, J. (1989). Engulfment of the Philadelphia strain of *Legionella pneumophila* within pseudopod coils in human phagocytes. Comparison with the other *Legionella* strains and species. *Acta Pathol. Microbiol. Immunol. Scand.[B]* **97**, 105–114.

Rodgers, F.G. and Gibson, F.C. III (1993). Opsonin-independent adherence and intracellular development of *Legionella pneumophila* within U-937 cells. *Can. J. Microbiol.* **39**, 718–722.

Rowbotham, T.J. (1980). Preliminary report on the pathogenicity of *Legionella pneumophila* for freshwater and soil amoebae. *J. Clin. Pathol.* **33**, 1179–1183.

Rowbotham, T.J. (1986). Current views on the relationships between amoebae, legionellae and man. *Isr. J. Med. Sci.* **22**, 678–689.

Roy, C.R. and Isberg, R.R. (1997). Topology of *Legionella pneumophila* DotA: an inner membrane protein required for replication in macrophages. *Infect. Immun.* **65**, 571–578.

Roy, C.R. and Tilney, L.G. (2002). The road less traveled: transport of *Legionella* to the endoplasmic reticulum. *J. Cell Biol.* **158**, 415–419.

Segal, G., Purcell, M., and Shuman, H.A. (1998). Host cell killing and bacterial conjugation require overlapping sets of genes within a 22-kb region of the *Legionella pneumophila* chromosome. *Proc. Natl. Acad. Sci. USA* **95**, 1669–1674.

Segal, G. and Shuman, H.A. (1998). Intracellular multiplication and human macrophage killing by *Legionella pneumophila* are inhibited by conjugal components of IncQ plasmid RSF1010. *Mol. Microbiol.* **30**, 197–208.

Segal, G., Russo, J.J., and Shuman, H.A. (1999). Relationships between a new type IV secretion system and the *icm/dot* virulence system of *Legionella pneumophila*. *Mol. Microbiol.* **34**, 799–809.

Segal, G. and Shuman, H.A. (1999). *Legionella pneumophila* utilizes the same genes to multiply within *Acanthamoeba castellanii* and human macrophages. *Infect. Immun.* **67**, 2117–2124.

Solomon, J.M. and Isberg, R.R. (2000). Growth of *Legionella pneumophila* in *Dictyostelium discoideum*: a novel system for genetic analysis of host-pathogen interactions. *Trends Microbiol.* **8**, 478–480.

Solomon, J.M., Rupper, A., Cardelli, J.A., and Isberg, R.R. (2000). Intracellular growth of *Legionella pneumophila* in *Dictyostelium discoideum*, a system for genetic analysis of host-pathogen interactions. *Infect. Immun.* **68**, 2939–2947.

Steinert, M., Emody, L., Amann, R., and Hacker, J. (1997). Resuscitation of viable but nonculturable *Legionella pneumophila* Philadelphia JR32 by *Acanthamoeba castellanii*. *Appl. Environ. Microbiol.* **63**, 2047–2053.

Stone, B.J., Brier, A., and Kwaik, Y.A. (1999). The *Legionella pneumophila prp* locus; required during infection of macrophages and amoebae. *Microb. Pathog.* **27**, 369–376.

Susa, M., Ticac, T., Rukavina, T., Doric, M., and Marre, R. (1998). *Legionella pneumophila* infection in intratracheally inoculated T cell depleted or non-depleted A/J mice. *J. Immunol.* **160**, 316–321.

Swanson, M.S. and Isberg, R.R. (1995a). Formation of the *Legionella pneumophila* replicative phagosome. *Infect. Agents Dis.* **2**, 269–271.

Swanson, M.S. and Isberg, R.R. (1995b). Association of *Legionella pneumophila* with the macrophage endoplasmic reticulum. *Infect. Immun.* **63**, 3609–3620.

Tateda, K., Moore, T.A., Deng, J.C., Newstead, M.W., Zeng, X., Matsukawa, A., Swanson, M.S., Yamaguchi, K., and Standiford, T.J. (2001a). Early recruitment of neutrophils determines subsequent T1/T2 host responses in a murine model of *Legionella pneumophila* pneumonia. *J. Immunol.* **166**, 3355–3361.

Tateda, K., Moore, T.A., Newstead, M.W., Tsai, W.C., Zeng, X., Deng, J.C., Chen, G., Reddy, R., Yamaguchi, K., and Standiford, T.J. (2001b). Chemokine-dependent neutrophil recruitment in a murine model of *Legionella pneumonia*: potential role of neutrophils as immunoregulatory cells. *Infect. Immun.* **69**, 2017–2024.

Tilney, L.G., Harb, O.S., Connelly, P.S., Robinson, C.G., and Roy, C.R. (2001). How the parasitic bacterium *Legionella pneumophila* modifies its phagosome and transforms it into rough ER: implications for conversion of plasma membrane to the ER membrane. *J. Cell Sci.* **114**, 4637–4650.

Venkataraman, C., Haack, B.J., Bondada, S., and Abu Kwaik, Y. (1997). Identification of a Gal/GalNAc lectin in the protozoan *Hartmannella vermiformis* as a potential receptor for attachment and invasion by the Legionnaires' disease bacterium, *Legionella pneumophila*. *J. Exp. Med.* **186**, 537–547.

Venkataraman, C., Gao, L.-Y., Bondada, S., and Abu Kwaik, Y. (1998). Identification of putative cytoskeletal protein homologues in the protozoan *Hartmannella vermiformis* as substrates for induced tyrosine phosphatase activity upon attachment to the Legionnaires' disease bacterium, *Legionella pneumophila*. *J. Exp. Med.* **188**, 505–514.

Vogel, J.P., Andrews, H.L., Wong, S.K., and Isberg, R.R. (1998). Conjugative transfer by the virulence system of *Legionella pneumophila*. *Science* **279**, 873–876.

Watarai, M., Andrews, H.L., and Isberg, R.R. (2001a). Formation of a fibrous structure on the surface of *Legionella pneumophila* associated with exposure of DotH and DotO proteins after intracellular growth. *Mol. Microbiol.* **39**, 313–330.

Watarai, M., Derre, I., Kirby, J., Growney, J.D., Dietrich, W.F., and Isberg, R.R. (2001b). *Legionella pneumophila* is internalized by a macropinocytotic uptake pathway controlled by the Dot/Icm system and the mouse *Lgn1* locus. *J. Exp. Med.* **194**, 1081–1096.

Weinbaum, D.L., Benner, R.R., Dowling, J.N., Alpern, A., Pasculle, A.W., and Donowitz, G.R. (1984). Interaction of *Legionella micdadei* with human monocytes. *Infect. Immun.* **46**, 68–73.

Winn, W.C. Jr. and Myerowitz, R.L. (1981). The pathology of the *Legionella* pneumonias. A review of 74 cases and the literature. *Hum. Pathol.* **12**, 401–422.

Yamamoto, H., Ezaki, T., Ikedo, M., and Yabuuchi, E. (1991). Effects of biocidal treatments to inhibit the growth of legionellae and other microorganisms in cooling towers. *Microbiol. Immunol.* **35**, 795–802.

Yamamoto, Y., Klein, T.W., and Friedman, H. (1992). Genetic control of macrophage susceptibility to infection by *Legionella pneumophila*. *FEMS Microbiol. Immunol.* **89**, 137–146.

Yamamoto, Y., Klein, T.W., and Friedman, H. (1996). Immunoregulatory role of nitric oxide in *Legionella pneumophila*-infected macrophages. *Cell Immunol.* **171**, 231–239.

Zink, S.D., Pedersen, L., Cianciotto, N.P., and Abu-Kwaik, Y. (2002). The Dot/Icm type IV secretion system of *Legionella pneumophila* is essential for the induction of apoptosis in human macrophages. *Infect. Immun.* **70**, 1657–1663.

Zuckman, D.M., Hung, J.B., and Roy, C.R. (1999). Pore-forming activity is not sufficient for *Legionella pneumophila* phagosome trafficking and intracellular replication. *Mol. Microbiol.* **32**, 990–1001.

Zusman, T., Gal-Mor, O., and Segal, G. (2002). Characterization of a *Legionella pneumophila relA* insertion mutant and toles of RelA and RpoS in virulence gene expression. *J. Bacteriol.* **184**, 67–75.

CHAPTER 6

Listeria monocytogenes invasion and intracellular growth

Kendy K.Y. Wong and Nancy E. Freitag

Studies focused on the Gram-positive, facultative intracellular bacterial pathogen *Listeria monocytogenes* have provided valuable insights into many facets of biology, including cell-mediated immunity, cell physiology, and bacterial pathogenesis. The bacterium invades and replicates within a wide variety of cell types, and is capable of infecting an astonishing diversity of hosts, including mammals, fish, and insects (Gray and Killinger, 1966). The well-established use of murine and tissue culture models of infection, the ease of growing *L. monocytogenes* within the laboratory, and the existence of numerous genetic tools for the generation and analysis of bacterial mutants have helped to make *L. monocytogenes* a powerful model system for the exploration of the molecular basis of host–pathogen interactions. Because of its ubiquity within the environment and robust survival skills, this important foodborne pathogen remains a constant concern for public health departments and the food industry.

Listeriae are noncapsulated, nonspore-forming, facultative anaerobic bacilli. They are 0.4 μm by 1 to 1.5 μm in size and are motile at 10°C to 25°C through the expression of polar flagellae (Farber and Peterkin, 1991; Lorber, 1997; Vazquez-Boland et al., 2001b). There are six species in the *Listeria* genus: *L. monocytogenes*, *L. ivanovii*, *L. innocua*, *L. seeligeri*, *L. welshimeri*, and *L. grayi* (Collins et al., 1991; Sheehan et al., 1994). Only *L. monocytogenes* and *L. ivanovii* are considered pathogenic and are responsible for the disease known as listeriosis. *L. ivanovii* is mainly an animal pathogen, whereas *L. monocytogenes* has been reported to cause infections in humans, other animals, birds, and fish (Gray and Killinger, 1966; McLauchlin, 1987; Alexander et al., 1992; Chand and Sadana, 1999; Ryser, 1999; Wesley, 1999). *L. monocytogenes* is the major listerial species pathogenic for humans; although rare, *L. ivanovii* has caused human infections, and at least one report of human listeriosis has been associated with *L. seeligeri* (Rocourt et al., 1986; Cummins et al., 1994; Lessing et al., 1994; Ramage et al., 1999). The *Listeria* spp. are widespread in nature and can be isolated from a variety of sources such as water, soil, food, and feces of animals (Schuchat et al., 1991; Fenlon, 1999). The natural habitat for *Listeria* organisms is believed to be decaying vegetation, but the two pathogenic species can invade and survive within a variety of hosts and host cell types. *L. monocytogenes* was first characterized in 1926 (Murray et al., 1926), but it was not until the early 1980s that a clear link was established between listeriosis and the ingestion of food contaminated with the organism (Schlech et al., 1983). Serious infections occur in immunocompromised individuals, neonates, the elderly, and pregnant women, usually through the ingestion of contaminated food such as dairy products and ready-to-eat deli meats (Gray and Killinger, 1966; Farber and Peterkin, 1991; Rocourt, 1996; Ryser, 1999). The infection of healthy individuals has been reported to cause gastroenteritis, a condition thought to be underdocumented because routine clinical testings do not search for listeriae (Hof, 2001). Although *L. monocytogenes* does not usually pose a high risk for healthy individuals, the mean mortality rate of human listeriosis is 20–30% (Farber and Peterkin, 1991; Schuchat et al., 1991; Rocourt, 1996). The ability of *L. monocytogenes* to tolerate high salt concentrations and low pH' and to multiply at refrigerated temperatures makes it a difficult and constant challenge for the food industry (Lammerding and Doyle, 1990; Lou and Yousef, 1999). This chapter briefly reviews the current knowledge regarding *L. monocytogenes* invasion of host cells and the resulting course of infection following host cell entry.

CLINICAL PRESENTATIONS OF LISTERIOSIS: AN OVERVIEW OF THE DIVERSITY OF CELL TYPES AND TISSUES THAT SUPPORT BACTERIAL REPLICATION

Among the first cells encountered by *L. monocytogenes* following ingestion of the bacterium by the host are the epithelial cells lining the gut. In the mouse, *L. monocytogenes* has been observed to translocate the intestine without causing gross lesions, and rapid listerial translocation occurs in rat ileal loop models of intestinal infection (Marco et al., 1992; Pron et al., 1998). On the basis of these findings, it has been suggested that bacterial replication in the intestinal mucosa is probably not required for the establishment of systemic infection, although *L. monocytogenes* is able to proliferate at this location. *L. monocytogenes* can cross the intestinal barrier by means of M cells of Peyer's patches, and the Peyer's patches appear to be preferential multiplication sites for the bacteria, which can then invade neighboring cells by direct cell-to-cell spread (Havell et al., 1999; Daniels et al., 2000; Hof, 2001). Dendritic cells are also early listerial targets in Peyer's patches (Pron et al., 2001). Gross intestinal lesions and gastroenteritis appear to occur only following the ingestion of large numbers of bacteria. The risk for gastroenteritis also appears to increase in the presence of small amounts of mucosal damage within the intestine; such damage may happen periodically in humans (MacDonald and Carter, 1980; Pron et al., 1998; Daniels et al., 2000). Once across the intestinal barrier, *L. monocytogenes* is translocated to the liver and spleen within minutes, where most of the organisms are then cleared (Conlan and North, 1991; Pron et al., 1998; Cousens and Wing, 2000; Daniels et al., 2000). In susceptible hosts, surviving *L. monocytogenes* multiplies, mainly in hepatocytes, and spreads through the liver parenchyma. Splenocytes are also target cells for *L. monocytogenes*, although they appear less permissive for bacterial replication than hepatocytes. In both organs, microabscess formation is observed and neutrophils are rapidly recruited to the infection foci; however, *L. monocytogenes*-infected spleen cells appear to be less susceptible to lysis by neutrophils during early infection (Conlan and North, 1994; Rogers et al., 1996). *L. monocytogenes* has been demonstrated to induce apoptosis in hepatocytes and Caco-2 cells (Rogers et al., 1996; Valenti et al., 1999), and in dendritic cells apoptosis was recently shown to be triggered by listeriolysin O (LLO; see Guzman et al., 1995). If *L. monocytogenes* infection is not resolved within the liver and spleen, the bacteria enter into circulation and disseminate to other tissues.

L. monocytogenes appears to have a tropism for the placenta of pregnant women and for the central nervous system (CNS). The organism has the ability to cross both the feto-placental barrier and the blood–brain barrier (Armstrong and Fung, 1993; Lorber, 1997; Greiffenberg et al., 1998; Parkassh et al., 1998). At the placenta, *L. monocytogenes* causes the formation of numerous microabscesses and the necrosis of placental villi (Parkassh et al., 1998; Vazquez-Boland et al., 2001b). *L. monocytogenes* infects the developing fetus and causes chorioamnionitis, which often leads to abortion or stillbirth. Alternatively, fetal listeriosis may lead to the birth of a baby with granulomatosis infantiseptica, characterized by microabscesses all over the body and internal organs (Klatt et al., 1986; Schuchat et al., 1991; Lorber, 1997). The mother may be asymptomatic or may only have mild flu-like symptoms, and CNS infection rarely occurs in pregnant women. For the establishment of CNS infections, active actin-based intra-axonal movement may facilitate *L. monocytogenes* spread (Otter and Blakemore, 1989; Antal et al., 2001).

Listeria-infected phagocytes and dendritic cells have also been shown to play a role in bacterial dissemination and in the initiation of CNS infection (Drevets et al., 1995, 2001; Guzman et al., 1995; Drevets, 1999; Pron et al., 2001). The growth of *L. monocytogenes* in brain tissues appears to be unrestricted, and the bacterium has been found to infect the epithelial cells of the choroids plexus, ependymal cells, microglial cells and neurons, and brain parenchymal tissue (Kirk, 1993; Schluter et al., 1999).

In nonpregnant individuals, meningitis and encephalitis are common manifestations of *L. monocytogenes* disease. Clinical features often seen with listeriosis include fever, headache, nausea, vomiting, movement disorders, altered consciousness, and seizures. Bacteremia is also frequently observed in adult listeriosis (Lorber, 1997; Vazquez-Boland et al., 2001b). Some unusual forms of listerial infections include endocarditis, hepatitis, myocarditis, arteritis, pneumonia, sinusitis, conjunctivitis, ophthalmitis, joint infection, and skin infection (Vazquez-Boland et al., 2001b and references therein). Febrile gastroenteritis has also been described in recent years (Salamina et al., 1996; Dalton et al., 1997; Miettinen et al., 1999; Aureli et al., 2000). This wide spectrum of *L. monocytogenes* disease manifestations reflects the diversity of cell types and tissues that can harbor the bacterium.

As *L. monocytogenes* can invade many types of cells and tissues within a host, a number of tissue culture cell model systems have been developed to study bacteria–host cell interactions, including epithelial cells, such as Caco-2, Henle 407, Vero, HeLa, Chinese hamster ovary (CHO) cells, and potoroo rat kidney (PtK2) cells; endothelial cells, such as human umbilical vein endothelial cells (HUVEC) and human brain microvascular endothelial cells (HBMEC); fibroblast cells; macrophages (primary and macrophage cell lines such as J774); hepatocytes; and dendritic cells (Gaillard et al., 1987; Sun et al., 1990; Dramsi et al., 1995; Drevets et al., 1995; Guzman et al., 1995; Mengaud et al., 1996; Greiffenberg et al., 1998; Parida et al., 1998; Kolb-Maurer et al., 2000; Suarez et al., 2001). The mechanisms governing the invasion of these various cell types by *L. monocytogenes* are the focus of the following section.

INVASION OF HOST CELLS BY *L. MONOCYTOGENES*

The intracellular life cycle of *L. monocytogenes* has been extensively studied. The overall process resulting in intracellular bacterial replication and cell-to-cell spread includes the following: (1) bacterial adherence or attachment to the host cells; (2) bacterial internalization by phagocytic cells, or bacterial induced-uptake into nonprofessional phagocytic cells; (3) bacterial escape

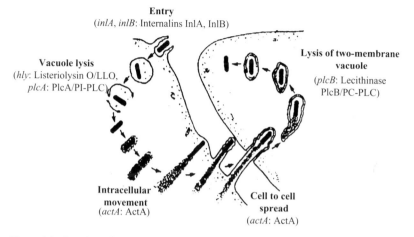

Entry
(*inlA, inlB*: Internalins InlA, InlB)

Vacuole lysis
(*hly*: Listeriolysin O/LLO,
plcA: PlcA/PI-PLC)

Lysis of two-membrane vacuole
(*plcB*: Lecithinase
PlcB/PC-PLC)

Intracellular movement
(*actA*: ActA)

Cell to cell spread
(*actA*: ActA)

Figure 6.1. Overview of *L. monocytogenes* infection of host cells. The stippled regions represent host actin filaments. The major gene products that contribute to intracellular growth and cell-to-cell spread are indicated. Adapted from Tilney and Portnoy (1989) with permission from the *Journal of Cell Biology*.

from the phagosome; (4) bacterial multiplication and actin polymerization-based movement within the host cell cytosol; and (5) spread of the bacteria to adjacent cells and escape from a double-membrane-bound vacuole (Fig. 6.1).

Adherence and internalization mechanisms for nonprofessional phagocytic cells

The first steps in *L. monocytogenes* infection following bacterial ingestion are bacterial adherence and invasion of the gastrointestinal epithelium. Following adherence, the internalization of *L. monocytogenes* proceeds by means of a zipper-like mechanism in which the plasma membrane of the host cell closely enwraps the bacterium until it is completely engulfed into a vacuole. This process is strikingly different from the trigger mechanism induced by *Salmonella* and *Shigella* spp., which leads to the formation of dramatic membrane ruffles in host cells during uptake (Swanson and Baer, 1995; Dramsi and Cossart, 1998; Kuhn and Goebel, 2000). *L. monocytogenes* has been found to enter polarized cells through the basolateral surface, and its internalization is increased following the disruption of intercellular junctions in the host cells (Gaillard and Finlay, 1996). Of the various listerial ligands described to participate in bacterial attachment and invasion of nonprofessional phagocytic host cells, the internalins (encoded by the *inl* genes) appear to be the major players and have been the best described.

The *inlAB* gene products

The *inlAB* locus was identified following the isolation of bacterial transposon-insertion mutants that were defective for invasion of Caco-2 cells (Gaillard et al., 1991). *L. monocytogenes inlAB* mutants were found to be severely impaired in host cell invasion, and the expression of InlA and InlB on the surface of latex beads or on the surface of normally noninvasive bacteria (such as *L. innocua* and *Enterococcus faecalis*) was sufficient to induce internalization (Lecuit et al., 1997; Braun et al., 1998).

L. monocytogenes InlA promotes entry into human intestinal epithelial cell lines such as Caco-2 cells and into other cells expressing the E-cadherin receptor ligand (Mengaud et al., 1996). E-cadherin is a 110-kDa transmembrane glycoprotein specific to epithelial tissues, such as those in the digestive tract and in the cells lining the choroid plexus and placental chorionic villi. The protein is predominantly expressed at the adherens junctions and on the basolateral face of cells, where it mediates calcium-dependent cell–cell adhesion and helps maintain tissue cohesion. It has been suggested that *L. monocytogenes* breaches the intestinal barrier through M cells at the Peyer's patch surface to gain access to the basolaterally located E-cadherins. Alternatively, the E-cadherin receptor might be transiently accessible when crypt cells migrate to the tips of intestinal villi for exfoliation (Mengaud et al., 1996; Vazquez-Boland et al., 2001b; Cossart, 2002). The cytoplasmic domain of E-cadherin interacts with proteins known as catenins, and these associations ultimately stimulate the cytoskeletal rearrangements within the host cell that result in *L. monocytogenes* uptake (Cossart and Lecuit, 1998; Kuhn and Goebel, 1998; Vazquez-Boland et al., 2001b).

InlA is an approximately 80-kDa protein that is targeted to the bacterial surface by an N-terminal signal sequence. The InlA C-terminus contains a hydrophobic stretch of 20 amino acids preceded by a cell wall anchor motif, LPXTG (Fig. 6.2). This motif is a signature of many Gram-positive surface proteins and is necessary for their covalent linkage to the peptidoglycan (Fischetti et al., 1990). A putative transamidase known as sortase, encoded by *srtA*, was recently identified and shown to be required for the cell wall anchoring of InlA (Bierne et al., 2002). Moreover, *L. monocytogenes* mutants lacking functional sortase were shown to be less invasive *in vitro* and attenuated in a mouse model of infection (Garandeau et al., 2002). Similar to other members of the internalin family, the N-terminus of InlA has a series of leucine-rich repeats (LRRs) consisting of a string of 22 amino

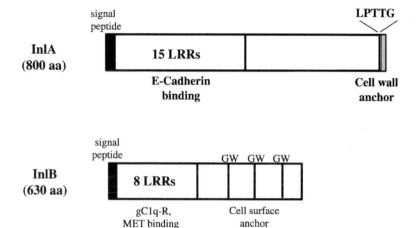

Figure 6.2. Schematic illustration of InlA and InlB. LRR represents leucine-rich repeats, and GW represents the GW repeats present in InlB that anchor the protein to the bacterial cell surface. The LPTTG motif is involved in cell wall anchoring of InlA. The receptors for InlA and InlB are indicated below the LRR binding regions.

acids containing leucine or isoleucine residues at seven fixed positions (at residues 3, 6, 9, 11, 16, 19, and 22; see Dramsi et al., 1997; Engelbrecht et al., 1998a, 1998b; Raffelsbauer et al., 1998). There are 15 successive LRR units in InlA; each LRR is composed of a beta strand that alternates with a more flexible antiparallel helix, and the repeats are connected by coils. An intact LRR region is necessary and sufficient to induce bacterial entry, whereas the inter-repeat region has been suggested to play a role in the proper folding and stabilization of the LRRs (Lecuit et al., 1997). The InlA protein is thought to interact by means of the LRR regions with the extracellular domain of E-cadherin, whose cytoplasmic domain then leads to rearrangement of the host cell actin cytoskeleton for bacterial internalization. The proline residue at position 16 of E-cadherin has been shown to be critical for InlA interaction. Interestingly, the mouse and rat E-cadherins have glycine in place of proline at this position and do not facilitate InlA-dependent internalization of *L. monocytogenes*. The position 16 proline of E-cadherin thus contributes some measure of host specificity for *L. monocytogenes* (Lecuit et al., 1999). InlA has also been shown to contribute to the attachment and internalization of *L. monocytogenes* by macrophages (Sawyer et al., 1996).

InlB is another *L. monocytogenes* surface protein that mediates bacterial invasion of a broader range of host cells, including CHO cells, Vero cells, hepatocytes, and several endothelial, epithelial, and fibroblast cell lines (Dramsi et al., 1995; Lingnau et al., 1995; Ireton et al., 1996; Greiffenberg et al., 1998;

Parida et al., 1998). Two host receptors have been identified for InlB. One, gC1q-R, is expressed in most tissues and is the receptor for the globular part of the complement component C1q (Braun et al., 2000; Galan, 2000). The gC1q-R protein has no transmembrane domain and no cytoplasmic tail; thus a co-receptor may be required to mediate the effects of InlB–gC1q-R interactions. The other receptor for InlB is Met, a 145-kDa tyrosine kinase that is also a high-affinity receptor for hepatocyte growth factor (Shen et al., 2000). InlB-mediated listerial entry into host cells leads to the stimulation of phosphoinositide 3-kinase (PI-3 kinase; see Ireton et al., 1996). In addition, the MEK-1, ERK-1, and ERK-2 protein kinases are activated during *L. monocytogenes* invasion of HeLa epithelial cells (Tang et al., 1998).

InlB is a 71-kDa member of the internalin family and has a signal sequence followed by 8 LRRs at the N-terminus. However, InlB does not possess the Gram-positive surface anchor LPXTG motif at its C-terminus. Instead, a "GW" motif, containing repeats led by the sequence GW, functions as the cell surface anchor domain (Fig. 6.2; also see Braun et al., 1997). InlB is only loosely associated with lipoteichoic acid in the cell wall by means of the GW motif, and some InlB is released into the supernatant (Jonquieres et al., 1999). The GW motif also binds cellular glycosaminoglycans, and this interaction appears to be required for efficient InlB-mediated bacterial invasion by enhancing the internalization triggered by the N-terminal LRRs (Jonquieres et al., 2001).

Besides *inlAB*, six additional genes encoding putative surface-anchored internalins have been identified in *L. monocytogenes*. Similar to InlA, these six members (InlC2, InlD, InlE, InlF, InlG, and InlH) all have signal sequences and various numbers of LRRs at the N-terminal region, plus the LPXTG motifs at the C-terminus (Dramsi et al., 1997; Engelbrecht et al., 1998a; Raffelsbauer et al., 1998). These internalins were recently shown to enhance InlA-mediated invasion of *L. monocytogenes* into Caco-2 cells and into HB-MEC. However, InlB-triggered bacterial internalization into these cell lines did not require any of the other internalin proteins (Bergmann et al., 2002).

InlC is an unusual member of the internalin protein family. It is small (~30 kDa) and is secreted by *L. monocytogenes* instead of being anchored to the cell wall. The potential role of InlC in *L. monocytogenes* pathogenesis is still unclear. It has been suggested to be involved in the dissemination of *L. monocytogenes* infection; however, cell-to-cell spread is not affected in *inlC* mutant strains (Engelbrecht et al., 1996; Lingnau et al., 1996; Domann et al., 1997). Southern blot and genome analysis has indicated the presence of internalin homologues in the nonpathogenic *L. innocua*, and it is possible that these homologues might have functions that are distinct from those required for invasion (Gaillard et al., 1991; Glaser et al., 2001).

Other possible *L. monocytogenes* adherence and internalization factors

In addition to the internalin protein family, a number of other *L. monocytogenes* surface proteins have been implicated in bacterial adherence and invasion of host cells (Kuhn and Goebel, 1989; Alvarez-Dominguez et al., 1997b). The *L. monocytogenes* surface protein ActA, which mediates the polymerization of host actin filaments required for bacterial motility within the host cytosol, has been shown to recognize host cell heparan sulfate proteoglycans (HSPG) and may contribute to epithelial cell invasion (Alvarez-Dominguez et al., 1997b; Suarez et al., 2001). A major extracellular protein, p60 (encoded by *iap*), has been reported to be required for bacterial attachment and invasion of mouse fibroblasts. However, it is possible that the decrease in invasion of *L. monocytogenes* p60 mutants was due to septation defects associated with the disruption of the p60 murein hydrolase activity (Bubert et al., 1992; Wuenscher et al., 1993). Bacterial adherence to Caco-2 cells seems to be independent of p60 and rather mediated by a 104-kDa surface protein known as the *Listeria* adhesion protein (Lap; see Pandiripally et al., 1999). There is also a fibronectin-binding protein that has been identified in *L. monocytogenes* (Gilot et al., 1999, 2000).

The completion of the *L. monocytogenes* genome sequence has revealed that, among the bacterial genomes currently sequenced, *L. monocytogenes* contains the largest number of proteins with the cell wall anchor motif LPXTG (Glaser et al., 2001). It is possible that some of these putative surface proteins are involved in adherence and internalization. In addition to protein ligands, other surface components, such as lipoteichoic acid, appear to be important for *L. monocytogenes* adhesion to various cell lines. The D-alanylation of the lipoteichoic acids, catalyzed by products expressed from the *dlt* operon, was recently shown to be required for adhesion to murine macrophages and hepatocytes as well as human Caco-2 epithelial cells (Abachin et al., 2002). A virulent *L. monocytogenes* strain possessing surface alpha-D-galactose residues was shown to bind through lectin-like interactions to a human hepatocarcinoma cell line, which expresses a well-characterized carbohydrate receptor for alpha-D-galactose at the surface (Cowart et al., 1990). Agglutination by lectins suggests the presence of other carbohydrate-binding ligands on the surface of *L. monocytogenes* (Cottin et al., 1990; Facinelli et al., 1998).

L. monocytogenes entry into professional phagocytes

Macrophages actively engulf *L. monocytogenes*, and several host cell components have been shown to participate in macrophage binding and uptake of the bacterium. The complement component C3b is involved in mediating listerial interactions with macrophages by means of CR3, the complement receptor type 3 (Drevets and Campbell, 1991; Croize et al., 1993; Drevets et al., 1993). *L. monocytogenes* also binds C1q in a specific, saturable, and dose-dependent manner. Enhanced uptake of C1q-bound *L. monocytogenes* was mediated by C1q complement receptors expressed on the macrophage surface (Alvarez-Dominguez et al., 1993). Nonopsonic interactions may also be involved, as *L. monocytogenes* uptake is efficient in the absence of serum (Pierce et al., 1996). The type I and type II macrophage scavenger receptors can recognize and bind a wide range of polyanions, and these receptors have been shown to bind listerial lipoteichoic acids (Dunne et al., 1994; Greenberg et al., 1996; Ishiguro et al., 2001). N-acetylneuraminic acid (NAcNeu) has also been suggested to play a role in the attachment of *L. monocytogenes* to murine macrophages (Maganti et al., 1998). Host nuclear factor κB (NF-κB) is activated when *L. monocytogenes* adheres to the surface of macrophage-like cell lines; long-lasting NF-κB activation appears to require LLO and products of the lecithinase operon (*mpl-actA-plcB*; see Hauf et al., 1994).

LIFE AFTER INVASION: THE PHAGOSOME IS NO PLACE FOR *L. MONOCYTOGENES* TO LIVE

Inside activated macrophages, most of the internalized *L. monocytogenes* are killed within minutes after uptake (Davies, 1983; Raybourne and Bunning, 1994). Nevertheless, *L. monocytogenes* appears well adapted to establish and maintain a niche inside various types of nonbactericidal host cells. *L. monocytogenes* mediates escape from the host cell phagosome, multiplies within the cytosol, and moves to infect adjacent cells by means of the expression of a number of protein products that function together to enable the bacterium to exploit the host cell environment.

L. monocytogenes escape from the phagosome

Host cell invasion ultimately provides access to a rich growth medium for *L. monocytogenes*: the host cytosol. Following host cell uptake of the bacterium, *L. monocytogenes* is enclosed in a primary phagosomal vacuole that becomes quickly acidified. *L. monocytogenes* prevents further maturation of this phagosome and enhances its own survival by escaping into the cytosol after disruption of the phagosomal membrane (Alvarez-Dominguez et al., 1997a). In mouse bone marrow-derived macrophages, bacterial-mediated perforation of the phagosome membrane is detectable within minutes

(Beauregard et al., 1997). Complete disruption of the phagosomal membrane within infected Caco-2 cells is observed in approximately 30 minutes (Gaillard et al., 1987). The major factor contributing to phagosomal escape is LLO, first described following the isolation of bacterial transposon-insertion mutants that had a nonhemolytic phenotype on sheep blood agar plates (Gaillard et al., 1986; Kathariou et al., 1987). LLO, encoded by *hly*, was the initial virulence factor to be identified as essential for *L. monocytogenes* virulence (Gaillard et al., 1986; Geoffroy et al., 1987; Kathariou et al., 1987; Mengaud et al., 1988; Cossart et al., 1989); bacterial mutants lacking LLO do not mediate phagoso-mal escape in most cell lines examined and are highly attenuated in mouse models of infection (Portnoy et al., 1992). The production of LLO by the non-pathogenic soil bacterium *Bacillus subtilis* is sufficient to mediate bacterial escape from macrophage phagosomes following uptake of *B. subtilis* into tissue culture cells (Bielecki et al., 1990).

LLO is a 58-kDa cytolysin belonging to a large family of cholesterol-dependent, pore-forming toxins that includes streptolysin O (SLO) and per-fringolysin O (PFO), secreted by *Streptococcus pyogenes* and *Clostridium per-fringens*, respectively (Palmer, 2001). Similar to other family members, LLO has a conserved tryptophan-rich undecapeptide (ECTGLAWEWWR) at its C-terminus. This region appears to be important for the cholesterol-dependent membrane binding of these cytolysins (Iwamoto et al., 1990; Boulnois et al., 1991; Sekino-Suzuki et al., 1996; Palmer, 2001). LLO and its family mem-bers mediate membrane disruption through the formation of oligomeric ring-shaped structures that form membrane pores that are approximately 20–30 nm in diameter (Palmer, 2001). The crystal structure of PFO has been solved, revealing four protein domains; this structure has provided valuable information regarding structure–function relationships within this family of proteins (Rossjohn et al., 1997). A recent study demonstrated that two vari-ants of LLO, comprising domains 1–3 or domain 4 alone, could be secreted efficiently and reassembled into a functionally active protein when the do-mains were expressed simultaneously (Dubail et al., 2001). Another study demonstrated that domains 1–3 of LLO were essential for cytokine induc-tion, whereas domain 4 was important for binding to membrane cholesterol (Kohda et al., 2002).

LLO is unique among the family members because of its acidic pH optimum (Geoffroy et al., 1987). LLO is active at pH 4.5 to 6.5, an ideal range for the acidic phagocytic vacuole, the pH of which is approximately 5. The cytolysin PFO, which does not have an acidic pH optimum, is toxic to host cells when expressed by *L. monocytogenes* Δ*hly* mutants, possibly as a result of PFO-mediated host cell plasma membrane damage (Jones and Portnoy,

1994). Mutations that result in single amino acid substitutions within PFO can apparently convert the protein into a vacuole-specific hemolysin similar to LLO, but *L. monocytogenes* strains expressing these PFO variant molecules are still highly attenuated in mouse models of infection (Jones et al., 1996). A single leucine located at position 461 within LLO has been recently identified to confer the acidic pH activity (Glomski et al., 2002). The substitution of leucine 461 with a threonine found at a similar position within PFO increased the activity of LLO at a neutral pH. *L. monocytogenes* mutants expressing the *hly* L461T variant were cytotoxic to infected host cells (Glomski et al., 2002). In addition, a PEST-like sequence in the N-terminus of LLO has been suggested to target LLO for rapid degradation within the host cytosol (Decatur and Portnoy, 2000). *L. monocytogenes* mutants producing LLO molecules that lack the PEST sequence accumulate the protein within the host cytosol, and the mutants are extremely cytotoxic for host cells and highly attenuated for virulence in mice. Moreover, the addition of the PEST sequence to PFO can convert it into a protein variant that is nontoxic to host cells (Decatur and Portnoy, 2000). An independent study has shown that *L. monocytogenes* strains that express an LLO variant with substitutions for all of the P, E, S, and T residues is also highly attenuated for virulence in mice (Lety et al., 2001). It thus appears that multiple mechanisms exist to restrict LLO activity within the host cytosol, thereby compartmentalizing the activity of the protein to the phagosome and preventing LLO-mediated damage to the host cell plasma membrane. The existence of multiple mechanisms that serve to regulate LLO activity highlights how extraordinarily suited *L. monocytogenes* has become for life within the host cytosol.

As previously discussed, LLO plays an important role in mediating the escape of *L. monocytogenes* from the phagosome of many cell types; however, its activity is not required for phagosome disruption within all cells. Non-hemolytic *L. monocytogenes* mutants lacking functional LLO are still capable of replicating within the cytosol of Henle 407 human epithelial cells, which suggests that other factors can contribute to phagosomal membrane disruption (Portnoy et al., 1988). *L. monocytogenes* encodes two phospholipases: a 33-kDa phosphatidylinositol-specific phospholipase C (PI-PLC) that is encoded by *plcA*, and a broad-specificity phospholipase that is encoded by *plcB* (Camilli et al., 1991; Geoffroy et al., 1991; Leimeister-Wachter et al., 1991; Mengaud et al., 1991; Vazquez-Boland et al., 1992).

PI-PLC can hydrolyze moieties that anchor many eukaryotic membrane proteins to the plasma membrane, and the secreted enzyme has an acidic pH optimum (Goldfine and Knob, 1992). PI-PLC is thought to be active in the phagocytic vacuole and to mediate lysis of the vacuolar membrane in conjunction with LLO (Goldfine and Knob, 1992). It has been suggested that the initial pores formed by LLO provide access to PI-PLC membrane substrates. Cooperation between the LLO hemolysin and the phospholipases was also suggested to increase the membrane affinity of LLO, thereby facilitating membrane lysis (Goldfine et al., 1995; Sibelius et al., 1996; Stachowiak and Bielecki, 2001). Recently, the combined activities of LLO and PI-PLC were shown to affect bacterial entry and vacuolar escape in J774 cells by influencing the translocation of protein kinase C and calcium signaling (Wadsworth and Goldfine, 2002).

The 29-kDa lecithinase or phosphatidylcholine-phospholipase C (PC-PLC) encoded by plcB is another factor that contributes to the escape of L. monocytogenes from cell phagosomes. This enzyme has a wide pH optima and a broad substrate spectrum (Geoffroy et al., 1991; Goldfine et al., 1993). PlcB is synthesized as an inactive proenzyme pro-PlcB, a regulatory step that has probably evolved to prevent PlcB-mediated bacterial membrane damage. The N-terminal propeptide of PlcB is cleaved by a zinc-dependent metalloprotease encoded by mpl (Poyart et al., 1993). Interestingly, L. monocytogenes appears to sequester pools of inactive pro-PlcB during bacterial replication within the host cytosol, and active enzyme is then rapidly released when pH decreases below 7.0 (Marquis and Hager, 2000). PlcB is required for the efficient escape of L. monocytogenes from the secondary double-membrane vacuoles formed during the process of cell-to-cell spread (discussed in the following section); however, the protein can functionally replace LLO to mediate escape from the primary phagosome in human epithelial cells (Marquis et al., 1995).

In addition to the factors mentioned herein, a previously unknown surface protein designated SvpA (surface virulence-associated protein) was recently suggested to promote phagosomal escape and thus intracellular survival of L. monocytogenes (Borezee et al., 2001). A svpA mutant was shown to be confined within the phagosomal vacuole and was attenuated in a mouse model of infection with decreased bacterial growth in the liver and spleen (Borezee et al., 2001). L. monocytogenes therefore appears to possess a variety of factors that may work in concert to facilitate efficient escape from the phagosome so that the bacterium can reach the promised land: the host cytosol.

THE GOOD LIFE: L. MONOCYTOGENES AT HOME IN THE CYTOSOL

The doubling time for L. monocytogenes within the host cytosol is very similar to the doubling time observed for bacterial cultures grown in rich broth media; therefore, the cytosol appears to be a very permissive environment for *L. monocytogenes* proliferation (Portnoy et al., 1988). Intracellular growth of *L. monocytogenes* does not require the induction of stress proteins, and several auxotrophic mutants of *L. monocytogenes* were able to replicate in the cytosol of the macrophage-like cell line J774 and the human epithelial cell line Henle 407 with growth rates similar to those of wild-type strains (Marquis et al., 1993; Hanawa et al., 1995). The expression of three genes involved in purine and pyrimidine biosynthesis (*purH*, *purD*, and *pyrE*), and a gene encoding an ATP-dependent arginine transporter (*arpJ*), is induced in bacteria located within host cells; however, *L. monocytogenes* strains carrying mutations in these genes were not affected for rates of intracellular growth and were fully virulent in a mouse model of infection (Klarsfeld et al., 1994). Other *in vivo*-induced genes that encode metabolic enzymes and nutrient transport systems have been identified through the use of a variety of genetic screens (Dubail et al., 2000; Gahan and Hill, 2000; Autret et al., 2001; Wilson et al., 2001). It has been suggested that some nutrients, such as nucleotides and certain amino acids, are not at limiting concentrations within the cytosol but are nonetheless at low concentration, and *L. monocytogenes* therefore induces the expression of genes required for the uptake and biosynthesis of various metabolites to facilitate efficient intracellular proliferation (Klarsfeld et al., 1994; Sheehan et al., 1994; Vazquez-Boland et al., 2001b).

Recent evidence has implicated Hpt, a *L. monocytogenes* homologue of the mammalian glucose-6-phosphate (G-6-P) translocase, in facilitating rapid intracellular proliferation of bacteria through utilization of hexose phosphates obtained from the host cytosol (Chico-Calero et al., 2002). The mammalian G-6-P translocase transports G-6-P from the cytosol into the endoplasmic reticulum. *L. monocytogenes* Hpt may function in an analogous manner to transport G-6-P from the cytosol into the bacterium; bacterial mutants lacking the G-6-P translocase were defective for intracellular proliferation and were attenuated for virulence in mice (Chico-Calero et al., 2002).

The acquisition of iron is important for *L. monocytogenes*, as it is for all other living cells. Iron is an essential cofactor for a wide variety of enzymatic processes. Iron stimulates *L. monocytogenes* growth *in vitro*, and it increases bacterial proliferation *in vivo* (Sword, 1966; Stelma et al., 1987; Payne, 1993). Increased concentrations of iron have been shown to upregulate the expression of *inlAB* and enhance *L. monocytogenes* invasion of Caco-2 cells (Conte et al., 1996). In macrophages, iron is required for antimicrobial responses involving the generation of reactive oxygen and nitrogen intermediates. However, humans with iron overload are more susceptible to listeriosis; thus a balance of iron must be maintained to appropriately respond to *L. monocytogenes* infection (Schuchat et al., 1991; Fleming and Campbell, 1997). In mammalian hosts, iron is normally sequestered by serum transferrin and is also complexed by intracellular heme compounds; it is therefore not freely available to *L. monocytogenes*. Unlike many pathogens, *L. monocytogenes* does not appear to secrete the high-affinity iron-binding siderophores; however, the organism can use exogenous siderophores by means of an extracellular ferric iron reductase. *L. monocytogenes* also possesses a citrate-inducible iron transport system and a cell-surface-associated transferrin-binding protein (Adams et al., 1990; Hartford et al., 1993; Deneer et al., 1995; Barchini and Cowart, 1996; Coulanges et al., 1997).

THE GRASS IS ALWAYS GREENER: SPREAD OF *L. MONOCYTOGENES* TO ADJACENT HOST CELLS

The cytosol may be a tasty smorgasbord for replicating *L. monocytogenes*, but it pays to move onto greener cell pastures. Each cell can only sustain a finite amount of bacterial replication, and evidence suggests that *L. monocytogenes* needs to spread to new host cells as a means of outrunning host cellular immune responses (Auerbuch et al., 2001). The necessity of cell-to-cell spread as a means of obtaining fresh supplies of nutrients and of avoiding humoral and cellular immune responses was discussed in a recent review (O'Riordan and Portnoy, 2002), in which the authors also noted that although there are not many cytosolic bacterial pathogens, most of them are capable of cell-to-cell spread.

L. monocytogenes-directed actin-based motility

L. monocytogenes cytosol motility and bacterial spread to adjacent cells is accomplished by the exploitation of host actin polymerization machinery. Observations of infected host cells by electron microscopy indicate that cytosolic *L. monocytogenes* is soon surrounded by a dense cloud of host-derived F-actin, which eventually forms a comet-tail-like structure at one bacterial pole (Tilney and Portnoy, 1989). Approximately 2 h after internalization, *L. monocytogenes* begins to move in the host cytosol with a velocity of 0.1–0.4 μm/s, a rate approximately proportional to the rate of actin polymerization (Theriot et al., 1992). The *L. monocytogenes* surface protein ActA is absolutely necessary for actin-based motility within the host cytosol.

ActA was identified following the isolation of *L. monocytogenes* mutants that lacked the ability to polymerize actin and spread to adjacent cells (Domann et al., 1992; Kocks et al., 1992). *L. monocytogenes actA* mutants invade host cells, escape from the phagosome, and replicate within the cytosol,

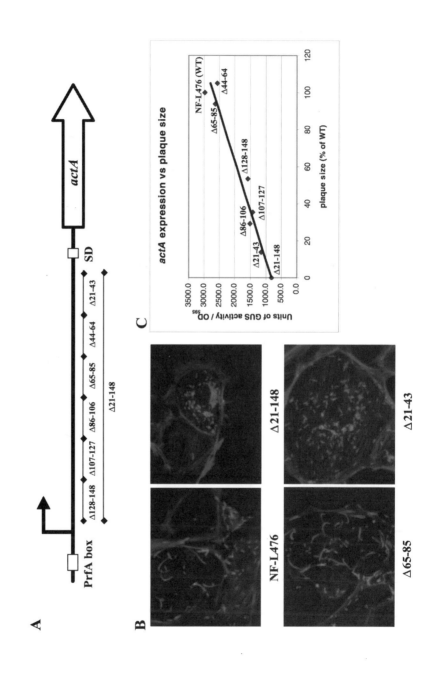

A

B

C

actA expression vs plaque size

NF-L476 (WT)

Δ44-64

Δ65-85

Δ128-148

Δ86-106

Δ107-127

Δ21-43

Δ21-148

plaque size (% of WT)

Units of GUS activity / OD$_{595}$

but the bacteria are unable to use actin polymerization as a motile force and thus accumulate as microcolonies within infected host cells. *L. monocytogenes actA* deletion mutants are 1000-fold less virulent in mice, and infection is restricted predominantly to the gut (Domann et al., 1992; Kocks et al., 1992; Manohar et al., 2001). ActA appears to be the only *L. monocytogenes* protein required to nucleate actin polymerization; actin-based motility and the formation of actin comet tails is observed in cell extracts for both streptococci and beads that have been asymmetrically coated with purified ActA (Smith et al., 1995; Cameron et al., 1999). *L. innocua*, a nonpathogenic *Listeria* species, does not polymerize actin when added to cell extracts, but the expression of the *actA* gene product on the *L. innocua* surface enables the bacteria to form actin tails and move (Kocks et al., 1995).

The length of *L. monocytogenes* actin tails reflects the rates of bacterial movement (Theriot et al., 1992; Sechi et al., 1997), and the rate of movement is likely to affect the efficiency of cell-to-cell spread. Interestingly, there seems to be an optimal window of ActA production required for normal movement and cell-to-cell spread. A single additional copy of the *actA* gene integrated within the *L. monocytogenes* chromosome can reduce the efficiency of cell-to-cell spread, apparently because of increased ActA production (Lauer et al., 2002). Conversely, recent research in our laboratory indicates that a reduction in *actA* expression appears to proportionally reduce cell-to-cell spread. A series of *actA* promoter mutants, demonstrated to express decreasing amounts of *actA* as measured through the use of *actA-gus* reporter gene fusions, produced correspondingly smaller plaques in monolayers of mouse L2 fibroblast cells (Fig. 6.3; K. Wong and N. Freitag, unpublished data). A linear relationship was found to exist between *actA* expression and plaque size. Decreasing *actA* expression was also associated with decreasing actin tail length as visualized by staining for filamentous actin in monolayers of infected PtK2 cells in tissue culture (Fig. 6.3). These data support the

Figure 6.3. (*facing page*). Correlation of *actA* expression levels with *L. monocytogenes* cell-to-cell spread. (A) Diagram of the *actA* promoter region. The dark arrow represents the transcript initiation site, SD indicates the ribosome binding site, and the PrfA binding site is indicated by the presence of the PrfA box. Targeted *actA* deletions introduced into the *L. monocytogenes* chromosome are shown, with the numbers representing the size of the deletion and the distance from the ATG start codon. (B) Actin staining of PtK2 epithelial cells infected with *L. monocytogenes* wild type (NF-L476) or the *actA* promoter mutant strains. Filamentous actin is shown in green, and bacteria are shown in red. (C) Correlation between levels of *actA* expression and cell-to-cell spread ability of *L. monocytogenes*. See color section.

premise that an optimal concentration of ActA at the bacterial surface is critical for function (Moors et al., 1999).

The mature ActA protein contains 610 amino acids after the cleavage of a 29-residue signal sequence from its N-terminus. ActA can be divided into three domains: an N-terminal domain (1–234) rich in cationic residues, a central domain (235–394) containing proline-rich repeats, and a C-terminal domain (395–610) containing a stretch of hydrophobic residues (585–606) that anchors the protein to the surface of *L. monocytogenes*. The N-terminal domain contains all the information necessary to mediate actin polymerization and support actin-based motility (Lasa et al., 1997). The cationic residues (129–153) within this domain appear to be critical for actin assembly; residues 21–97 are responsible for the continuity of the assembly process; and residues 117–121 contribute to tail formation (Lasa et al., 1997).

A key function of the ActA N-terminal domain is the recruitment of the host Arp2/3 complex, which is critical for the nucleation step initiating actin filament formation (Welch et al., 1997). The N-terminus of ActA shares homology with the Wiskott-Aldrich syndrome protein (WASP) family in mammalian cells (Skoble et al., 2000; Boujemaa-Paterski et al., 2001), which have been shown to stimulate actin polymerization through activation of Arp2/3 and to connect the cytoskeleton with signaling pathways that mediate actin rearrangement (Welch et al., 1997; Machesky and Insall, 1998; Goldberg, 2001). ActA binds two actin monomers and three of the seven subunits of the Arp2/3 complex (Zalevsky et al., 2001). A mutational analysis of the ActA N-terminus indicates that the actin-binding region spans 40 residues (31–72); a separate region of ActA (136–231) contributes to the actin cloud-to-tail transition that affects intracellular motility rates (Lauer et al., 2001).

The central domain of ActA contains four short proline-rich repeats (PRRs) with consensus sequence DFPPPPTDEEL. These PRRs are similar to PRRs present in mammalian cytoskeletal proteins, and they serve as binding sites for the host tetrameric vasodilator phospho-protein (VASP; see Reinhard et al., 1995; Niebuhr et al., 1997; Dramsi and Cossart, 1998). VASP binds host profilin, an actin-sequestering protein that recruits actin monomers for the elongation of the actin filaments. Both VASP and profilin have been shown to co-localize at the interface between *L. monocytogenes* and the actin tails (Theriot et al., 1994; Chakraborty et al., 1995). Interestingly, *L. monocytogenes* is still capable of actin recruitment and intracellular movement in the absence of the PRRs, VASP, or profilin, although the organism moves at a significantly lower rate (Lasa et al., 1995; Smith et al., 1996; Niebuhr et al., 1997; Loisel

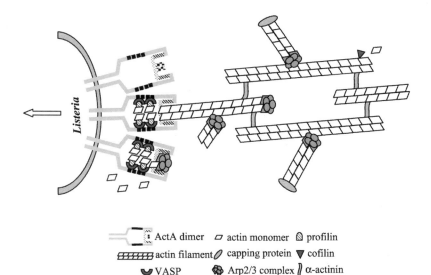

ActA dimer ◻ actin monomer ⬧ profilin
actin filament ⬙ capping protein ▼ cofilin
⬺ VASP ⬡ Arp2/3 complex ⫽ α-actinin

Figure 6.4. Model of actin nucleation and polymerization as directed by *L. monocytogenes* (based on previous models described by Cossart and Bierne, 2001 and Skoble et al., 2000). VASP binds to the four proline-rich repeats present in the central domain of ActA. The N-terminal domain of ActA contains three regions that contribute to actin binding and the activation of the Arp2/3 complex.

et al., 1999). This suggests a stimulatory role for the ActA PRRs and VASP in actin-based motility.

Several other host proteins are involved in *L. monocytogenes*-directed actin polymerization. Alpha-actinin is distributed over the actin tail and is responsible for stabilizing the tail by cross-linking actin filaments (Dabiri et al., 1990; Loisel et al., 1999). Although alpha-actinin is not required for *L. monocytogenes* actin-based motility, the tails are less rigid and the bacteria tend to drift in its absence. Capping protein blocks actin assembly by binding to barbed ends of actin filaments. This protein was shown to be required in the reconstitution of *L. monocytogenes* motility *in vitro*. The absence of this protein or its presence at low concentrations results in a fishbone-like web of actin filaments, which is inefficient at providing motile force (Marchand et al., 1995; Loisel et al., 1999; Pantaloni et al., 2000). Cofilin depolymerizes actin and may play a role in generating pools of actin monomers (Carlier et al., 1997); it is also required for reconstituting bacterial movement *in vitro* (Loisel et al., 1999). Taken together, a model depicting *L. monocytogenes* actin-based movement is shown in Fig. 6.4.

L. monocytogenes invasion of adjacent cells

As *L. monocytogenes* moves through the cell and contacts the plasma membrane, it appears to push out the membrane and form a pseudopod-like (or *Listeria*-pod-like) structure that is somehow recognized or taken up by the neighboring cell. The bacterium is then enclosed in a double-membrane-bound secondary vacuole, and lysis of this compartment is facilitated by the phospholipase PC-PLC (Vazquez-Boland et al., 1992). The *plcA* gene product (PI-PLC) appears to function synergistically with PC-PLC, as *L. monocytogenes* mutants lacking both enzymes have a more serious defect in cell-to-cell spread than either of the single mutants (Marquis et al., 1995). LLO is also required for escape from the double-membrane vacuoles (Gedde et al., 2000). Through the coordinated activities of the *hly*, *plcA*, and *plcB* gene products, *L. monocytogenes* disrupts the double-membrane vacuole and makes a fresh home within the cytosol of the new cell victim. In this way, the bacterium gains new pastures without exposure to the humoral immune responses of the host.

COORDINATING THE BACTERIAL DANCE STEPS OF INVASION, REPLICATION, AND CELL-TO-CELL SPREAD

L. monocytogenes is similar to other bacterial pathogens in that many of the gene products that contribute to infection are expressed preferentially within host cells. The expression of virulence determinants by *L. monocytogenes* appears to be tightly regulated and can be triggered by specific host cell compartment environments, such as the phagosome or cytosol. The process of *L. monocytogenes* virulence gene regulation in response to host cell environments is a complex and fascinating story, with many of the players still waiting to be discovered.

Most of the virulence genes identified thus far in *L. monocytogenes* are located within a 10-kb chromosomal region known as the *Listeria* pathogenicity island 1 (LIPI-1; see Vazquez-Boland et al., 2001a). In addition to the aforementioned virulence genes *plcA*, *hly*, *mpl*, *actA*, and *plcB*, a positive transcriptional regulator known as *prfA* is located within this cluster, and LIPI-1 is commonly referred to as the PrfA regulon (Fig. 6.5). The LIPI-1 is only present in pathogenic *Listeria* species, although an apparently inactivated form is also present in *L. seeligeri*. This region is located between *prs* and *ldh*, two housekeeping genes coding for phosphoribosyl-pyrophosphate synthetase and lactate dehydrogenase, respectively. DNA sequences in the *prs–ldh* intergenic region of the nonpathogenic listeriae were shown to have

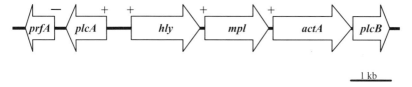

Figure 6.5. The *L. monocytogenes* LIPI−1, also referred to as the PrfA regulon; + designates promoters whose expression is increased by PrfA binding, and − designates promoters for which PrfA appears to decrease expression.

homology with integration consensus sequences for a transposon, supporting the hypothesis that horizontal gene transfer might be involved in the evolution of this gene cluster (Gouin et al., 1994; Cai and Wiedmann, 2001).

PrfA is essential for *L. monocytogenes* virulence, and it is required for the expression of all the other genes within the LIPI-1. The *prfA* gene was first identified during the characterization of a spontaneous nonhemolytic *L. monocytogenes* mutant that was found to contain a small deletion in an open reading frame downstream of *plcA*. Complementation of the mutation with the complete open reading frame led to increased transcription from genes located within LIPI-1 and demonstrated the pleiotropic effects of the *prfA* gene product (Leimeister-Wachter et al., 1989, 1990). The expressions of several other virulence genes, such as *inlA*, *inlB*, *hpt*, and *bsh* (encoding a bile salt hydrolase), have also subsequently been shown to be completely or partially dependent on PrfA (Kreft and Vazquez-Boland, 2001; Chico-Calero et al., 2002; Dussurget et al., 2002). Other *L. monocytogenes* genes, such as *clpC*, whose gene product contributes to stress responses, and *flaA*, encoding the flagellin subunit, appear to be negatively regulated by PrfA (Ripio et al., 1997, 1998). In contrast, there are virulence genes that are PrfA-independent, such as *svpA* and the *inlGHE* operon (Lingnau et al., 1995; Ripio et al., 1997; Raffelsbauer et al., 1998; Borezee et al., 2001).

Although the expression of many *L. monocytogenes* virulence genes is dependent on PrfA, the patterns of gene expression vary for individual genes during bacterial growth and during the course of host cell infection. *inlA* and *inlB* expression is readily detectable during bacterial growth outside of host cells, whereas *hly*, *plcA*, and *plcB* have been reported to be induced within the phagosome (Bubert et al., 1999). The expression of *mpl* and *actA* is highly upregulated within the host cytosol, whereas *actA* expression alone appears downregulated within the secondary vacuoles that are formed following the spread of *L. monocytogenes* to adjacent cells (Bohne et al., 1994; Bubert et al., 1999; Freitag and Jacobs, 1999). The patterns of virulence gene expression observed appear to correspond perfectly with the functional roles ascribed to the different products. However, it is unclear why *plcB* transcripts were detected in the phagosome in the absence of detectable *actA* mRNA, as the two genes are cotranscribed with no apparent intergenic promoter (Vazquez-Boland et al., 1992). It has been suggested that the *actA* mRNA region and the *plcB* mRNA region may differ in message stability.

The 27-kDa PrfA protein is a member of the Crp/Fnr family of transcriptional activators, most members of which have been identified in Gram-negative bacteria (Lampidis et al., 1994; Ripio et al., 1997). In *Escherichia coli*, Crp (also known as Cap) and Fnr are involved in carbon utilization and anaerobic growth responses, respectively (Spiro and Guest, 1990; Kolb et al., 1993). The PrfA sequence is only approximately 20% identical and 30% similar to that of Crp; however, the two proteins appear to share significant structural similarities (based on the reported Crp structure and on sequence structure predictions for PrfA). Both proteins contain an N-terminal β-roll structure and a C-terminal DNA-binding helix-turn-helix (HTH) motif. The activation regions that are involved in the interaction of Crp with RNA polymerase may also be conserved in PrfA. Perhaps most intriguing in regard to the similarities between Crp and PrfA is the existence of mutationally activated forms of the proteins, known as Crp* and PrfA*. Crp requires the presence of a cofactor, cAMP, for full activity, and two mutations within Crp have been described, G145S and A144T, that lead to cofactor-independent Crp activation (Crp* mutants; see Garges and Adhya, 1988). Ripio et al. (1997) found that *L. monocytogenes* isolates that expressed high amounts of LLO or PC-PLC contained a single amino acid substitution within PrfA located in a region (G145S) similar to the mutations found to confer the Crp* phenotype to Crp. The PrfA* G145S substitution causes constitutively high expression of all PrfA-regulated genes, suggesting that, like Crp, a cofactor (which does not appear to be cAMP) may be required for PrfA activation (Lampidis et al., 1994; Ripio et al., 1997; Goebel et al., 2000).

Our laboratory has recently identified two additional mutations within PrfA that appear to confer a PrfA* phenotype (Shetron-Rama et al., 2003). *L. monocytogenes* PrfA E77K contains a glutamate-to-lysine substitution within the β-roll structure of PrfA, whereas PrfA G155S contains a glycine-to-serine substitution near the location of the original PrfA* mutation (G145S). Both mutations confer high-level expression of PrfA-regulated genes, and strains containing either mutation appear hyper-invasive for epithelial cell lines. Interestingly, the PrfA G155S mutant strain is approximately fivefold more virulent in a mouse model of infection than the wild-type strain, suggesting that constitutive activation of PrfA is not detrimental to virulence.

PrfA is a site-specific DNA-binding protein that recognizes a 14-bp palindrome known as the "PrfA-box" within the –40 region of PrfA-dependent promoters. The binding affinity of PrfA for its target promoters is influenced by the nucleotide sequence of the PrfA box. PrfA binds with high affinity to the perfectly symmetrical PrfA box (TTAACA-NN-TGTTAA) shared by the divergently transcribed *hly* and *plcA* promoters. The PrfA boxes of the *mpl* and *actA* promoters contain single base pair asymmetries, and the *inlA* PrfA box has two base pair asymmetries. These PrfA box asymmetries have been postulated to cause PrfA to bind with lower affinity to these target promoters, and thus it has been speculated that the relative amount of PrfA present at any time within the cell determines the level of expression of the various virulence genes (Freitag et al., 1993; Goebel et al., 2000; Kreft and Vazquez-Boland, 2001). Recent studies have shown that although the expression of some genes appears to follow the PrfA binding affinity rule (*hly*, *mpl*), other PrfA-dependent promoters, such as the *actA* promoter, do not (Williams et al., 2000), and that even the increased expression of the mutationally activated *prfA** allele is not sufficient to raise levels of *actA* expression to those observed for bacteria located within the host cell cytosol (Shetron-Rama et al., 2002; Greene and Freitag, 2003). Our laboratory has isolated *L. monocytogenes* mutants that express high levels of *actA* during growth in broth culture, and these mutations map at least 40 kb outside of the PrfA regulon, indicating the participation of additional factors in the induction of *actA* expression (Shetron-Rama et al., 2003).

Environmental signals that influence virulence gene expression

Multiple conditions have been reported to influence *L. monocytogenes* virulence gene expression in culture, including temperature, available carbon sources, nutrient deprivation, and iron (Kreft and Vazquez-Boland, 2001). Virulence gene expression has been reported to be optimally activated at 37°C, the body temperature of mammalian hosts (Leimeister-Wachter et al., 1992), and the expression of the key regulatory protein PrfA appears to be under the control of a thermosensitive mechanism of translation in which optimal translation occurs at 37°C (Johansson et al., 2002). However, *L. monocytogenes* virulence gene expression has been observed within infected insect cell lines maintained at room temperature (Mansfield et al., 2003 (in press)), indicating that the appropriate cell environment can stimulate bacterial gene expression even at low temperatures. Iron and available carbon sources influence the expression of *L. monocytogenes* virulence genes, and bacterial gene products involved in iron uptake or in the utilization of hexose sugars are upregulated by intracellular *L. monocytogenes* (Sword, 1966; Payne, 1993; Klarsfeld et al., 1994; Borezee et al., 2001; Chico-Calero et al., 2002). Interestingly, the proximal *prf*AP2 promoter was recently demonstrated to be regulated by the stress-responsive alternative sigma factor σ^B; however, the loss of σ^B function only modestly reduces the virulence of *L. monocytogenes* (Nadon et al., 2002). As *L. monocytogenes* is a bacterium that survives in a multitude of environments, regulation of gene expression has probably evolved in order to respond to a variety of signals that tell the bacterium whether its location lies inside or outside of a multicellular organism. When the bacterium determines that it is indeed inside of a prospective host organism, additional signals may then serve to direct the bacterium to its replicative niche within the cytosol.

REFERENCES

Abachin, E., Poyart, C., Pellegrini, E., Milohanic, E., Fiedler, F., Berche, P., and Trieu-Cuot, P. (2002). Formation of D-alanyl-lipoteichoic acid is required for adhesion and virulence of *Listeria monocytogenes*. *Mol. Microbiol.* **43**, 1–14.

Adams, T.J., Vartivarian, S., and Cowart, R.E. (1990). Iron acquisition systems of *Listeria monocytogenes*. *Infect. Immun.* **58**, 2715–2718.

Alexander, A.V., Walker, R.L., Johnson, B.J., Charlton, B.R., and Woods, L.W. (1992). Bovine abortions attributable to *Listeria ivanovii*: four cases (1988–1990). *J. Am. Vet. Med. Assoc.* **200**, 711–714.

Alvarez-Dominguez, C., Carrasco-Marin, E., and Leyva-Cobian, F. (1993). Role of complement component C1q in phagocytosis of *Listeria monocytogenes* by murine macrophage-like cell lines. *Infect. Immun.* **61**, 3664–3672.

Alvarez-Dominguez, C., Roberts, R., and Stahl, P.D. (1997a). Internalized *Listeria monocytogenes* modulates intracellular trafficking and delays maturation of the phagosome. *J. Cell Sci.* **110**, 731–743.

Alvarez-Dominguez, C., Vazquez-Boland, J.A., Carrasco-Marin, E., Lopez-Mato, P., and Leyva-Cobian, F. (1997b). Host cell heparan sulfate proteoglycans mediate attachment and entry of *Listeria monocytogenes*, and the listerial surface protein ActA is involved in heparan sulfate receptor recognition. *Infect. Immun.* **65**, 78–88.

Antal, E.A., Loberg, E.M., Bracht, P., Melby, K.K., and Maehlen, J. (2001). Evidence for intraaxonal spread of *Listeria monocytogenes* from the periphery to the central nervous system. *Brain Pathol.* **11**, 432–438.

Armstrong, R.W. and Fung, P.C. (1993). Brainstem encephalitis (rhomben-cephalitis) due to *Listeria monocytogenes*: case report and review. *Clin. Infect. Dis.* **16**, 689–702.

Auerbuch, V., Lenz, L.L., and Portnoy, D.A. (2001). Development of a competitive index assay to evaluate the virulence of *Listeria monocytogenes actA* mutants during primary and secondary infection of mice. *Infect. Immun.* **69**, 5953–5957.

Aureli, P., Fiorucci, G.C., Caroli, D., Marchiaro, G., Novara, O., Leone, L., and Salmaso, S. (2000). An outbreak of febrile gastroenteritis associated with corn contaminated by *Listeria monocytogenes*. *N. Engl. J. Med.* **342**, 1236–1241.

Autret, N., Dubail, I., Trieu-Cuot, P., Berche, P., and Charbit, A. (2001). Identi-fication of new genes involved in the virulence of *Listeria monocytogenes* by signature-tagged transposon mutagenesis. *Infect. Immun.* **69**, 2054–2065.

Barchini, E. and Cowart, R.E. (1996). Extracellular iron reductase activity produced by *Listeria monocytogenes*. *Arch. Microbiol.* **166**, 51–57.

Beauregard, K.E., Lee, K.D., Collier, R.J., and Swanson, J.A. (1997). pH-dependent perforation of macrophage phagosomes by listeriolysin O from *Listeria mono-cytogenes*. *J. Exp. Med.* **186**, 1159–1163.

Bergmann, B., Raffelsbauer, D., Kuhn, M., Goetz, M., Hom, S., and Goebel, W. (2002). InlA- but not InlB-mediated internalization of *Listeria monocytogenes* by non-phagocytic mammalian cells needs the support of other internalins. *Mol. Microbiol.* **43**, 557–570.

Bielecki, J., Youngman, P., Connelly, P., and Portnoy, D.A. (1990). *Bacillus sub-tilis* expressing a haemolysin gene from *Listeria monocytogenes* can grow in mammalian cells. *Nature* **345**, 175–176.

Bierne, H., Mazmanian, S.K., Trost, M., Pucciarelli, M.G., Liu, G., Dehoux, P., Jansch, L., Garcia-del Portillo, F., Schneewind, O., and Cossart, P. (2002). Inactivation of the *srtA* gene in *Listeria monocytogenes* inhibits anchoring of surface proteins and affects virulence. *Mol. Microbiol.* **43**, 869–881.

Bohne, J., Sokolovic, Z., and Goebel, W. (1994). Transcriptional regulation of *prfA* and PrfA-regulated virulence genes in *Listeria monocytogenes*. *Mol. Microbiol.* **11**, 1141–1150.

Borezee, E., Pellegrini, E., Beretti, J.L., and Berche, P. (2001). SvpA, a novel sur-face virulence-associated protein required for intracellular survival of *Listeria monocytogenes*. *Microbiology* **147**, 2913–2923.

Boujemaa-Paterski, R., Gouin, E., Hansen, G., Samarin, S., Le Clainche, C., Didry, D., Dehoux, P., Cossart, P., Kocks, C., Carlier, M.F., and Pantalovi, D. (2001). *Listeria* protein ActA mimics WASP family proteins: it activates filament barbed end branching by Arp2/3 complex. *Biochemistry* **40**, 11,390–11, 404.

Boulnois, G.J., Paton, J.C., Mitchell, T.J., and Andrew, P.W. (1991). Structure and function of pneumolysin, the multifunctional, thiol-activated toxin of *Streptococcus pneumoniae. Mol. Microbiol.* **5**, 2611–2616.

Braun, L., Dramsi, S., Dehoux, P., Bierne, H., Lindahl, G., and Cossart, P. (1997). InlB: an invasion protein of *Listeria monocytogenes* with a novel type of surface association. *Mol. Microbiol.* **25**, 285–294.

Braun, L., Ohayon, H., and Cossart, P. (1998). The InlB protein of *Listeria monocytogenes* is sufficient to promote entry into mammalian cells. *Mol. Microbiol.* **27**, 1077–1087.

Braun, L., Ghebrehiwet, B., and Cossart, P. (2000). gC1q-R/p32, a C1q-binding protein, is a receptor for the InlB invasion protein of *Listeria monocytogenes. EMBO J.* **19**, 1458–1466.

Bubert, A., Kuhn, M., Goebel, W., and Kohler, S. (1992). Structural and functional properties of the p60 proteins from different *Listeria* species. *J. Bacteriol.* **174**, 8166–8171.

Bubert, A., Sokolovic, Z., Chun, S.K., Papatheodorou, L., Simm, A., and Goebel, W. (1999). Differential expression of *Listeria monocytogenes* virulence genes in mammalian host cells. *Mol. Gen. Genet.* **261**, 323–336.

Cai, S. and Wiedmann, M. (2001). Characterization of the *prfA* virulence gene cluster insertion site in non-hemolytic *Listeria* spp.: probing the evolution of the *Listeria* virulence gene island. *Curr. Microbiol.* **43**, 271–277.

Cameron, L.A., Footer, M.J., van Oudenaarden, A., and Theriot, J.A. (1999). Motility of ActA protein-coated microspheres driven by actin polymerization. *Proc. Natl. Acad. Sci. USA* **96**, 4908–4913.

Camilli, A., Goldfine, H., and Portnoy, D.A. (1991). *Listeria monocytogenes* mutants lacking phosphatidylinositol-specific phospholipase C are avirulent. *J. Exp. Med.* **173**, 751–754.

Carlier, M.F., Laurent, V., Santolini, J., Melki, R., Didry, D., Xia, G.X., Hong, Y., Chua, N.H., and Pantaloni, D. (1997). Actin depolymerizing factor (ADF/cofilin) enhances the rate of filament turnover: implication in actin-based motility. *J. Cell Biol.* **136**, 1307–1322.

Chakraborty, T., Ebel, F., Domann, E., Niebuhr, K., Gerstel, B., Pistor, S., Temm-Grove, C.J., Jockusch, B.M., Reinhard, M., and Walter, U. (1995). A focal adhesion factor directly linking intracellularly motile *Listeria monocytogenes* and *Listeria ivanovii* to the actin-based cytoskeleton of mammalian cells. *EMBO J.* **14**, 1314–1321.

Chand, P. and Sadana, J.R. (1999). Outbreak of *Listeria ivanovii* abortion in sheep in India. *Vet. Rec.* **145**, 83–84.

Chico-Calero, I., Suarez, M., Gonzalez-Zorn, B., Scortti, M., Slaghuis, J., Goebel, W., and Vazquez-Boland, J.A. (2002). Hpt, a bacterial homolog of the microsomal glucose-6-phosphate translocase, mediates rapid intracellular proliferation in *Listeria*. *Proc. Natl. Acad. Sci. USA* **99**, 431–436.

Collins, M.D., Wallbanks, S., Lane, D.J., Shah, J., Nietupski, R., Smida, J., Dorsch, M., and Stackebrandt, E. (1991). Phylogenetic analysis of the genus *Listeria* based on reverse transcriptase sequencing of 16S rRNA. *Int. J. Syst. Bacteriol.* **41**, 240–246.

Conlan, J.W. and North, R.J. (1991). Neutrophil-mediated dissolution of infected host cells as a defense strategy against a facultative intracellular bacterium. *J. Exp. Med.* **174**, 741–744.

Conlan, J.W. and North, R.J. (1994). Neutrophils are essential for early anti-*Listeria* defense in the liver, but not in the spleen or peritoneal cavity, as revealed by a granulocyte-depleting monoclonal antibody. *J. Exp. Med.* **179**, 259–268.

Conte, M.P., Longhi, C., Polidoro, M., Petrone, G., Buonfiglio, V., Di Santo, S., Papi, E., Seganti, L., Visca, P., and Valenti, P. (1996). Iron availability affects entry of *Listeria monocytogenes* into the enterocytelike cell line Caco-2. *Infect. Immun.* **64**, 3925–3929.

Cossart, P., Vicente, M.F., Mengaud, J., Baquero, F., Perez-Diaz, J.C., and Berche, P. (1989). Listeriolysin O is essential for virulence of *Listeria monocytogenes*: direct evidence obtained by gene complementation. *Infect. Immun.* **57**, 3629–3636.

Cossart, P. and Lecuit, M. (1998). Interactions of *Listeria monocytogenes* with mammalian cells during entry and actin-based movement: bacterial factors, cellular ligands and signaling. *EMBO J.* **17**, 3797–3806.

Cossart, P. and Bierne, H. (2001). The use of host cell machinery in the pathogenesis of *Listeria monocytogenes*. *Curr. Opin. Immunol.* **13**, 96–103.

Cossart, P. (2002). Molecular and cellular basis of the infection by *Listeria monocytogenes*: an overview. *Int. J. Med. Microbiol.* **291**, 401–409.

Cottin, J., Loiseau, O., Robert, R., Mahaza, C., Carbonnelle, B., and Senet, J.M. (1990). Surface *Listeria monocytogenes* carbohydrate-binding components revealed by agglutination with neoglycoproteins. *FEMS Microbiol. Lett.* **56**, 301–305.

Coulanges, V., Andre, P., Ziegler, O., Buchheit, L., and Vidon, D.J. (1997). Utilization of iron-catecholamine complexes involving ferric reductase activity in *Listeria monocytogenes*. *Infect. Immun.* **65**, 2778–2785.

Cousens, L.P. and Wing, E.J. (2000). Innate defenses in the liver during *Listeria* infection. *Immunol. Rev.* **174**, 150–159.

Cowart, R.E., Lashmet, J., McIntosh, M.E., and Adams, T.J. (1990). Adherence of a virulent strain of *Listeria monocytogenes* to the surface of a hepatocarcinoma cell line via lectin-substrate interaction. *Arch. Microbiol.* **153**, 282–286.

Croize, J., Arvieux, J., Berche, P., and Colomb, M.G. (1993). Activation of the human complement alternative pathway by *Listeria monocytogenes*: evidence for direct binding and proteolysis of the C3 component on bacteria. *Infect. Immun.* **61**, 5134–5139.

Cummins, A.J., Fielding, A.K., and McLauchlin, J. (1994). *Listeria ivanovii* infection in a patient with AIDS. *J. Infect.* **28**, 89–91.

Dabiri, G.A., Sanger, J.M., Portnoy, D.A., and Southwick, F.S. (1990). *Listeria monocytogenes* moves rapidly through the host-cell cytoplasm by inducing directional actin assembly. *Proc. Natl. Acad. Sci. USA* **87**, 6068–6072.

Dalton, C.B., Austin, C.C., Sobel, J., Hayes, P.S., Bibb, W.F., Graves, L.M., Swaminathan, B., Proctor, M.E., and Griffin, P.M. (1997). An outbreak of gastroenteritis and fever due to *Listeria monocytogenes* in milk. *N. Engl. J. Med.* **336**, 100–105.

Daniels, J.J., Autenrieth, I.B., and Goebel, W. (2000). Interaction of *Listeria monocytogenes* with the intestinal epithelium. *FEMS Microbiol. Lett.* **190**, 323–328.

Davies, W.A. (1983). Kinetics of killing *Listeria monocytogenes* by macrophages: rapid killing accompanying phagocytosis. *J. Reticuloendothel. Soc.* **34**, 131–141.

Decatur, A.L. and Portnoy, D.A. (2000). A PEST-like sequence in listeriolysin O essential for *Listeria monocytogenes* pathogenicity. *Science* **290**, 992–995.

Deneer, H.G., Healey, V., and Boychuk, I. (1995). Reduction of exogenous ferric iron by a surface-associated ferric reductase of *Listeria* spp. *Microbiology* **141**, 1985–1992.

Domann, E., Wehland, J., Rohde, M., Pistor, S., Hartl, M., Goebel, W., Leimeister-Wachter, M., Wuenscher, M., and Chakraborty, T. (1992). A novel bacterial virulence gene in *Listeria monocytogenes* required for host cell microfilament interaction with homology to the proline-rich region of vinculin. *EMBO J.* **11**, 1981–1990.

Domann, E., Zechel, S., Lingnau, A., Hain, T., Darji, A., Nichterlein, T., Wehland, J., and Chakraborty, T. (1997). Identification and characterization of a novel PrfA-regulated gene in *Listeria monocytogenes* whose product, IrpA, is highly homologous to internalin proteins, which contain leucine-rich repeats. *Infect. Immun.* **65**, 101–109.

Dramsi, S., Biswas, I., Maguin, E., Braun, L., Mastroeni, P., and Cossart, P. (1995). Entry of *Listeria monocytogenes* into hepatocytes requires expression of *inIB*, a surface protein of the internalin multigene family. *Mol. Microbiol.* **16**, 251–261.

Dramsi, S., Dehoux, P., Lebrun, M., Goossens, P.L., and Cossart, P. (1997). Identification of four new members of the internalin multigene family of *Listeria monocytogenes* EGD. *Infect. Immun.* **65**, 1615–1625.

Dramsi, S. and Cossart, P. (1998). Intracellular pathogens and the actin cytoskeleton. *Annu. Rev. Cell Dev. Biol.* **14**, 137–166.

Drevets, D.A. and Campbell, P.A. (1991). Roles of complement and complement receptor type 3 in phagocytosis of *Listeria monocytogenes* by inflammatory mouse peritoneal macrophages. *Infect. Immun.* **59**, 2645–2652.

Drevets, D.A., Leenen, P.J., and Campbell, P.A. (1993). Complement receptor type 3 (CD11b/CD18) involvement is essential for killing of *Listeria monocytogenes* by mouse macrophages. *J. Immunol.* **151**, 5431–5439.

Drevets, D.A., Sawyer, R.T., Potter, T.A., and Campbell, P.A. (1995). *Listeria monocytogenes* infects human endothelial cells by two distinct mechanisms. *Infect. Immun.* **63**, 4268–4276.

Drevets, D.A. (1999). Dissemination of *Listeria monocytogenes* by infected phagocytes. *Infect. Immun.* **67**, 3512–3517.

Drevets, D.A., Jelinek, T.A., and Freitag, N.E. (2001). *Listeria monocytogenes*-infected phagocytes can initiate central nervous system infection in mice. *Infect. Immun.* **69**, 1344–1350.

Dubail, I., Berche, P., and Charbit, A. (2000). Listeriolysin O as a reporter to identify constitutive and *in vivo*-inducible promoters in the pathogen *Listeria monocytogenes*. *Infect. Immun.* **68**, 3242–3250.

Dubail, I., Autret, N., Beretti, J.L., Kayal, S., Berche, P., and Charbit, A. (2001). Functional assembly of two membrane-binding domains in listeriolysin O, the cytolysin of *Listeria monocytogenes*. *Microbiology* **147**, 2679–2688.

Dunne, D.W., Resnick, D., Greenberg, J., Krieger, M., and Joiner, K.A. (1994). The type I macrophage scavenger receptor binds to gram-positive bacteria and recognizes lipoteichoic acid. *Proc. Natl. Acad. Sci. USA* **91**, 1863–1867.

Dussurget, O., Cabanes, D., Dehoux, P., Lecuit, M., Buchrieser, C., Glaser, P., and Cossart, P. (2002). *Listeria monocytogenes* bile salt hydrolase is a PrfA-regulated virulence factor involved in the intestinal and hepatic phases of listeriosis. *Mol. Microbiol.* **45**, 1095–1106.

Engelbrecht, F., Chun, S.K., Ochs, C., Hess, J., Lottspeich, F., Goebel, W., and Sokolovic, Z. (1996). A new PrfA-regulated gene of *Listeria monocytogenes* encoding a small, secreted protein which belongs to the family of internalins. *Mol. Microbiol.* **21**, 823–837.

Engelbrecht, F., Dickneite, C., Lampidis, R., Gotz, M., DasGupta, U., and Goebel, W. (1998a). Sequence comparison of the chromosomal regions encompassing the internalin C genes (*inlC*) of *Listeria monocytogenes* and *L. ivanovii*. *Mol. Gen. Genet.* **257**, 186–197.

Engelbrecht, F., Dominguez-Bernal, G., Hess, J., Dickneite, C., Greiffenberg, L., Lampidis, R., Raffelsbauer, D., Daniels, J.J., Kreft, J., Kaufmann, S.H., Vazquez-Boland, J.A., and Goebel W. (1998b). A novel PrfA-regulated chromosomal locus, which is specific for *Listeria ivanovii*, encodes two small, secreted internalins and contributes to virulence in mice. *Mol. Microbiol.* **30**, 405–417.

Facinelli, B., Giovanetti, E., Magi, G., Biavasco, F., and Varaldo, P.E. (1998). Lectin reactivity and virulence among strains of *Listeria monocytogenes* determined *in vitro* using the enterocyte-like cell line Caco-2. *Microbiology* **144**, 109–118.

Farber, J.M. and Peterkin, P.I. (1991). *Listeria monocytogenes*, a food-borne pathogen. *Microbiol. Rev.* **55**, 476–511.

Fenlon, D.R. (1999). *Listeria monocytogenes* in the natural environment. In *Listeria, Listeriosis, and Food Safety*, ed. E.T. Ryser and E.H. Marth, pp. 21–37. New York: Marcel Dekker.

Fischetti, V.A., Pancholi, V., and Schneewind, O. (1990). Conservation of a hexapeptide sequence in the anchor region of surface proteins from gram-positive cocci. *Mol. Microbiol.* **4**, 1603–1605.

Fleming, S.D. and Campbell, P.A. (1997). Some macrophages kill *Listeria monocytogenes* while others do not. *Immunol. Rev.* **158**, 69–77.

Freitag, N.E., Rong, L., and Portnoy, D.A. (1993). Regulation of the *prfA* transcriptional activator of *Listeria monocytogenes*: multiple promoter elements contribute to intracellular growth and cell-to-cell spread. *Infect. Immun.* **61**, 2537–2544.

Freitag, N.E. and Jacobs, K.E. (1999). Examination of *Listeria monocytogenes* intracellular gene expression by using the green fluorescent protein of *Aequorea victoria*. *Infect. Immun.* **67**, 1844–1852.

Gahan, C.G. and Hill, C. (2000). The use of listeriolysin to identify *in vivo* induced genes in the gram-positive intracellular pathogen *Listeria monocytogenes*. *Mol. Microbiol.* **36**, 498–507.

Gaillard, J.L., Berche, P., and Sansonetti, P. (1986). Transposon mutagenesis as a tool to study the role of hemolysin in the virulence of *Listeria monocytogenes*. *Infect. Immun.* **52**, 50–55.

Gaillard, J.L., Berche, P., Mounier, J., Richard, S., and Sansonetti, P. (1987). *In vitro* model of penetration and intracellular growth of *Listeria monocytogenes* in the human enterocyte-like cell line Caco-2. *Infect. Immun.* **55**, 2822–2829.

Gaillard, J.L., Berche, P., Frehel, C., Gouin, E., and Cossart, P. (1991). Entry of *L. monocytogenes* into cells is mediated by internalin, a repeat protein reminiscent of surface antigens from gram-positive cocci. *Cell* **65**, 1127–1141.

Gaillard, J.L. and Finlay, B.B. (1996). Effect of cell polarization and differentiation on entry of *Listeria monocytogenes* into the enterocyte-like Caco-2 cell line. *Infect. Immun.* **64**, 1299–1308.

Galan, J.E. (2000). Alternative strategies for becoming an insider: lessons from the bacterial world. *Cell* **103**, 363–366.

Garandeau, C., Reglier-Poupet, H., Dubail, I., Beretti, J.L., Berche, P., and Charbit, A. (2002). The sortase SrtA of *Listeria monocytogenes* is involved in processing of internalin and in virulence. *Infect. Immun.* **70**, 1382–1390.

Garges, S. and Adhya, S. (1988). Cyclic AMP-induced conformational change of cyclic AMP receptor protein (CRP): intragenic suppressors of cyclic AMP-independent CRP mutations. *J. Bacteriol.* **170**, 1417–1422.

Gedde, M.M., Higgins, D.E., Tilney, L.G., and Portnoy, D.A. (2000). Role of listeriolysin O in cell-to-cell spread of *Listeria monocytogenes*. *Infect. Immun.* **68**, 999–1003.

Geoffroy, C., Gaillard, J.L., Alouf, J.E., and Berche, P. (1987). Purification, characterization, and toxicity of the sulfhydryl-activated hemolysin listeriolysin O from *Listeria monocytogenes*. *Infect. Immun.* **55**, 1641–1646.

Geoffroy, C., Raveneau, J., Beretti, J.L., Lecroisey, A., Vazquez-Boland, J.A., Alouf, J.E., and Berche, P. (1991). Purification and characterization of an extracellular 29-kilodalton phospholipase C from *Listeria monocytogenes*. *Infect. Immun.* **59**, 2382–2388.

Gilot, P., Andre, P., and Content, J. (1999). *Listeria monocytogenes* possesses adhesins for fibronectin. *Infect. Immun.* **67**, 6698–6701.

Gilot, P., Jossin, Y., and Content, J. (2000). Cloning, sequencing and characterisation of a *Listeria monocytogenes* gene encoding a fibronectin-binding protein. *J. Med. Microbiol.* **49**, 887–896.

Glaser, P., Frangeul, L., Buchrieser, C., Rusniok, C., Amend, A., Baquero, F., Berche, P., Bloecker, H., Brandt, P., Chakraborty, T., Charbit, A., Chetovani, F., Couvre, E., de Daruvar, A., Dehoux, P., Domann, E., Domingvez-Bernal, G., Duchard, E., Durant, L., Dussurget, O., Entian, K.D., Fsihi, H., Portillo, F.G., Garrido, P., Gautier, L., Goebel, W., Gomez-Lopez, N., Hain, T., Hauf, J., Jackson, D., Jones, L.M., Kaerst, U., Kreft, J., Kuhn, M., Kunst, F., Kurapkat, G., Madveno, E., Maitdurnam, A., Vicente, J.M., Ng, E., Nedjari, H., Nordsiek, G., Novella, S., de Pablos, B., Perez-Diaz, J.C., Purcell, R., Remmel, B., Rose, M., Schlueter, T., Simoes, N., Tierrez, A., Vazquez-Boland, J.A., Voss, H., Wehland, J., and Cossart, P. (2001). Comparative genomics of *Listeria* species. *Science* **294**, 849–852.

Glomski, I.J., Gedde, M.M., Tsang, A.W., Swanson, J.A., and Portnoy, D.A. (2002). The *Listeria monocytogenes* hemolysin has an acidic pH optimum to compartmentalize activity and prevent damage to infected host cells. *J. Cell Biol.* **156**, 1029–1038.

Goebel, W., Kreft, J., and Bockmann, R. (2000). Regulation of virulence genes in pathogenic *Listeria*. In *Gram-Positive Pathogens*, ed. V.A. Fischetti, R.P. Novick, J.J. Ferretti, D.A. Portnoy, and J.I. Rood, pp. 449–506. Washington, DC: American Society for Microbiology.

Goldberg, M.B. (2001). Actin-based motility of intracellular microbial pathogens. *Microbiol. Mol. Biol. Rev.* **65**, 595–626.

Goldfine, H. and Knob, C. (1992). Purification and characterization of *Listeria monocytogenes* phosphatidylinositol-specific phospholipase C. *Infect. Immun.* **60**, 4059–4067.

Goldfine, H., Johnston, N.C., and Knob, C. (1993). Nonspecific phospholipase C of *Listeria monocytogenes*: activity on phospholipids in Triton X-100-mixed micelles and in biological membranes. *J. Bacteriol.* **175**, 4298–4306.

Goldfine, H., Knob, C., Alford, D., and Bentz, J. (1995). Membrane permeabilization by *Listeria monocytogenes* phosphatidylinositol-specific phospholipase C is independent of phospholipid hydrolysis and cooperative with listeriolysin O. *Proc. Natl. Acad. Sci. USA* **92**, 2979–2983.

Gouin, E., Mengaud, J., and Cossart, P. (1994). The virulence gene cluster of *Listeria monocytogenes* is also present in *Listeria ivanovii*, an animal pathogen, and *Listeria seeligeri*, a nonpathogenic species. *Infect. Immun.* **62**, 3550–3553.

Gray, M.L. and Killinger, A.H. (1966). *Listeria monocytogenes* and listeric infections. *Bacteriol. Rev.* **30**, 309–382.

Greenberg, J.W., Fischer, W., and Joiner, K.A. (1996). Influence of lipoteichoic acid structure on recognition by the macrophage scavenger receptor. *Infect. Immun.* **64**, 3318–3325.

Greene, S.L. and Freitag, N.E. (2003). Negative regulation of PrfA, the key activator of *Listeria monocytogenes* virulence gene expression, is dispensable for bacterial pathogenesis. Microbiol. **149**, 111–120.

Greiffenberg, L., Goebel, W., Kim, K.S., Weiglein, I., Bubert, A., Engelbrecht, F., Stins, M., and Kuhn, M. (1998). Interaction of *Listeria monocytogenes* with human brain microvascular endothelial cells: InlB-dependent invasion, long-term intracellular growth, and spread from macrophages to endothelial cells. *Infect. Immun.* **66**, 5260–5267.

Guzman, C.A., Rohde, M., Chakraborty, T., Domann, E., Hudel, M., Wehland, J., and Timmis, K.N. (1995). Interaction of *Listeria monocytogenes* with mouse dendritic cells. *Infect. Immun.* **63**, 3665–3673.

Hanawa, T., Yamamoto, T., and Kamiya, S. (1995). *Listeria monocytogenes* can grow in macrophages without the aid of proteins induced by environmental stresses. *Infect. Immun.* **63**, 4595–4599.

Hartford, T., O'Brien, S., Andrew, P.W., Jones, D., and Roberts, I.S. (1993). Utilization of transferrin-bound iron by *Listeria monocytogenes. FEMS Microbiol. Lett.* **108**, 311–318.

Hauf, N., Goebel, W., Serfling, E., and Kuhn, M. (1994). *Listeria monocytogenes* infection enhances transcription factor NF-kappa B in P388D$_1$ macrophage-like cells. *Infect. Immun.* **62**, 2740–2747.

Havell, E.A., Beretich, G.R. Jr., and Carter, P.B. (1999). The mucosal phase of *Listeria* infection. *Immunobiology* **201**, 164–177.

Hof, H. (2001). *Listeria monocytogenes*: a causative agent of gastroenteritis? *Eur. J. Clin. Microbiol. Infect. Dis.* **20**, 369–373.

Ireton, K., Payrastre, B., Chap, H., Ogawa, W., Sakaue, H., Kasuga, M., and Cossart, P. (1996). A role for phosphoinositide 3-kinase in bacterial invasion. *Science* **274**, 780–782.

Ishiguro, T., Naito, M., Yamamoto, T., Hasegawa, G., Gejyo, F., Mitsuyama, M., Suzuki, H., and Kodama, T. (2001). Role of macrophage scavenger receptors in response to *Listeria monocytogenes* infection in mice. *Am. J. Pathol.* **158**, 179–188.

Iwamoto, M., Ohno-Iwashita, Y., and Ando, S. (1990). Effect of isolated C-terminal fragment of theta-toxin (perfringolysin O) on toxin assembly and membrane lysis. *Eur. J. Biochem.* **194**, 25–31.

Johansson, J., Mandin, P., Renzoni, A., Chiaruttini, C., Springer, M., and Cossart, P. (2002). An RNA thermosensor controls expression of virulence genes in *Listeria monocytogenes. Cell* **110**, 551–561.

Jones, S. and Portnoy, D.A. (1994). Characterization of *Listeria monocytogenes* pathogenesis in a strain expressing perfringolysin O in place of listeriolysin O. *Infect. Immun.* **62**, 5608–5613.

Jones, S., Preiter, K., and Portnoy, D.A. (1996). Conversion of an extracellular cytolysin into a phagosome-specific lysin which supports the growth of an intracellular pathogen. *Mol. Microbiol.* **21**, 1219–1225.

Jonquieres, R., Bierne, H., Fiedler, F., Gounon, P., and Cossart, P. (1999). Interaction between the protein InlB of *Listeria monocytogenes* and lipoteichoic acid: a novel mechanism of protein association at the surface of gram-positive bacteria. *Mol. Microbiol.* **34**, 902–914.

Jonquieres, R., Pizarro-Cerda, J., and Cossart, P. (2001). Synergy between the N- and C-terminal domains of InlB for efficient invasion of non-phagocytic cells by *Listeria monocytogenes. Mol. Microbiol.* **42**, 955–965.

Kathariou, S., Metz, P., Hof, H., and Goebel, W. (1987). Tn916-induced mutations in the hemolysin determinant affecting virulence of *Listeria monocytogenes. J. Bacteriol.* **169**, 1291–1297.

Kirk, J. (1993). Diagnostic ultrastructure of *Listeria monocytogenes* in human central nervous tissue. *Ultrastruct. Pathol.* **17**, 583–592.

Klarsfeld, A.D., Goossens, P.L., and Cossart, P. (1994). Five *Listeria monocytogenes* genes preferentially expressed in infected mammalian cells: *plcA, purH, purD, pyrE* and an arginine ABC transporter gene, *arpJ. Mol. Microbiol.* **13**, 585–597.

Klatt, E.C., Pavlova, Z., Teberg, A.J., and Yonekura, M.L. (1986). Epidemic perinatal listeriosis at autopsy. *Hum. Pathol.* **17**, 1278–1281.

Kocks, C., Gouin, E., Tabouret, M., Berche, P., Ohayon, H., and Cossart, P. (1992). *L. monocytogenes*-induced actin assembly requires the *actA* gene product, a surface protein. *Cell* **68**, 521–531.

Kocks, C., Marchand, J.B., Gouin, E., d'Hauteville, H., Sansonetti, P.J., Carlier, M.F., and Cossart, P. (1995). The unrelated surface proteins ActA of *Listeria monocytogenes* and IcsA of *Shigella flexneri* are sufficient to confer actin-based motility on *Listeria innocua* and *Escherichia coli* respectively. *Mol. Microbiol.* **18**, 413–423.

Kohda, C., Kawamura, I., Baba, H., Nomura, T., Ito, Y., Kimoto, T., Watanabe, I., and Mitsuyama, M. (2002). Dissociated linkage of cytokine-inducing activity and cytotoxicity to different domains of listeriolysin O from *Listeria monocytogenes. Infect. Immun.* **70**, 1334–1341.

Kolb, A., Busby, S., Buc, H., Garges, S., and Adhya, S. (1993). Transcriptional regulation by cAMP and its receptor protein. *Annu. Rev. Biochem.* **62**, 749–795.

Kolb-Maurer, A., Gentschev, I., Fries, H.W., Fiedler, F., Brocker, E.B., Kampgen, E., and Goebel, W. (2000). *Listeria monocytogenes*-infected human dendritic cells: uptake and host cell response. *Infect. Immun.* **68**, 3680–3688.

Kreft, J. and Vazquez-Boland, J.A. (2001). Regulation of virulence genes in *Listeria. Int. J. Med. Microbiol.* **291**, 145–157.

Kuhn, M. and Goebel, W. (1989). Identification of an extracellular protein of *Listeria monocytogenes* possibly involved in intracellular uptake by mammalian cells. *Infect. Immun.* **57**, 55–61.

Kuhn, M. and Goebel, W. (1998). Host cell signaling during *Listeria monocytogenes* infection. *Trends Microbiol.* **6**, 11–15.

Kuhn, M. and Goebel, W. (2000). Internalization of *Listeria monocytogenes* by nonprofessional and professional phagocytes. *Subcell Biochem.* **33**, 411–436.

Lammerding, A.M. and Doyle, M.P. (1990). Stability of *Listeria monocytogenes* by non-thermal processing conditions. In *Foodborne Listeriosis*, ed. A.J. Miller, J.L. Smith, and G.A. Somkuti, pp. 195–202. New York: Elsevier.

Lampidis, R., Gross, R., Sokolovic, Z., Goebel, W., and Kreft, J. (1994). The virulence regulator protein of *Listeria ivanovii* is highly homologous to PrfA from *Listeria monocytogenes* and both belong to the Crp-Fnr family of transcription regulators. *Mol. Microbiol.* **13**, 141–151.

Lasa, I., David, V., Gouin, E., Marchand, J.B., and Cossart, P. (1995). The amino-terminal part of ActA is critical for the actin-based motility of *Listeria monocytogenes*; the central proline-rich region acts as a stimulator. *Mol. Microbiol.* **18**, 425–436.

Lasa, I., Gouin, E., Goethals, M., Vancompernolle, K., David, V., Vandekerckhove, J., and Cossart, P. (1997). Identification of two regions in the N-terminal domain of ActA involved in the actin comet tail formation by *Listeria monocytogenes*. *EMBO J.* **16**, 1531–1540.

Lauer, P., Theriot, J.A., Skoble, J., Welch, M.D., and Portnoy, D.A. (2001). Systematic mutational analysis of the amino-terminal domain of the *Listeria monocytogenes* ActA protein reveals novel functions in actin-based motility. *Mol. Microbiol.* **42**, 1163–1177.

Lauer, P., Chow, M.Y., Loessner, M.J., Portnoy, D.A., and Calendar, R. (2002). Construction, characterization, and use of two *Listeria monocytogenes* site-specific phage integration vectors. *J. Bacteriol.* **184**, 4177–4186.

Lecuit, M., Ohayon, H., Braun, L., Mengaud, J., and Cossart, P. (1997). Internalin of *Listeria monocytogenes* with an intact leucine-rich repeat region is sufficient to promote internalization. *Infect. Immun.* **65**, 5309–5319.

Lecuit, M., Dramsi, S., Gottardi, C., Fedor-Chaiken, M., Gumbiner, B., and Cossart, P. (1999). A single amino acid in E-cadherin responsible for host specificity towards the human pathogen *Listeria monocytogenes*. *EMBO J.* **18**, 3956–3963.

Leimeister-Wachter, M., Goebel, W., and Chakraborty, T. (1989). Mutations affecting hemolysin production in *Listeria monocytogenes* located outside the listeriolysin gene. *FEMS Microbiol. Lett.* **53**, 23–29.

Leimeister-Wachter, M., Haffner, C., Domann, E., Goebel, W., and Chakraborty, T. (1990). Identification of a gene that positively regulates expression of listeriolysin, the major virulence factor of *Listeria monocytogenes*. *Proc. Natl. Acad. Sci. USA* **87**, 8336–8340.

Leimeister-Wachter, M., Domann, E., and Chakraborty, T. (1991). Detection of a gene encoding a phosphatidylinositol-specific phospholipase C that is coordinately expressed with listeriolysin in *Listeria monocytogenes*. *Mol. Microbiol.* **5**, 361–366.

Leimeister-Wachter, M., Domann, E., and Chakraborty, T. (1992). The expression of virulence genes in *Listeria monocytogenes* is thermoregulated. *J. Bacteriol.* **174**, 947–952.

Lessing, M.P., Curtis, G.D., and Bowler, I.C. (1994). *Listeria ivanovii* infection. *J. Infect.* **29**, 230–231.

Lety, M.A., Frehel, C., Dubail, I., Beretti, J.L., Kayal, S., Berche, P., and Charbit, A. (2001). Identification of a PEST-like motif in listeriolysin O required for phagosomal escape and for virulence in *Listeria monocytogenes*. *Mol. Microbiol.* **39**, 1124–1139.

Lingnau, A., Domann, E., Hudel, M., Bock, M., Nichterlein, T., Wehland, J., and Chakraborty, T. (1995). Expression of the *Listeria monocytogenes* EGD inlA and *inlB* genes, whose products mediate bacterial entry into tissue culture cell lines, by PrfA-dependent and -independent mechanisms. *Infect. Immun.* **63**, 3896–3903.

Lingnau, A., Chakraborty, T., Niebuhr, K., Domann, E., and Wehland, J. (1996). Identification and purification of novel internalin-related proteins in *Listeria monocytogenes* and *Listeria ivanovii. Infect. Immun.* **64**, 1002–1006.

Loisel, T.P., Boujemaa, R., Pantaloni, D., and Carlier, M.F. (1999). Reconstitution of actin-based motility of *Listeria* and *Shigella* using pure proteins. *Nature* **401**, 613–616.

Lorber, B. (1997). Listeriosis. *Clin. Infect. Dis.* **24**, 1–9.

Lou, Y. and Yousef, A.E. (1999). Characteristics of *Listeria monocytogenes* important to food processors. In *Listeria, Listeriosis, and Food safety*, ed. E.T. Ryser and E.H. Marth, pp. 131–224. New York: Marcel Dekker.

MacDonald, T.T. and Carter, P.B. (1980). Cell-mediated immunity to intestinal infection. *Infect. Immun.* **28**, 516–523.

Machesky, L.M. and Insall, R.H. (1998). Scar1 and the related Wiskott-Aldrich syndrome protein, WASP, regulate the actin cytoskeleton through the Arp2/3 complex. *Curr. Biol.* **8**, 1347–1356.

Maganti, S., Pierce, M.M., Hoffmaster, A., and Rodgers, F.G. (1998). The role of sialic acid in opsonin-dependent and opsonin-independent adhesion of *Listeria monocytogenes* to murine peritoneal macrophages. *Infect. Immun.* **66**, 620–626.

Manohar, M., Baumann, D.O., Bos, N.A., and Cebra, J.J. (2001). Gut colonization of mice with *actA*-negative mutant of *Listeria monocytogenes* can stimulate a humoral mucosal immune response. *Infect. Immun.* **69**, 3542–3549.

Mansfield, B., Dionne, M., Schneider, D., and Freitag, N.E. (2003). *Drosophila* as a model host for *Listeria monocytogenes* infections. *Cell. Microbiol.*, in press.

Marchand, J.B., Moreau, P., Paoletti, A., Cossart, P., Carlier, M.F., and Pantaloni, D. (1995). Actin-based movement of *Listeria monocytogenes*: actin assembly results from the local maintenance of uncapped filament barbed ends at the bacterium surface. *J. Cell Biol.* **130**, 331–343.

Marco, A.J., Prats, N., Ramos, J.A., Briones, V., Blanco, M., Dominguez, L., and Domingo, M. (1992). A microbiological, histopathological and immunohistological study of the intragastric inoculation of *Listeria monocytogenes* in mice. *J. Comp. Pathol.* **107**, 1–9.

Marquis, H., Bouwer, H.G., Hinrichs, D.J., and Portnoy, D.A. (1993). Intracytoplasmic growth and virulence of *Listeria monocytogenes* auxotrophic mutants. *Infect. Immun.* **61**, 3756–3760.

Marquis, H., Doshi, V., and Portnoy, D.A. (1995). The broad-range phospholipase C and a metalloprotease mediate listeriolysin O-independent escape of *Listeria monocytogenes* from a primary vacuole in human epithelial cells. *Infect. Immun.* **63**, 4531–4534.

Marquis, H. and Hager, E.J. (2000). pH-regulated activation and release of a bacteria-associated phospholipase C during intracellular infection by *Listeria monocytogenes*. *Mol. Microbiol.* **35**, 289–298.

McLauchlin, J. (1987). *Listeria monocytogenes*, recent advances in the taxonomy and epidemiology of listeriosis in humans. *J. Appl. Bacteriol.* **63**, 1–11.

Mengaud, J., Vicente, M.F., Chenevert, J., Pereira, J.M., Geoffroy, C., Gicquel-Sanzey, B., Baquero, F., Perez-Diaz, J.C., and Cossart, P. (1988). Expression in *Escherichia coli* and sequence analysis of the listeriolysin O determinant of *Listeria monocytogenes*. *Infect. Immun.* **56**, 766–772.

Mengaud, J., Braun-Breton, C., and Cossart, P. (1991). Identification of phosphatidylinositol-specific phospholipase C activity in *Listeria monocytogenes*: a novel type of virulence factor? *Mol. Microbiol.* **5**, 367–372.

Mengaud, J., Ohayon, H., Gounon, P., Mege, R.M., and Cossart, P. (1996). E-cadherin is the receptor for internalin, a surface protein required for entry of *L. monocytogenes* into epithelial cells. *Cell* **84**, 923–932.

Miettinen, M.K., Siitonen, A., Heiskanen, P., Haajanen, H., Bjorkroth, K.J., and Korkeala, H.J. (1999). Molecular epidemiology of an outbreak of febrile gastroenteritis caused by *Listeria monocytogenes* in cold-smoked rainbow trout. *J. Clin. Microbiol.* **37**, 2358–2360.

Moors, M.A., Auerbuch, V., and Portnoy, D.A. (1999). Stability of the *Listeria monocytogenes* ActA protein in mammalian cells is regulated by the N-end rule pathway. *Cell. Microbiol.* **1**, 249–257.

Murray, E.G.D., Webb, R.E., and Swann, M.B.R. (1926). A disease of rabbits characterized by a large mononuclear leucocytosis, caused by a hitherto undescribed bacillus *Bacterium monocytogenes* (n. sp.). *J. Pathol. Bacteriol.* **29**, 407–439.

Nadon, C.A., Bowen, B.M., Wiedmann, M., and Boor, K.J. (2002). Sigma B contributes to PrfA-mediated virulence in *Listeria monocytogenes*. *Infect. Immun.* **70**, 3948–3952.

Niebuhr, K., Ebel, F., Frank, R., Reinhard, M., Domann, E., Carl, U.D., Walter, U., Gertler, F.B., Wehland, J., and Chakraborty, T. (1997). A novel proline-rich motif present in ActA of *Listeria monocytogenes* and cytoskeletal proteins is the ligand for the EVH1 domain, a protein module present in the Ena/VASP family. *EMBO J.* **16**, 5433–5444.

O'Riordan, M. and Portnoy, D. (2002). The host cytosol: front-line or home front? *Trends Microbiol.* **10**, 361.

Otter, A. and Blakemore, W.F. (1989). Observation on the presence of *Listeria monocytogenes* in axons. *Acta Microbiol. Hung.* **36**, 125–131.

Palmer, M. (2001). The family of thiol-activated, cholesterol-binding cytolysins. *Toxicon* **39**, 1681–1689.

Pandiripally, V.K., Westbrook, D.G., Sunki, G.R., and Bhunia, A.K. (1999). Surface protein p104 is involved in adhesion of *Listeria monocytogenes* to human intestinal cell line, Caco-2. *J. Med. Microbiol.* **48**, 117–124.

Pantaloni, D., Boujemaa, R., Didry, D., Gounon, P., and Carlier, M.F. (2000). The Arp2/3 complex branches filament barbed ends: functional antagonism with capping proteins. *Nat. Cell Biol.* **2**, 385–391.

Parida, S.K., Domann, E., Rohde, M., Muller, S., Darji, A., Hain, T., Wehland, J., and Chakraborty, T. (1998). Internalin B is essential for adhesion and mediates the invasion of *Listeria monocytogenes* into human endothelial cells. *Mol. Microbiol.* **28**, 81–93.

Parkassh, V., Morotti, R.A., Joshi, V., Cartun, R., Rauch, C.A., and West, A.B. (1998). Immunohistochemical detection of *Listeria* antigen in the placenta in perinatal listeriosis. *Int. J. Gynecol. Pathol.* **17**, 343–350.

Payne, S.M. (1993). Iron acquisition in microbial pathogenesis. *Trends Microbiol.* **1**, 66–69.

Pierce, M.M., Gibson, R.E., and Rodgers, F.G. (1996). Opsonin-independent adherence and phagocytosis of *Listeria monocytogenes* by murine peritoneal macrophages. *J. Med. Microbiol.* **45**, 258–262.

Portnoy, D.A., Jacks, P.S., and Hinrichs, D.J. (1988). Role of hemolysin for the intracellular growth of *Listeria monocytogenes*. *J. Exp. Med.* **167**, 1459–1471.

Portnoy, D.A., Chakraborty, T., Goebel, W., and Cossart, P. (1992). Molecular determinants of *Listeria monocytogenes* pathogenesis. *Infect. Immun.* **60**, 1263–1267.

Poyart, C., Abachin, E., Razafimanantsoa, I., and Berche, P. (1993). The zinc metalloprotease of *Listeria monocytogenes* is required for maturation of phosphatidylcholine phospholipase C: direct evidence obtained by gene complementation. *Infect. Immun.* **61**, 1576–1580.

Pron, B., Boumaila, C., Jaubert, F., Sarnacki, S., Monnet, J.P., Berche, P., and Gaillard, J.L. (1998). Comprehensive study of the intestinal stage of listeriosis in a rat ligated ileal loop system. *Infect. Immun.* **66**, 747–755.

Pron, B., Boumaila, C., Jaubert, F., Berche, P., Milon, G., Geissmann, F., and Gaillard, J.L. (2001). Dendritic cells are early cellular targets of *Listeria monocytogenes* after intestinal delivery and are involved in bacterial spread in the host. *Cell. Microbiol.* **3**, 331–340.

Raffelsbauer, D., Bubert, A., Engelbrecht, F., Scheinpflug, J., Simm, A., Hess, J., Kaufmann, S.H., and Goebel, W. (1998). The gene cluster *inlC2DE* of *Listeria monocytogenes* contains additional new internalin genes and is important for virulence in mice. *Mol. Gen. Genet.* **260**, 144–158.

Ramage, C.P., Low, J.C., McLauchlin, J., and Donachie, W. (1999). Characterisation of *Listeria ivanovii* isolates from the UK using pulsed-field gel electrophoresis. *FEMS Microbiol. Lett.* **170**, 349–353.

Raybourne, R.B. and Bunning, V.K. (1994). Bacterium-host cell interactions at the cellular level: fluorescent labeling of bacteria and analysis of short-term bacterium-phagocyte interaction by flow cytometry. *Infect. Immun.* **62**, 665–672.

Reinhard, M., Jouvenal, K., Tripier, D., and Walter, U. (1995). Identification, purification, and characterization of a zyxin-related protein that binds the focal adhesion and microfilament protein VASP (vasodilator-stimulated phosphoprotein). *Proc. Natl. Acad. Sci. USA* **92**, 7956–7960.

Ripio, M.T., Dominguez-Bernal, G., Lara, M., Suarez, M., and Vazquez-Boland, J.A. (1997). A Gly145Ser substitution in the transcriptional activator PrfA causes constitutive overexpression of virulence factors in *Listeria monocytogenes. J. Bacteriol.* **179**, 1533–1540.

Ripio, M.T., Vazquez-Boland, J.A., Vega, Y., Nair, S., and Berche, P. (1998). Evidence for expressional crosstalk between the central virulence regulator PrfA and the stress response mediator ClpC in *Listeria monocytogenes. FEMS Microbiol. Lett.* **158**, 45–50.

Rocourt, J., Hof, H., Schrettenbrunner, A., Malinverni, R., and Bille, J. (1986). Acute purulent *Listeria seeligeri* meningitis in an immunocompetent adult. *Schweiz. Med. Wochenschr.* **116**, 248–251.

Rocourt, J. (1996). Risk factors for listeriosis. *Food Control* **7**, 195–202.

Rogers, H.W., Callery, M.P., Deck, B., and Unanue, E.R. (1996). *Listeria monocytogenes* induces apoptosis of infected hepatocytes. *J. Immunol.* **156**, 679–684.

Rossjohn, J., Feil, S.C., McKinstry, W.J., Tweten, R.K., and Parker, M.W. (1997). Structure of a cholesterol-binding, thiol-activated cytolysin and a model of its membrane form. *Cell* **89**, 685–692.

Ryser, E.T. (1999). Foodborne listeriosis. In *Listeria, Listeriosis, and Food Safety*, ed. E.T. Ryser and E.H. Marth, pp. 299–358. New York: Marcel Dekker.

Salamina, G., Dalle Donne, E., Niccolini, A., Poda, G., Cesaroni, D., Bucci, M., Fini, R., Maldini, M., Schuchat, A., Swaminathan, B., Bibb, W., Rocourt, J., Binkin, N., and Salmaso, S. (1996). A foodborne outbreak of gastroenteritis involving *Listeria monocytogenes. Epidemiol. Infect.* **117**, 429–436.

Sawyer, R.T., Drevets, D.A., Campbell, P.A., and Potter, T.A. (1996). Internalin A can mediate phagocytosis of *Listeria monocytogenes* by mouse macrophage cell lines. *J. Leukoc. Biol.* **60**, 603–610.

Schlech, W.F., Lavigne, P.M., Bortolussi, R.A., Allen, A.C., Haldane, E.V., Wort, A.J., Hightower, A.W., Johnson, S.E., King, S.H., Nicholls, E.S., and Broome, C.V. (1983). Epidemic listeriosis – evidence for transmission by food. *N. Engl. J. Med.* **308**, 203–206.

Schluter, D., Buck, C., Reiter, S., Meyer, T., Hof, H., and Deckert-Schluter, M. (1999). Immune reactions to *Listeria monocytogenes* in the brain. *Immunobiology* **201**, 188–195.

Schuchat, A., Swaminathan, B., and Broome, C.V. (1991). Epidemiology of human listeriosis. *Clin. Microbiol. Rev.* **4**, 169–183.

Sechi, A.S., Wehland, J., and Small, J.V. (1997). The isolated comet tail pseudopodium of *Listeria monocytogenes*: a tail of two actin filament populations, long and axial and short and random. *J. Cell Biol.* **137**, 155–167.

Sekino-Suzuki, N., Nakamura, M., Mitsui, K.I., and Ohno-Iwashita, Y. (1996). Contribution of individual tryptophan residues to the structure and activity of theta-toxin (perfringolysin O), a cholesterol-binding cytolysin. *Eur. J. Biochem.* **241**, 941–947.

Sheehan, B., Kocks, C., Dramsi, S., Gouin, E., Klarsfeld, A.D., Mengaud, J., and Cossart, P. (1994). Molecular and genetic determinants of the *Listeria monocytogenes* infectious process. *Curr. Topics Microbiol.* **192**, 187–216.

Shen, Y., Naujokas, M., Park, M., and Ireton, K. (2000). InIB-dependent internalization of *Listeria* is mediated by the Met receptor tyrosine kinase. *Cell* **103**, 501–510.

Shetron-Rama, L.M., Marquis, H., Bouwer, H.G., and Freitag, N.E. (2002). Intracellular induction of *Listeria monocytogenes actA* expression. *Infect. Immun.* **70**, 1087–1096.

Shetron-Rama, L.M., Mueller, K., Bravo, J.M., Bouwer, H.G., and Freitag, N.E. (2003). Isolation of *Listeria monocytogenes* mutants with high level *in vitro* expression of host cytosol-induced gene products. *Mol. Microbiol.*, **48**, 1537–1551.

Sibelius, U., Chakraborty, T., Krogel, B., Wolf, J., Rose, F., Schmidt, R., Wehland, J., Seeger, W., and Grimminger, F. (1996). The listerial exotoxins listeriolysin and phosphatidylinositol-specific phospholipase C synergize to elicit endothelial cell phosphoinositide metabolism. *J. Immunol.* **157**, 4055–4060.

Skoble, J., Portnoy, D.A., and Welch, M.D. (2000). Three regions within ActA promote Arp2/3 complex-mediated actin nucleation and *Listeria monocytogenes* motility. *J. Cell Biol.* **150**, 527–538.

Smith, G.A., Portnoy, D.A., and Theriot, J.A. (1995). Asymmetric distribution of the *Listeria monocytogenes* ActA protein is required and sufficient to direct actin-based motility. *Mol. Microbiol.* **17**, 945–951.

Smith, G.A., Theriot, J.A., and Portnoy, D.A. (1996). The tandem repeat domain in the *Listeria monocytogenes* ActA protein controls the rate of actin-based motility, the percentage of moving bacteria, and the localization of vasodilator-stimulated phosphoprotein and profilin. *J. Cell Biol.* **135**, 647–660.

Spiro, S. and Guest, J.R. (1990). FNR and its role in oxygen-regulated gene expression in *Escherichia coli*. *FEMS Microbiol. Rev.* **6**, 399–428.

Stachowiak, R. and Bielecki, J. (2001). Contribution of hemolysin and phospholipase activity to cytolytic properties and viability of *Listeria monocytogenes*. *Acta Microbiol. Pol.* **50**, 243–250.

Stelma, G.N. Jr., Reyes, A.L., Peeler, J.T., Francis, D.W., Hunt, J.M., Spaulding, P.L., Johnson, C.H., and Lovett, J. (1987). Pathogenicity test for *Listeria monocytogenes* using immunocompromised mice. *J. Clin. Microbiol.* **25**, 2085–2089.

Suarez, M., Gonzalez-Zorn, B., Vega, Y., Chico-Calero, I., and Vazquez-Boland, J.A. (2001). A role for ActA in epithelial cell invasion by *Listeria monocytogenes*. *Cell. Microbiol.* **3**, 853–864.

Sun, A.N., Camilli, A., and Portnoy, D.A. (1990). Isolation of *Listeria monocytogenes* small-plaque mutants defective for intracellular growth and cell-to-cell spread. *Infect. Immun.* **58**, 3770–3778.

Swanson, J.A. and Baer, S.C. (1995). Phagocytosis by zippers and triggers. *Trends Cell Biol.* **5**, 89–93.

Sword, C.P. (1966). Mechanisms of pathogenesis in *Listeria monocytogenes* infection. I. Influence of iron. *J. Bacteriol.* **92**, 536–542.

Tang, P., Sutherland, C.L., Gold, M.R., and Finlay, B.B. (1998). *Listeria monocytogenes* invasion of epithelial cells requires the MEK-1/ERK-2 mitogen-activated protein kinase pathway. *Infect. Immun.* **66**, 1106–1112.

Theriot, J.A., Mitchison, T.J., Tilney, L.G., and Portnoy, D.A. (1992). The rate of actin-based motility of intracellular *Listeria monocytogenes* equals the rate of actin polymerization. *Nature* **357**, 257–260.

Theriot, J.A., Rosenblatt, J., Portnoy, D.A., Goldschmidt-Clermont, P.J., and Mitchison, T.J. (1994). Involvement of profilin in the actin-based motility of *L. monocytogenes* in cells and in cell-free extracts. *Cell* **76**, 505–517.

Tilney, L.G. and Portnoy, D.A. (1989). Actin filaments and the growth, movement, and spread of the intracellular bacterial parasite, *Listeria monocytogenes*. *J. Cell Biol.* **109**, 1597–1608.

Valenti, P., Greco, R., Pitari, G., Rossi, P., Ajello, M., Melino, G., and Antonini, G. (1999). Apoptosis of Caco-2 intestinal cells invaded by *Listeria monocytogenes*: protective effect of lactoferrin. *Exp. Cell Res.* **250**, 197–202.

Vazquez-Boland, J.A., Kocks, C., Dramsi, S., Ohayon, H., Geoffroy, C., Mengaud, J., and Cossart, P. (1992). Nucleotide sequence of the lecithinase operon of

Listeria monocytogenes and possible role of lecithinase in cell-to-cell spread. *Infect. Immun.* **60**, 219–230.

Vazquez-Boland, J.A., Dominguez-Bernal, G., Gonzalez-Zorn, B., Kreft, J., and Goebel, W. (2001a). Pathogenicity islands and virulence evolution in *Listeria*. *Microbes Infect.* **3**, 571–584.

Vazquez-Boland, J.A., Kuhn, M., Berche, P., Chakraborty, T., Dominguez-Bernal, G., Goebel, W., Gonzalez-Zorn, B., Wehland, J., and Kreft, J. (2001b). *Listeria* pathogenesis and molecular virulence determinants. *Clin. Microbiol. Rev.* **14**, 584–640.

Wadsworth, S.J. and Goldfine, H. (2002). Mobilization of protein kinase C in macrophages induced by *Listeria monocytogenes* affects its internalization and escape from the phagosome. *Infect. Immun.* **70**, 4650–4660.

Welch, M.D., Iwamatsu, A., and Mitchison, T.J. (1997). Actin polymerization is induced by Arp2/3 protein complex at the surface of *Listeria monocytogenes*. *Nature* **385**, 265–269.

Wesley, I.V. (1999). Listeriosis in animals. In *Listeria, Listeriosis, and Food Safety*, ed. E.T. Ryser and E.H. Marth, pp. 39–73. New York: Marcel Dekker.

Williams, J.R., Thayyullathil, C., and Freitag, N.E. (2000). Sequence variations within PrfA DNA binding sites and effects on *Listeria monocytogenes* virulence gene expression. *J. Bacteriol.* **182**, 837–841.

Wilson, R.L., Tvinnereim, A.R., Jones, B.D., and Harty, J.T. (2001). Identification of *Listeria monocytogenes* in *in vivo*-induced genes by fluorescence-activated cell sorting. *Infect. Immun.* **69**, 5016–5024.

Wuenscher, M.D., Kohler, S., Bubert, A., Gerike, U., and Goebel, W. (1993). The *iap* gene of *Listeria monocytogenes* is essential for cell viability, and its gene product, p60, has bacteriolytic activity. *J. Bacteriol.* **175**, 3491–3501.

Zalevsky, J., Grigorova, I., and Mullins, R.D. (2001). Activation of the Arp2/3 complex by the *Listeria* ActA protein. ActA binds two actin monomers and three subunits of the Arp2/3 complex. *J. Biol. Chem.* **276**, 3468–3475.

CHAPTER 7

N. gonorrhoeae: The varying mechanism of pathogenesis in males and females

Jennifer L. Edwards, Hillery A. Harvey, and Michael A. Apicella

Neisseria gonorrhoeae, the gonococcus, is the causative agent of gonorrhea, one of the oldest human diseases on record. Biblical references to gonorrhea in Leviticus (15:1–15:19) showed that the infectious nature of the disease was recognized even at that time. Probably, the best description of gonorrhea in a man in the preantibiotic era can be found in the writings of Boswell, who described in detail each of his 19 episodes of infection (Ober, 1970). These descriptions also allude to the asymptomatic nature of the disease in women, as many of the contacts from whom he acquired the infection were without symptoms of disease. Today, it is estimated that greater than 1 million cases of *N. gonorrhoeae* infection occur in the United States, and 60 million cases are reported annually worldwide. Hence, *N. gonorrhoeae* infection remains prevalent in the general population, despite the fact that antibiotic therapy is readily available. The high incidence of this disease remains a major concern in lower socioeconomic groups in the United States; however, the highest incidence of infection and of complications resulting from infection occur in developing countries. In underdeveloped nations it has been shown that patients with gonorrhea are at much higher risk for contracting human immunodeficiency virus (HIV). It is proposed that this increased susceptibility to HIV infection results from the inflammatory response generated by infecting gonococci with the subsequent disruption and shedding of the mucosal epithelium.

The gonococcus was classically considered an extracellular pathogen that colonized the epithelial surfaces of the human genital tract. This was in spite of the fact that over the past three decades the pathogenesis of gonococci in eukaryotic cells has been studied extensively in patient exudates and in tissue and organ culture. Electron microscopic studies by Evans (1977) of biopsies of infected females show that *N. gonorrhoeae* colonize and invade epithelial

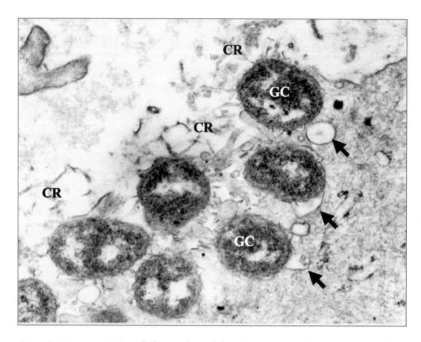

Figure 7.1. A cross section of a biopsy derived from the squamocolumnar junction of the uterine cervix of an infected female demonstrates the invasive nature of *N. gonorrhoeae*. Extensive cytoskeletal rearrangements (CR; e.g., filopodia, lamellipodia, and ruffles) of the mucosal epithelium are elicited by gonococcal infection and invasion and are visible on the apical cell surface. Gonococci (GC) are internalized in spacious vacuoles, which are denoted by arrows.

cells of the cervical squamocolumnar junction (Fig. 7.1). Studies by Apicella and coworkers (1996) show that urethral epithelial cells in exudates from infected men contain numerous intracellular organisms (Fig. 7.2). It is now understood that the ability of this organism to gain entry into host cells is crucial to its capability to establish and sustain infection.

The gonococcus can infect a number of different epithelial surfaces, thereby causing an array of clinical syndromes. Anorectal and pharyngeal infection usually results in an asymptomatic or a subclinical disease state. Conjunctival infection generally results in a severe conjunctivitis with corneal ulceration, although mild disease is occasionally reported. The gonococcus can also cause disseminated infection. However, disseminated gonococcal infection (DGI) occurs less frequently (1–3%) in both men and women. DGI results from gonococcal bacteremia and is commonly associated with untreated asymptomatic gonococcal infection or, less frequently (less than 3%),

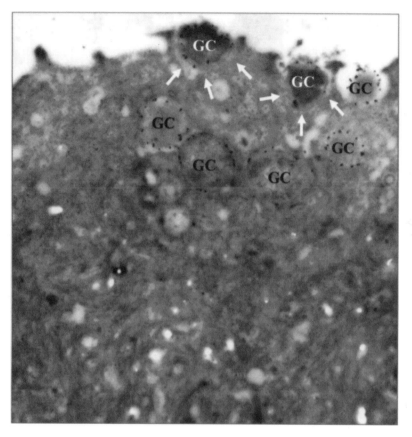

Figure 7.2. An exfoliated urethral epithelial cell obtained from the exudate of a male with documented gonococcal urethritis. Gonococci (GC) are labeled with a gold-bead antibody-conjugate that is specific for lipooligosaccharide. Arrows denote the intimate association of *N. gonorrhoeae* with the apical surface of the urethral epithelial cell and pedestal formation.

with complement (C′) deficiency. The majority of DGIs are found in women, in which the onset of clinical symptoms is often associated with menses. Arthritis–dermatitis syndrome is the most common manifestation of DGI, but severe cases can lead to endocarditis and meningitis.

In men, uncomplicated gonococcal infection is usually presented as an acute urethritis but can result in an acute epididymitis. Symptoms of disease primarily result as the consequence of the inflammatory response directed at the invasive gonococcus. However, experimental gonococcal infections in male human volunteers show that there is a delay between infection and

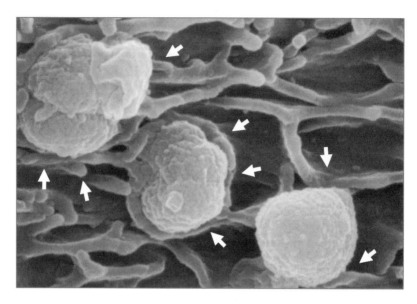

Figure 7.3. A 30-min experimental infection of a primary urethral epithelial cell with *N. gonorrhoeae* demonstrates filopodia extension induced by the gonococcus. The diplococcus arrangement, characteristic of *N. gonorrhoeae*, is visible at the upper left. Arrows denote filopodia engulfing bacteria.

the onset of clinical symptoms (Schneider et al., 1995). During this time, gonococci cannot be cultured from the urethra. Microscopy studies of urethral exudates from men with gonorrhea demonstrate that *N. gonorrhoeae* is an intracellular pathogen (Apicella et al., 1996). These studies also reveal that intracellular invasion by the gonococcus occurs through the intimate association between the gonococcus and the urethral cell membrane (Fig. 7.3), followed by internalization of gonococci within vacuoles. These observations, in conjunction with data obtained from experimental infections of men, suggest that viable gonococci are later released from the epithelial cells back into the lumen of the urethra. The organism's ability to gain access to and to survive the intracellular environment and the mechanisms that allow its subsequent release to the extracellular environment are poorly understood.

If left untreated, gonococcal infection of most men will resolve over a period of several weeks, but resolution may be mistaken for an asymptomatic carrier state. The occurrence of asymptomatic infection in men, however, is very low. Less than 3% of (infected) men exhibit asymptomatic gonococcal infection. In striking contrast to gonococcal infection of men, 50–80% of women exhibit asymptomatic genital tract infection, and 70–90% of women with DGI lack symptoms of genital tract involvement (Densen et al., 1982). Additionally, many women exhibit subclinical manifestations of *N. gonorrhoeae* infection; consequently, medical treatment is often not sought, or gonococcal infection is undiagnosed.

The high incidence of asymptomatic gonorrhea in women contributes substantially to the prevalence of *N. gonorrhoeae* in the general population and leads to acute and chronic complications in these women. Women with gonorrhea often exhibit multiple sites of *N. gonorrhoeae* colonization, which is usually attributed to contamination by gonococci-containing cervical secretions. Infection in women is further compounded by coinfection with other sexually transmittable organisms and a number of host physiological factors that, cooperatively, complicate disease (Bolan et al., 1999). Of the women who exhibit symptomatic gonorrhea, the severity of clinical presentation of disease is widely variable. Ascending gonococcal infection occurs in up to 45% of infected women and can result in pelvic inflammatory disease (PID), which can cause permanent fallopian tube scarring and blockage with subsequent infertility and ectopic pregnancies. One in 10 women suffer from PID (Aral et al., 1991), of which *N. gonorrhoeae* is the etiological agent in 40% of all reported cases (Sweet et al., 1986). Greater than 25% of women with PID will exhibit long-term complications of disease, 20% will suffer chronic pelvic pain, and 20% will become infertile. In contrast, men with a history of gonorrhea very rarely become infertile because of *N. gonorrhoeae* infection. An additional 10% of women with PID will have an ectopic pregnancy, which is the leading cause of death of women in their first trimester of pregnancy. Pregnancy is also considered to put a woman at an increased risk for acquiring DGI. The association of DGI with asymptomatic infection is reflected by the disproportionate percentage (80%) of DGIs that occur in women.

As with other bacterial pathogens, the widespread emergence of antibiotic-resistant strains of *N. gonorrhoeae* has increased the urgency of vaccine development. The hope of a vaccine, however, requires a greater understanding of both the human and the bacterial factors that contribute to a diseased state. Previous attempts at vaccine development have met with failure or with limited success. The propensity of gonococcal surface proteins to undergo both phase and antigenic variation has contributed to the ill success of vaccine development. Recent analysis of the *N. gonorrhoeae* genome has identified 100 gonococcal genes that exhibit phase variation (Snyder et al., 2001). Gonococcal infection was more recently shown to elicit only a mild humoral immune response, and no immunological memory has been observed. This suggests gonococcal subversion of the immune response by an as yet undefined mechanism. These difficulties may potentially be overcome by the use of genetically engineered hybrid proteins.

N. GONORRHOEAE CONSTITUENTS IMPLICATED IN PATHOGENESIS

Pili

Pili (or fimbriae) are filamentous appendages that extend from the surface of a bacterial cell. *N. gonorrhoeae* express the type IV-A class of bacterial pili, which are characterized by the presence of a N-methyl-phenylalanine amino acid residue at the first position in the mature pilin protein. An individual pilus fiber exists as a polymer composed of individual subunits (i.e., pilin) of approximately 18–22 kDa. Individual pilin subunits contain conserved and semivariable (SV) and hypervariable (HV) regions. Pilus assembly requires a repertoire of proteins that are highly conserved between bacterial species. Several levels of control regulate the expression of type IV pilus in the gonococcus. Multiple copies of the pilin gene can be found, scattered, throughout the gonococcal genome. Usually only one copy (but sometimes two copies), termed *pilE* (E for expressed), contains a promoter region and, thus, is expressed. The remaining copies, termed *pilS* (S for silent), provide genetic material for nonreciprocal recombination events, which leads to antigenic variation between strains of gonococci.

Pili are also subject to phase variation that can result from alternative cleavage of the N-terminal hydrophobic region of PilE to produce soluble S-pilin or from the production of aberrant or excessively long (L-)pilin (Forest and Tainer, 1997). Phase and antigenic variation provide mechanisms of immune avoidance, which allows the gonococcus to proliferate within a particular microenvironment of its human host. The importance of pili to *N. gonorrhoeae* colonization of epithelial cells has been demonstrated by several laboratories, and it is generally accepted that pili facilitate adherence by allowing the gonococcus to overcome electrostatic repulsion that occurs with the host cell. PilC, a pilus-associated protein, and which is also associated with the bacterial outer membrane (Rahman et al., 1977), may facilitate pilus-mediated adherence as a pilus tip-adhesin. Phase variation of PilC expression occurs by frameshift mutations within a polyguanine region upstream of the two alleles (*pilC1* and *pilC2*) encoding PilC (Dehio et al., 2000).

The absence of pilus does not appear to influence adhesion of the gonococcus to polymorphonuclear cells (PMNs), and it is suggested that the presence of pili may confer resistance to phagocytosis by these cells (Dilworth et al., 1975). Pili also mediate genetic transformation and agglutination, and they are postulated to provide a mechanism by which bacteria are able to move across and colonize mucosal epithelia through twitching motility. The gonococcal pilus is covalently modified with an O-linked,

N-acetylglucosamine(α1-3) galactose (Galα1-3GlcNac) disaccharide, which (to date) makes it unique among all prokaryotic proteins (Parge et al., 1995). A trisaccharide, galactose(β1-4) galactose(α1-3)2,4-diacetimido-2,4,6-trideoxyhexose (Galβ1-4Galα1-3-DATDH; see Stimson et al., 1995) is present in this same position of meningococcal pilin (Marceau et al., 1998). Removal of the meningococcal glycosyl residue confers a hyperadherent phenotype to mutant organisms (Marceau et al., 1998), suggesting that pilin glycosylation modulates pilin function and its interaction with epithelial cells. The role of phosphate (Forest et al., 1999) and phase-variable phosphorylcholine (Weiser et al., 1998) posttranslational modifications to pilus function are, currently, not known.

The role of pili in gonococcal pathogenesis is implicated *in vivo* by examination of gonococcal isolates from patients with gonorrhea. In addition, multiple *in vitro* studies have demonstrated the significance of pili in gonococcal adherence to epithelial cells. Complement receptor type 3 (CR3; see Edwards et al., 2001) and the C' regulatory protein, CD46 (or membrane cofactor receptor; see Källström et al., 1997), can serve as receptors for gonococcal pilus. CD46 is a human-specific transmembrane protein that is expressed by all nucleated cells. The association of pili with CD46 results in a rapid calcium influx within the host cell (Källström et al., 1998). This suggests that pili may modulate host cell signaling mechanisms to aid gonococcal epithelial invasion. However, Tobiason and Seifert (2001) demonstrated that the presence of CD46 on several epithelial cell lines is inversely related to gonococcal association with these cells. Additionally, CD46 does not appear to play a role in the CR3-mediated initial colonization of the uterine cervix (Edwards et al., 2002). Consequently, discernment of the role of CD46 in *N. gonorrhoeae* colonization will require further examination.

Porin

Porin (P.I; 32–39 kDa) accounts for greater than 60% of the total weight of proteins present in the outer membrane of the gonococcus, making it the most abundant outer membrane protein. Gonococcal nomenclature regarding porin is based on homology to meningococcal PorB. A gonococcal homologue of *N. meningitidis* PorA was not thought to exist; however, a gene showing homology to the PorA gene sequence has recently been identified (Dehio et al., 2000). Unlike pili, opacity-associated outer membrane proteins (Opa), and lipooligosaccharide (LOS), all of which exhibit phase or antigenic variation or both, porin expression is stable within a given strain of *N. gonorrhoeae*. Antigenic variation between strains, however, has provided one system of

N. gonorrhoeae serotyping. Two isotypes of porin are found in gonococci, P.IA and P.IB. Isotypes P.IA and P.IB can be further differentiated into serovars (classes 1–9) by their ability to react with a panel of monoclonal antibodies. For any *N. gonorrhoeae*, only one porin isotype is produced; however, anomalous P.IA/P.IB hybrid porins have been clinically isolated on rare occasion (Cooke et al., 1998). Gonococci exhibiting a P.IA isotype are associated with an increased risk for DGI and are thought to exhibit increased resistance to killing by normal human serum (i.e., serum-resistance).

Gonococcal porin resembles porins expressed by other Gram-negative bacteria in that it forms a water-filled channel in the bacterial outer membrane through which small solutes may cross. By analogy to *Escherichia coli* porins, gonococcal porin is suggested to exist in the bacterial outer membrane as a homotrimer, with each monomer forming a channel. Each porin monomer possesses 16 membrane-spanning regions, resulting in the exposure of eight antigenically variable loops in the extracellular environment. The longest of these extracellular loops express the greatest antigenic variability; they are immunodominant and serve as docking sites for bactericidal antibodies. Although porin is highly immunogenic, its intimate association with reduction-modifiable protein (Rmp or P.III) and with LOS within the outer membrane makes it inaccessible to immunological attack in the presence of LOS- or Rmp-directed blocking antibody.

In contrast to other Gram-negative porins, which exhibit cation selectivity, gonococcal porins resemble mitochondrial porins in their anion selectivity. Another distinguishing property of gonococcal and mitochondrial porins is their ability to interact with the adenosine and guanosine triphosphates (ATP and GTP, respectively) as a means to regulate pore size, voltage-dependent gating, and ion selectivity. Also unique among Gram-negative porins is the ability of gonococcal porin to translocate into eukaryotic host cell membranes. Within the eukaryotic cell membrane, porin it is thought to form a voltage-gated channel that is modulated by the host cell ATP and GTP.

The role of porin in potentiating disease by *N. gonorrhoeae* is thought to be multifactorial. Porin appears to modulate host cell function. Additional diverse functions of porin suggest that it is an important virulence factor for the gonococcus. Studies using PMNs have indicated that porin changes the membrane potential of these cells upon PMN membrane insertion (Haines et al., 1988). A change in membrane potential results in degranulation inhibition without altering the respiratory burst associated with oxidative killing. In contrast, gonococcal porin increases formyl-methionine-leucine-phenylalanine (fMLP)-mediated hydrogen peroxide production in PMN (Haines et al., 1991).

Further modulation of host cell function includes the ability of porin to inhibit phagosome maturation and downregulation of immunologically important cell surface receptors, such as immunoglobulin G (IgG), Fc receptors II (Fcγ RII) and III (Fcγ RIII), and complement receptor 1 (CR1) and CR3, in professional phagocytic cells.

Although porin appears to inhibit PMN actin polymerization in response to chemoattractants (Bjerknes et al., 1995), in epithelial cells porin acts as an actin-nucleating protein (Wen et al., 2000). In this respect porin may facilitate the cytoskeletal rearrangements required for actin-mediated entry of the gonococcus into its target host cell. Porin also modulates apoptosis of epithelial cells by inducing a calcium influx and, consequently, calpain and caspase activity within these cells (Dehio et al., 2000). Recent data have suggested that the ability of porin to induce apoptosis in epithelial cells may play a role in the cytotoxicity observed in fallopian tube organ culture (FTOC) and in shedding of epithelial cells (Dehio et al., 2000), which occurs *in vivo* during mucosal infection. Porin may also aid Opa–heparin sulfate proteoglycan (HSPG)-mediated invasion and Opa-independent invasion of Chang cells in the absence of phosphate (van Putten et al., 1998). In cooperation with gonococcus-bound iC3b and gonococcal pilus, porin also acts as an adhesin, mediating adherence to the cervical epithelium (Edwards et al., 2002).

Opacity-associated outer membrane proteins

Opa (P.II or class 5) proteins comprise a family of closely related integral outer membrane proteins. Opa proteins were originally identified by their ability to confer opacity and color changes to colonies of *N. gonorrhoeae* (grown on translucent agar) when viewed under diffused light with a stereomicroscope. Subsequent studies revealed that opacity is the result of Opa–LOS interactions that occur between adjacent bacteria. Opa proteins are heat modifiable, exhibiting altered electrophoretic mobility subsequent to heating at 100°C. Molecular mass (M_r) for individual Opa proteins ranges from 24 kDa to 30 kDa at 37°C and from 30 kDa to 32 kDa at 100°C. Gel filtration suggests that purified Opa proteins are trimers or tetramers; however, Opa isolation has proven difficult because of its high (one third) content of hydrophobic amino acids and, thus, its insolubility in the absence of detergents.

Structural models predict eight membrane-spanning regions with four membrane loops exposed on the extracellular face of the bacterium. The first three loops correspond to the SV, HV1, and HV2 regions; the fourth (C-terminal) loop is highly conserved. There are 11 or 12 *opa* loci that occur throughout the chromosome of a given *N. gonorrhoeae* strain. Expression of an individual *opa* gene occurs independently of other *opa* genes; consequently, a single gonococcus may exhibit none, one, or more than one Opa protein simultaneously. Each *opa* locus is subject to antigenic and phase variation; therefore, a given strain of *N. gonorrhoeae* may be highly heterogeneous with respect to Opa expression. Opa phase variation results from slipped-strand mispairing during replication in which alteration in the number of pentameric (CTCTT) repeat elements causes translational frameshifts (Lemon and Sparling, 1999).

A correlation has been made between the presence (and absence) of Opa proteins on *N. gonorrhoeae* clinical isolates with the site of isolation. In one human volunteer study, gonococci recovered from males infected with Opa$^-$ organisms had shifted to Opa$^+$ (Schneider et al., 1995). Clinical isolates obtained from men tend to express Opa proteins as do isolates recovered from human volunteer studies. Similarly, cervical isolates obtained from women at the time of ovulation (i.e., midcycle) also exhibit Opa expression. Translucent, Opa$^-$ organisms predominate in cervical isolates obtained during menses; in the fallopian tube; in genital, blood, and joint fluid obtained from patients with DGI; and in asymptomatic men. Opa proteins facilitate adherence to PMNs and, consequently, may play a role in potentiating gonococcal urethritis. A recent study has proposed a novel role for Opa proteins. Williams et al. (1998) suggested that Opa proteins may contribute to the intracellular survival of gonococci by binding host cell pyruvate kinase in order to acquire intracellular pyruvate, which is required for growth.

Lipooligosaccharide

LOS is an amphipathic glycolipid that constitutes a significant proportion of the *N. gonorrhoeae* outer membranes. In addition to differences in biosynthesis genes, the major difference between LOS and lipopolysaccharide (LPS), which is prevalent among Gram-negative bacteria, is its lack of a repeating O-antigen. LOS is composed of a lipid A moiety, a biantennary or triantennary core region composed of two L-glycero-D-manno-heptopyranose (heptose) and two 2-keto-3-deoxyoctulosonic acid (KDO) residues, and variable oligosaccharide side chains (Mandrell and Apicella, 1993; also see Fig. 7.4). The lipid A moiety anchors the LOS molecule within the outer membrane and consists of a dihexosamine backbone linked to four hydroxymeristic acids. These are substituted on the 3′ position by lauric acid. KDO molecules are juxtaposed to the lipid A and one KDO serves as the site of the diheptose addition. The heptose molecules serve as docking sites for short (6–10 sugar moieties) oligosaccharide addition(s). Additionally, the second heptose

$$\text{NANA } \alpha\, 2 \rightarrow 3\text{GlcNAc } \beta 1 \rightarrow 3\text{Gal } \beta 1 \rightarrow 4\text{Glc1} \rightarrow 4\text{Hep1} \rightarrow 5\text{KDO}_2 \rightarrow \text{Lipid A}$$

$$
\begin{array}{c}
3\\
\uparrow\\
1\\
\text{GlcNAc1} \rightarrow 2\text{Hep}\\
3\\
\uparrow\\
\text{PEA}
\end{array}
$$

Figure 7.4. A model of the sialylated LOS glycoform from *N. gonorrhoeae* strain 1291 is shown. This glycoform is represented as containing the sialyl-lactosamine found on the organism in the presence of CMP-NANA. Over 97% of gonococcal isolates express this lactosamine-containing glycoform that is important as a ligand to the human asialoglycoprotein receptor.

(213)

(which is not directly liked to KDO) bears an N-acetylglucosamine residue. Oligosaccharide substitutions exhibit interstrain and intrastrain variability.

Interconversion of LOS oligosaccharides occurs spontaneously and is dependent on the presence or absence of available substrates for, and the enzymes involved in, LOS biosynthesis. Several genes involved in LOS biosynthesis have been identified. The *lgt* locus, which encodes the genes for synthesis of lacto-N-neotetraose (LNnT) and digalactoside moieties, has recently been described. Several of these genes contain a polyguanine region that, in a manner similar to PilC phase variation, can result in frameshift mutations with concurrent loss (or gain) of certain oligosaccharide moieties available for LOS assembly. Phosphate, phosphoethanolamine or pyrophosphoethanolamine, and O-acetyl substitutions of the core region along with the addition of sialic acid to terminal galactose residues confer further variation to the LOS structure.

LOS oligosaccharide side chains terminate in epitopes that mimic oligosaccharide moieties of mammalian glycosphingolipids (including P^k, *i*, and paragloboside), regardless of oligosaccharide chain length. This form of molecular mimicry not only provides the bacterium with a method of immune avoidance but also allows the bacterium to use host-derived molecules that normally associate with the mimicked structure. The presence of a terminal LNnT epitope (Galβ1-4GlcNacβ1-3Galβ1-4Glc) on LOS, which mimics human paragloboside, allows the gonococcus to invade the urethral epithelium of men by adherence to the asialoglycoprotein receptor (ASGP-R; see Harvey et al., 2001), and this may facilitate disease transmission by adherence to the ASGP-R on human sperm (Harvey et al., 2000).

An analysis of *N. gonorrhoeae* strains demonstrates the predominance of the LNnT epitope among gonococci (Campagnari et al., 1990). These observations were confirmed *in situ* in that the LNnT epitope is selected for in men in human volunteer studies and in men with naturally acquired gonococcal urethritis. Schneider et al. (1995) further demonstrated that selection of the paragloboside epitope significantly enhances the ability of the gonococcus to cause disease in human volunteers. The predominance of a paragloboside moiety may enhance gonococcal survival in two ways: first, by avoidance of immune recognition by molecular mimicry of the LNnT, and second, by an increased serum-resistance by sialylation of the terminal galactose residue of the LNnT epitope. Thus, a lower infectious dose is required to establish disease, because a greater proportion of the inoculum would survive and proliferate.

The prevalence of the paragloboside moiety among *N. gonorrhoeae* is consistent with the finding that most clinically isolated gonococci initially exhibit serum-resistance; however, this property is lost with subsequent subculture. This unstable form of serum-resistance is attributed to sialylation of the terminal lactosamine of the LNnT moiety present on LOS. LOS and LPS stimulate cytokine production, which is associated with cytotoxicity. Stimulation of cytokine production by gonococcal LOS may allow this bacterium to access subepithelial tissues while increased serum-resistance and the predominance of human-like epitopes on the bacterium surface enhances survival. Sialylation of the gonococcus surface inhibits its ability to be phagocytosed by PMNs and may increase the survival of those gonococci that are phagocytosed. Sialylation of the gonococcus, however, significantly inhibits its ability to invade urethral epithelial cells (Harvey et al., 2001) and to initiate disease in human volunteer studies (Schneider et al., 1996).

Studies that explored the importance of LOS sialylation during infection and that used human volunteers and *in vitro* tissue culture models led to the hypothesis that although LOS sialylation may protect extracellular organisms from complement-mediated killing, these organisms become desialylated prior to internalization by host cells. Immunoelectron microscopy studies of urethral exudates from men with gonococcal urethritis show that *in vivo* sialylation of gonococcal LOS Galβ1-4-GlcNAc residue occurs during human infection and that although the majority of the LOS of intracellular gonococci is sialylated, approximately 10% of LNnT-terminal LOS remain unsialylated (Apicella et al., 1990). Studies by other groups later showed evidence that gonococci with sialylated LOS are less invasive for immortalized tissue culture epithelial cells than unsialylated gonococci (van Putten and Robertson, 1995). Conversion (by phase variation) of LOS to epitopes unable to be sialylated allows the gonococcus to adapt to environmental changes associated with progressive disease. LOS significantly contributes to the pathogenicity of the gonococcus in that it not only functions as an adhesin but also facilitates disease transmission, modulates bacterial invasion, and modulates disease progression.

Additional virulence factors

Through coevolution with its sole human host, the gonococcus has developed an impressive repertoire of gene products that allow it to exert redundant mechanisms by which it is able to infect, colonize, and proliferate within the hostile and benign microenvironments of the human body. Completion of the *N. gonorrhoeae* (strain FA1090) genome project and continued research will undoubtedly result in the identification of (as yet unknown) additional factors that contribute to gonococcal virulence. Several researchers have described factors that may contribute to virulence; however, their actual role in human disease is less well defined.

Many mucosal pathogens escape immunological killing by their ability to produce, and secrete, proteases that inactivate IgA1 and its secretory counterpart (sIgA1), the principal antibody isotype of the mucosa and mucosal secretions. Cleavage of these antibodies within their hinge region yields Fab$_\alpha$ monomeric fragments that are separate from their effector Fc$_\alpha$ counterparts; consequently, they are rendered inactive. Serine, cysteine, and metallo IgA proteases have been identified among those mucosal pathogens that produce these endopeptidases. *N. gonorrhoeae, N. meningitidis*, and *Haemophilus influenzae* produce serine proteases that cleave the IgA hinge region between proline and serine residues (i.e., type 1 activity) or between proline and threonine residues (i.e., type 2 activity). All *N. gonorrhoeae* and *N. meningitidis* strains examined to date have been found to constitutively express either a type 1 or a type 2 IgA1 protease. In gonococci the type of IgA1 protease produced correlates with auxotype and the serovar of porin produced. Gonococci that possess a nutritional requirement for arginine, hypoxanthine, and uracil (i.e., AHU auxotype) and a P.IA class 1 or 2 porin serovar produce a type 1 IgA1 protease. Type 2 IgA1 protease activity is prominent in gonococci that are excluded from the AHU/P.IA1/2 serotype.

The role of neisserial IgA1 protease activity in disease processes is undetermined. It is generally presumed that inactivation of potentially bactericidal IgA1 by a mucosal pathogen potentiates colonization and, therefore, potentiates disease caused by IgA protease-producing microbes. However, infection with *N. gonorrhoeae* IgA1 null mutants in human volunteer studies resulted in infection and disease indistinguishable from that observed with the parental wild-type strain. Analysis of cervical secretions yielded similar findings in that the level of IgA1 protease activity in samples obtained from women with gonococcal cervicitis were comparable with those obtained from women who were not infected. These observations suggest that IgA1 protease does not significantly influence colonization of the (uro)genital epithelia.

In epithelial cells, cleavage of lysosome-associated membrane proteins 1 (LAMP1) with concurrent phagosome modulation has been proposed as an alternative function for IgA1 protease activity. However, LAMP1 of professional phagocytes are resistant to (gonococcal) IgA1 protease activity. Within professional phagocytes LAMP1 are heavily glycoslyated; consequently, it is thought the hinge region is protected from protease cleavage. These studies suggest that IgA1 protease activity is limited to epithelial cells. This apparent conundrum pertaining to gonococcal IgA1 protease activity has to be examined further before a role for IgA1 protease activity in gonococcal pathogenesis can be resolved.

The ability of pathogenic organisms to sequester iron from their host animal is well established to contribute significantly to a diseased state. The gonococcus is no exception, and gonococci possess multiple mechanisms by which they are able to scavenge iron (Schryvers and Stojiljkovic, 1999). *Neisseria* do not produce siderophores; however, they do produce siderophore receptors that are homologous to siderophore receptors of other bacterial species. Expression of siderophore receptors allows *N. gonorrhoeae* to use siderophores produced by other bacterial species with which they may share an ecological niche. In addition to siderophore receptors, most pathogenic *Neisseria* also produce receptors for transferrin (Tf), lactoferrin (Lf), hemoglobin (Hb), heme, and haptoglobin-hemoglobin (Hp-Hb). Lf is found in higher concentration on mucosal surfaces than is Tf and, therefore, was presumed to be the preferred iron source for pathogenic *Neisseria* spp. However, some gonococci do not produce receptors for Lf, and mutants defective in Lf acquisition retain their virulence in human volunteer studies (Cornelissen et al., 1998).

In contrast, however, gonococci deficient in Tf acquisition are avirulent in human volunteer studies (Cornelissen et al., 1998), suggesting that Tf is required for gonococcal colonization of (at least) the male urethral epithelium. The ability of HmbR (the heme and Hb receptor), HpuAB (the Hb and Hp-Hb receptor), and TonB mutants to grow in the presence of heme as a sole iron source has presented the idea that heme may passively diffuse across the gonococcal outer membranes as a result of its hydrophobic character. Alternatively, porin may facilitate heme uptake (Schryvers and Stojiljkovic, 1999). However, recent data indicate that PilQ may serve a dual function, allowing pilin transport (with subsequent pilus assembly) out of, and heme passage into, the bacterial cell body (Chen et al., 2002).

The ability of gonococci to use heme, Hb, and Hp-Hb has been postulated to be responsible for the increased risk observed in women to develop PID and DGI during menses. However, fluctuations in hormone levels and associated changes to the female reproductive tract may also be responsible for the increased risk of complicated gonococcal infection with menses. Environmental fluctuations may allow gonococci of a particular isotype to gain access to subepithelial tissues or ascend the female genital tract and, thereby, initiate DGI or PID, respectively.

Interaction of the gonococcus with the lutropin receptor (LHr) is suggested to confer a contact-inducible phenotype to this bacterium that increases its invasive character, thereby allowing it to invade epithelia by adherence to the LHr (Spence et al., 1997). Identification of a LHr–gonococcus interaction involved the use of several immortal cell lines. Because LHr is upregulated in immortal cells and because ligand binding downregulates LHr surface expression, the significance of a LHr–gonococcus interaction *in vivo* within the lower female genital tract (e.g., the ectocervix and endocervix and the distal endometrium) remains to be determined. Gorby et al. (1991) demonstrated that gonococci bind to the LHr on nonciliated cells in FTOC, suggesting that the LHr may be paramount to colonization of the fallopian epithelia. The presence of the LHr on human uterus, placenta, decidua, fetal membrane, and fallopian tube tissues, in conjunction with increased expression of this receptor during menses, suggests that a gonococcus–LHr interaction could facilitate ascending infection in women.

Although speculative, a gonococcus–LHr interaction occurring on decidua and placental membranes could potentially result in severe complications and may, in part, contribute to the increased risk of spontaneous abortion associated with *N. gonorrhoeae* infection. The inhibitory effect of human chorionic gonadotropin (hCG), a ligand for LHr, on the observed LHr–gonococcus interaction suggests that gonococci possess an unidentified surface molecule that mimics hCG; however, the gonococcal LHr ligand has yet to be identified. The interaction of the gonococcus with the LHr appears to be a significant factor in colonization of fallopian tube tissue and may pose a serious risk to a developing fetus in pregnant women.

Several gonococcal gene products appear to be induced under conditions of limited oxygen. Because the oxygen tension is presumed to be low in some sites of gonococcal infection (and under certain circumstances), the contribution of these gene products, such as AniA (or Pan1; see Householder et al., 1999), Pan2, and Pan3 (Clark et al., 1987) to progressive infection and disease cannot be overlooked. Currently, however, there is little or no evidence to confirm or negate the contribution of these anaerobically induced proteins to the pathogenesis of *N. gonorrhoeae* in males or females.

COMPLEMENT AND PATHOGENIC *NEISSERIA*

Coevolution of the human C′ system with potentially pathogenic micro-organisms has allowed several human pathogens to adapt mechanisms by which they are able to avoid C′-mediated eradication (Vogel and Frosch, 1999). Methods evolved by microorganisms that allow them to escape the potentially lethal effects of C′ activation include the following: (1) development of an an-tiphagocytic capsule; (2) the development of surface structures that inhibit C3-convertase activity; (3) evasion of the lytic activity of the membrane at-tack complex (MAC), mediated by cell surface structures; (4) mimicry of C′ components; and (5) exploitation of C′ components to facilitate infection. The pathogenic *Neisseria*, that is, *N. meningitidis* and *N. gonorrhoeae*, are strict human pathogens that remain prevalent in the general population, in part because of their exquisite capability to avoid host defense mechanisms. In addition to the generalized defense mechanisms of phase and antigenic varia-tion, pathogenic *Neisseria* have evolved highly specialized immune avoidance mechanisms that allow them to persist within their (sole) human host.

N. meningitidis (the meningococcus) produces a capsule composed pri-marily, or entirely, of sialic acid. The presence of an extracellular capsule prohibits the interaction between cell-wall-deposited opsonins and host cells by providing a physical barrier impermeable to the phagocyte; that is, it is antiphagocytic. Meningococci of serogroup B possess a sialic acid capsule that is structurally similar to neural cell adhesion molecules (NCAM). This NCAM-like capsule confers further protection to serogroup B meningococci by rendering these bacteria nonimmunogenic. Sialic acid is widely distributed upon host cells and thus is recognized as an indigenous (or nonactivating) surface. In the absence of an effective antibody response (as already noted), direct recognition of a potential pathogen becomes critical to bacterial clear-ance. Within the alternative pathway (AP) of the C′ system, the presence of sialic acid on a target surface favors binding of factor H (fH) (with C′ inactiva-tion) over binding of factor B (fB), which would otherwise result in activation of the C′ system.

Although *N. gonorrhoeae* does not possess a capsule, it does share with the meningococcus the ability to sialylate its LOS. Gonococci that bear sialic acid moieties on their LOS have been found in cell-free systems to be resis-tant to C′-mediated killing; that is, they are serum-resistant. Serum-resistance conferred by sialylated LOS is attributed to the ability of LOS to bind a specific site within short consensus repeats (SCRs) 16–20 of fH (Ram et al., 1998). This results in C′ inactivation through the cofactor activity of fH in factor I (fI)-mediated inactivation of C3b. Because gonococci are not capable of producing cytosine 5′-monophosphate N-acetylneuraminic acid (CMP-NANA, the precursor for sialic acid synthesis), sialylation of a gonococcal LOS requires the use of an exogenous (i.e., host derived) substrate. In contrast, *N. meningitidis* is capable of synthesizing its own sialic acid and, therefore, incorporates endogenous sialic acid into its capsular and LOS structures. Significantly less fH is deposited upon capsulated meningococci when compared with strains that lack a capsule (Estabrook et al., 1997), suggesting that other factors may play a role in fH deposition on encapsulated organisms. An additional role for sialic acid has also been elucidated. The presence of sialic acid moieties can lead to a decrease in C9 deposition (Jarvis, 1995) upon the gonococcal surface, thereby prohibiting MAC assembly and, thus, its associated ability to lyse these cells.

Serum resistance of gonococci is also attributed to the expression of one of two isotypes of porin, PI.A and PI.B. Isotype PI.A is found more frequently on gonococci in people with DGI. Expression of PI.A by the gonococcus is attributed to increased serum resistance in comparison with gonococcal strains that express the PI.B isotype. Although porin is intimately associated with gonococcal LOS, binding of porin to fH occurs independently of LOS, at a site distinct from that defined for sialic acid (Ram et al., 1998). Within the classical pathway (CP) of the C′ system, C4bp functions in an analogous manner to fH of the AP. C4bp does not bind to the surface of sialylated gonococci; however, C4bp can bind gonococci when it is not sialylated (Ram et al., 2001). Binding of C4bp occurs on gonococci that express a PI.B isotype. Several studies have indicated that gonococcal killing is mediated primarily through the action of the CP; consequently, the ability of certain gonococcal strains to directly bind C4bp, which allows fI-mediated inactivation of C4b deposited on the gonococcal surface, may be critical to their survival within their human host.

In vitro N. gonorrhoeae infection studies and an examination of clinically isolated *N. gonorrhoeae* reveal that inactive C3b (i.e., iC3b, in comparison with active C3b) is predominant on the surface of this bacterium. iC3b opsonization allows the gonococcus to invade primary cervical epithelial cells in a process mediated by the action of gonococcal porin and pilus (Edwards et al., 2002), and which occurs independently of the sialylation status of the bacterium (Edwards and Apicella, 2002). Because endocytosis that is mediated by CR3 occurs independently of a proinflammatory response, internalization of the gonococcus within cervical epithelial cells may contribute to the asymptomatic nature associated with gonococcal infection in women.

The membrane-associated fI co-factor, CD46, may serve as a receptor for gonococcal pilus on some cell lines (Källström et al., 1997). Although the physiological relevance of a CD46–pilus interaction on the urogenital epithelium remains controversial, pilus binding to CD46 present on PMNs could facilitate fI-mediated inactivation of C3b or C4b on an opsonized bacterium.

Another family of surface molecules expressed by the pathogenic *Neisseria* is the Opa proteins. Opa proteins serve as ligands for members of the carcinoembryonic antigen-related family of cell adhesion molecules (CEACAM or CD66). CEACAM molecules can serve as coreceptors for the leukocyte integrins, including CR3 and CR4. Interaction of CEACAM with CD18 (the beta subunit shared by CR3 and CR4) is thought to inhibit the respiratory burst that occurs with ligand binding and phagocytosis by PMNs. This may promote the intracellular survival of PMN phagocytosed *Neisseria*.

An additional role for Opa proteins relies on their ability to bind heparin. Gonococcal binding of heparin by means of their Opa proteins increases the serum-resistance observed by these bacteria (Chen et al., 1995). Although the mechanism by which serum-resistance is conferred has yet to be defined, the ability of heparin to augment the regulatory role of vitronectin in the inhibition of MAC assembly provides one explanation for the observed increased serum-resistance.

Despite the redundant ability of the pathogenic *Neisseria* to evade C′-mediated killing, the C′ system is still capable of eliciting an efficient antimicrobial defense. This becomes most evident in individuals with various C′ deficiencies that are prone to recurrent bacterial infections, most notably with *Neisseria* spp. Individuals deficient in C′ components of the MAC exhibit an increased incidence of neisserial infections. In particular, deficiencies of C5, C6, C7, and C8 lead to a much greater risk for recurrent, and commonly systemic, infection with *Neisseria* spp. Despite the increased risk for systemic neisserial infections, a decreased mortality associated with infection is also observed and is thought to be attributable to a decrease in endotoxin released in the absence of a functional MAC.

People who are deficient in C9 are also at a much greater risk for *Neisseria* spp. infection. This has led to the suggestion that the bactericidal action of normal human serum is critical to controlling *Neisseria* spp. infection. Similarly, greater than 50% of people who lack the AP components fB, factor D (fD), and properdin exhibit recurrent neisserial infections. Studies focusing on neisserial serum resistance have primarily involved cell-free systems in which serum-resistance has been measured as the ability of these organisms to survive the lytic action of pooled human serum. However, mucosal membranes serve as the primary site for neisserial infection. Increasing evidence suggests that AP C′ components are produced by epithelial cells, albeit at a much lower level than is observed in human serum. Therefore, the ability of the pathogenic *Neisseria* to survive within their human host may reside in their ability to evade AP C′-mediated killing at the level of the mucosal epithelium with a subsequent transient carrier-like state. Progressive infection with dissemination into areas of the human body where specific antibody and C′ serum concentrations would far exceed those observed at the mucosal epithelium may lead to C′-mediated *Neisseria* eradication.

THE GONOCOCCUS AS AN INTRACELLULAR PATHOGEN

Despite the historic prevalence of *N. gonorrhoeae* within the human population and the considerable severity and frequency with which adverse complications accompany disease (most notably in women), it has only been in recent years that we have begun to understand gonococcal pathogenesis as it occurs within its (sole) human host. Discernment of gonococcal pathogenesis has been hindered by the lack of a good animal model by which to study the pathology of *N. gonorrhoeae* infection. Although several animal models have been described, no animal model exists that mirrors the full spectrum of human disease caused by the gonococcus. Consequently, researchers have relied on the use of human volunteers, tissue and organ cultures, and immortalized or malignant tissue culture cell lines by which to study gonococcal pathogenesis. Microscopic analyses of clinical biopsies and male urethral exudates have allowed extrapolation of successful gonococcus infection as it occurs *in vivo* with the realization that the gonococcus exists as an intracellular pathogen. However, more recent studies indicate that the mechanisms used by the gonococcus to obtain an intracellular status vary with the particular host cell target and microenvironment.

Infection of the male urethra

The development of a primary, human, male urethral epithelial cell culture system (Harvey et al., 1997) has greatly enhanced analysis of *N. gonorrhoeae* infection of the urethral epithelium. Within the urethral epithelium of men, infection is thought to occur as a sequential process that is initiated by attachment to and an intimate association with the targeted host cell that results in pedestal formation of the urethral cell membrane beneath an adherent

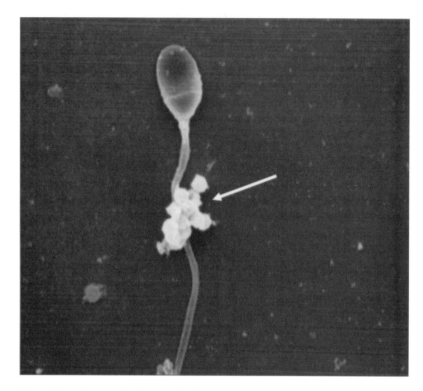

Figure 7.5. Polystyrene beads coated with LOS isolated from *N. gonorrhoeae* bind to the ASGP-R on human sperm as denoted by the arrow. This LOS glycoform harbors a terminal lactosamine as shown in Fig. 7.4. The LOS–ASGP-R interaction between bacteria and sperm may facilitate disease transmission.

gonococcus. Gonococci are internalized within vacuoles by actin-dependent (Giardina et al., 1998) and clathrin-dependent (Harvey et al., 1997) processes, where they replicate prior to their release to the extracellular milieu. Epithelial cells are subsequently shed. The intimate association observed between the gonococcus and the urethral cell surface is mediated by the ASGP-R (Harvey et al., 2001). This receptor binds to the terminal galactose of the gonococcus LOS. Consequently, LOS sialylation greatly impairs the ability of gonococci to associate with the male urethral epithelium. However, the presence or absence of sialic acid on gonococcal LOS or the presence of the ASGP-R on the cervical epithelium does not appear to contribute to lower female genital tract colonization. The ASGP-R is also present on human sperm (Harvey et al., 2000), suggesting that an LOS–ASGP-R interaction may also facilitate disease transmission (Fig. 7.5).

Studies in fallopian tube explants

Similar processes to those observed with infection of the male urethra have been demonstrated to occur in FTOC. In this model, however, gonococcal adherence occurs selectively on the nonciliated cells of fallopian tube tissue (McGee et al., 1981). Adherence is mediated by the LHr in a contact-inducible manner (Gorby et al., 1991). Following endocytosis of gonococci, bacteria-containing vacuoles are trancytosed to the basolateral surface of the infected epithelial cell, where they are released to the extracellular space. In this manner invasive bacteria are able to access the subepithelial tissues. Although the gonococcus selectively invades nonciliated cells, it is the adjacent ciliated cells that exhibit gonococcus-mediated cytotoxicity and ultimately results in the complete loss of ciliary action. The release of LOS and peptidoglycan by the gonococcus is thought to facilitate cytotoxicty of ciliated cells. Cytotoxicity occurs either directly or indirectly by the induction of increased production of the inflammatory cytokine, tumor necrosis factor (TNF).

Analysis of clinical *N. gonorrhoeae* isolates obtained from the fallopian tube as well as FTOC infection studies suggest that gonococcal pili also facilitate fallopian tube colonization. Additionally, gonococci that exhibit a translucent phenotype (Tr), and, therefore, lack Opa outer membrane proteins, are more invasive in FTOC. The prevalence of Tr gonococci in clinical isolates from the fallopian tube and in genital, blood, and joint fluid of DGI patients supports this observation. Expression of the LHr increases in an ascending manner from the endometrium to the fallopian tubes. Conversely, surface level expression of CR3 decreases progressively from the ectocervix to the upper female genital tract. Although it is currently not known if the expression of CR3 within the cervical epithelia exhibits cyclic variation, synthesis of C3 and surface expression of the LHr is upregulated at the time of menses. In a concerted action these molecules may contribute to an increased risk of complicated gonococcal infection.

Infection studies of the uterine cervix

In contrast to the pathology of *N. gonorrhoeae* infection observed for the male urethra and the upper female genital tract, lower genital tract (i.e., the uterine cervix) infection does not provoke a proinflammatory response. Recent data suggest that this subclinical or asymptomatic condition may be the result of the ability of the gonococcus to subvert the AP of the C' system. Primary cervical cells produce all of the alternative pathway C' proteins

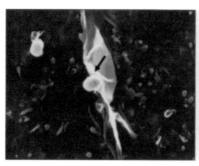

Figure 7.6. Gonococci (denoted by arrows) elicit membrane ruffles upon their association with the cervical epithelium. Membrane ruffling is the result of extensive host cytoskeletal rearrangements and results in bacterial internalization within macropinosomes. Membrane ruffles induced on ectocervical cells (left) exhibit a long ribbon-like morphology; ruffles formed by endocervical cells (right) are spherical in nature.

required for C′ activation and its inactivation. C′ protein C3 is activated upon *N. gonorrhoeae* infection to form C3b, which is deposited upon the lipid A portion of gonococcal LOS and is rapidly inactivated to form iC3b (Edwards and Apicella, 2002). The presence of iC3b on the gonococcal surface allows binding to CR3 in a cooperative manner with gonococcal pilus and porin (Edwards et al., 2002). Several studies demonstrate that ligand binding to the I-(or inserted) domain of CR3 does not invoke a proinflammatory response in immune cells. Therefore, the cooperative binding of iC3b, pilus, and porin to this region of CR3 may contribute to the asymptomatic nature of gonococcal cervicitis. Once the gonococcus adheres to CR3, a complex signal transduction cascade is initiated that results in extensive cervical cell cytoskeletal rearrangement. Large protrusions of the cervical cell membrane, termed membrane ruffles (Fig. 7.6), engulf the bacterium, resulting in gonococci internalization within macropinosomes (Fig. 7.7) in an actin-dependent manner (Edwards et al., 2000).

The interaction of the gonococcus with human neutrophils

The gonococcus, being a strict human pathogen, exquisitely senses its particular microenvironment (within its sole human host) and alters its physiological response to that environment in such a manner as to promote its own survival. This is evident by the finding that the interaction of the gonococcus with PMNs, which express CR3, occurs independently of a CR3 association (Farrell and Rest, 1990). The gonococcus–PMN association requires

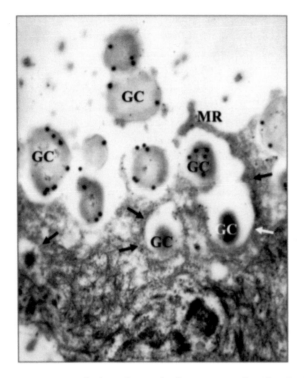

Figure 7.7. A cross section of polarized cervical cells experimentally infected with *N. gonorrhoeae* reveals that gonococci (GC) reside within spacious macropinosomes (denoted by arrows). Gonococci have been labeled with a gold-bead antibody-conjugate to gonococcal porin. A large membrane protrusion, indicative of a membrane ruffle (MR), can be seen on the right.

the presence of Opa outer membrane proteins; however, pilin proteins are not required. In contrast, gonococcal adherence to primary cervical and male urethral epithelial cells occurs independently of Opa proteins but requires gonococcal pilin.

Opa proteins are thought to confer distinct cellular trophisms to individual gonococci by the ability of different Opa proteins to differentially recognize host cell surface molecules. Two broad classes of Opa proteins exist that are represented by Opa_{50} (i.e., Opa proteins that recognize host cell HSPG) and Opa_{52} (i.e., Opa proteins that recognize CEACAM). The interaction of Opa proteins with HSPG is dependent on the presence of vitronectin or fibronectin (van Putten et al., 1998), which functions as a bridging molecule between the gonococcus and its target receptor(s). Association with an integrin coreceptor ($\alpha_v\beta_3$ or $\alpha_v\beta_5$ for vitronectin-mediated adherence or $\alpha_v\beta_1$

for fibronectin-mediated adherence) triggers a signaling cascade within the target cell that is dependent on the activation of protein kinase C (PKC).

A second, distinct, interaction between Opa_{50} and HSPG is demonstrated in Chang conjunctiva epithelial cells. Interaction of Opa_{50} with HSPG initiates a signaling cascade that activates phosphatidylcholine-dependent phospholipase C (PC-PLC). Generation of diacylglycerol (DAG) in turn activates acidic spingomyelinase (ASM), which results in ceramide production from spingomyelin. Ceramide is speculated to modulate the cytoskeletal rearrangements required for endocytosis of the cell-associated gonococcus.

Variable CEACAM isoforms are differentially expressed by a variety of different cell types, including PMNs and epithelial cells. Activation of the host cell Src tyrosine kinases, Hck and Fgr, results in the subsequent activation of Rac, which results in epithelial cytoskeletal rearrangement and internalization of the gonococcus. Although a single, primary receptor has not been identified that modulates the gonococcus–PMN interaction, the presence of the CEACAM, integrin, and HSPG receptors on PMNs suggests that any or all of these molecules may play a role in gonococcus phagocytosis. Additionally, binding of specific antibody to the gonococcus may facilitate phagocytosis by means of $Fc\gamma$ receptors.

A COMPREHENSIVE MODEL OF *N. GONORRHOEAE* PATHOGENESIS IN MEN AND WOMEN

The male urethra and the female uterine cervix are the primary sites of the majority of gonococcal infections. These sites are illustrative of the varying environments the organism must face in its pathogenesis, as these are entirely different epithelial surfaces that are derived from distinct embryological origins. The urethra in the male and female is derived from the initial division of the internal cloaca to form the urogenital sinus. The intermediate section of this sinus becomes the pars pelvina, which ultimately becomes the female urethra, a portion of the vestibule and the distal vagina in the female, and the prostatic urethra in the male. The caudalmost portion of the urogenital sinus becomes the membranous and penile portions of the urethra in the male and a portion of the vestibule in the female. In contrast, the uterine cervix is derived from the paramesonephric ducts. It is, therefore, not surprising that the surface receptors interacting with the organism on epithelial surfaces of these structures are different. The gonococcus has adapted very well to these different conditions; consequently, it is not surprising that distinct pathogenic mechanisms occur with *N. gonorrhoeae* infection in men and women.

N. gonorrhoeae possesses an impressive array of mechanisms that allow it to gain entry into host cells and to evade the host immune response. The importance of virulence factors such as pilus and Opa adhesins in gonococcal association with host cells is well established both *in vitro* and *in vivo*. In addition, although gonococcal LOS is implicated in the inflammatory response associated with gonococcal infection, clinical studies support *in vitro* evidence that gonococcal LOS plays a role in gonococcal adherence to and invasion of host cells. LOS and LPS play a role in invasion of host cells by other pathogenic bacteria. All of these virulence factors, which are located on the surface of the gonococcus, are capable of phase or antigenic variation or both. Because an association between the gonococcus and the host cell surface is a prerequisite for entry, it is believed that pili, Opa, LOS, and porin play an important role in the initiation of disease. Multiple studies show that the process of adherence to, followed by the invasion of, host cells involves the interaction of a repertoire of bacterial and host constituents. Given that gonorrhea is a sexually transmitted disease and reflecting upon what is known about gonococcal pathogenesis, assembling a model in the context of the male and female urogenital tracts may promote an understanding of this process at the level of human disease (Fig. 7.8).

Gonococci enter and survive within cervical epithelial cells. Because the intracellular environment is a source of CMP-NANA, the model shown in Fig. 7.8 presumes that gonococci with sialylated LOS are released from cervical epithelial cells to the extracellular environment. LOS sialylation is mediated by gonococcus-encoded sialyltransferase, but with host-derived CMP-NANA. Cooperativity between the gonococcus and other microbes that produce sialidases in the vaginal microflora results in desialylation of gonococcal LOS. Loss of the sialic acid moiety on gonococcal LOS, while in residence within the lower female genital tract, may prime the gonococcus for infection of the male urethra (where the presence of a sialic acid moiety would inhibit urethral interaction with ASGP-R). Upon transfer to the male partner, these desialylated gonococci can interact with the ASGP-R through the now available lactosamine residue on the nonreducing end of the LOS oligosaccharide. This initiates clathrin-mediated endocytosis of the organism, facilitating entry into the urethral epithelial cell.

Experiments examining the mechanism of gonococcal internalization involving LOS and the ASGP-R show that, although the number of ASGP-Rs at the epithelial cell surface increases after 4 h following infection with terminal lactosamine-expressing *N. gonorrhoeae*, no difference in the levels of total cellular ASGP-R protein or ASGP-R mRNA could be detected. This suggests cellular ASGP-R traffics to the cell surface during infection rather

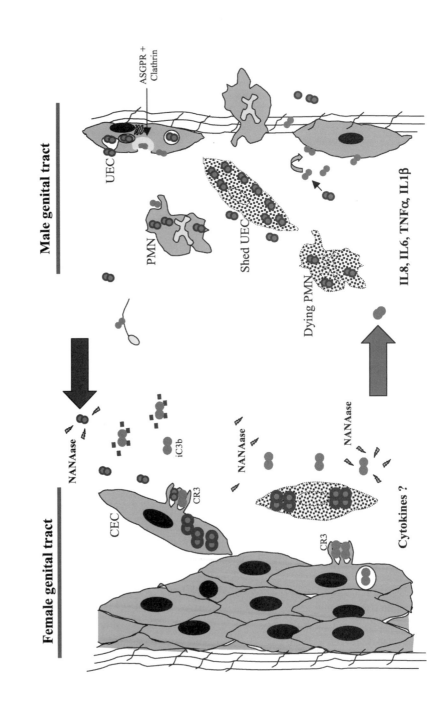

Female genital tract

Male genital tract

CEC

CR3

iC3b

NANAase

CR3

NANAase

NANAase

Cytokines ?

UEC

ASGPR + Clathrin

PMN

Shed UEC

Dying PMN

IL8, IL6, TNFα, IL1β

than being newly synthesized. Considering these findings together with what is known about the ASGP-R, one can propose a model in which extracellular gonococci bind ASGP-R and are internalized in clathrin-coated vesicles.

Clathrin-coated vesicles, containing receptor–ligand complexes, fuse with other clathrin-coated vesicles. In this early endosome, the clathrin coat is shed in an ATP-dependent process, and a pH drop results in uncoupling of the receptor–ligand complex. The ASGP-R is then recycled back to the cell surface where it is again available for binding ligand. In addition to the recycling back of ASGP-R to the cell surface in the absence of ligand, evidence is presented that a large fraction of ASGP-R is recycled to the cell surface prior to ligand dissociation, a process termed *diacytosis*, which may play a role in potentiating gonococcal urethritis.

Signaling mechanisms that are required for recruitment of cellular ASGP-R to the cell surface, whether or not fusion of gonococci-containing vesicles occurs *in vivo*, and the intracellular fate of the gonococcus remain to be determined. Human volunteer studies, *in vitro* invasion assays, and microscopy studies suggest that gonococci survive intracellularly. Lysosomes are considered the endpoint of ASGP-R-mediated endocytosis, but it is not known if gonococci internalized by means of ASGP-R escape lysosomal degradation. Pathogenic *Neisseria* species secrete an IgA protease, which cleaves host cell LAMP1. LAMP1 is involved in maturation of the endosome; consequently, the ability of the gonococcus to cleave this molecule is believed to contribute to gonococcal survival in the intracellular environment.

However, recent studies with a gonococcal IgA1 protease mutant showed that, compared to the wild type, this mutant was not compromised in its ability

---

Figure 7.8. (*facing page*). The different pathogenic interactions of the gonococcus in men and women. In women, infection of the cervical epithelial cells CEC occurs by membrane ruffling, which is an actin-dependent process. Once within the epithelial cell, the organism acquires the substrate for sialylation and sialylates its LOS (•). The inflammatory response is minimal, and as cells are shed they are released into the vaginal environment, which is rich in sialidases (🔨) produced by endogenous microflora that results in a desialylation of the LOS (•). Organisms passed to men are predominately desialylated and capable of entering urethral epithelial cells (UEC) by the interaction of the terminal lactosamine on the LOS and the asialoglycoprotein receptor on the urethral cell by a clathrin-dependent receptor-mediated endocytosis. Once within the cell, the organisms become resialylated. As cells are shed, organisms are released and phagocytosed by PMNs. Infected epithelial cells and PMNs are transferred to the female in the ejaculate. Uptake by the cervical cell through CR3 is unaffected by the sialylation state of the organism. Nonsialylated gonococci can also bind to human sperm and may aid in infection spread. See color section.

to cause urethritis in human volunteers. Whether the organism can proliferate within the host cell is still a matter for study, although exfoliated urethral epithelial cells in exudates from infected men show numerous organisms. This suggests that gonococcus replication very likely occurs within host (epithelial) cells. Within urethral epithelial cells, host-derived CMP-NANA is again available as a substrate for sialylation of gonococcal LOS. By an as yet undefined mechanism, gonococci traffic to the surface of the epithelial cell and are released from the surface. Once on the surface, these sialylated gonococci cannot reenter urethral epithelial cells or interact with the ASGP-R on sperm until the bacteria have divided in this CMP-NANA-free environment and the terminal lactosamine of their LOS becomes free of sialic acid.

In males, gonococcal infection likely results in inflammatory cytokine release by urethral epithelial cells. Few studies have addressed the question of epithelial cell signal transduction resulting from gonococcal adherence and invasion of these cells. However, *in vitro* studies using primary and immortal epithelial cells demonstrate that infection with *N. gonorrhoeae* elicits the production of the cytokines, TNF-α, interleukin (IL)-6, and IL-1β and the chemokine, IL-8, which triggers PMN influx and significantly contributes to local inflammation within the male urethra. It is also shown that, in experimental gonococcal infection in men, the cytokines, IL-6, IL-8, and TNF-α, are released at the local site of infection, the urethral mucosa. This is in contrast with findings observed in women with gonococcal cervicitis where local levels of IL-1, IL-6, and IL-8 are not elevated (Hedges et al., 1998). Therefore, the release of inflammatory cytokines by the cervical epithelium in response to *N. gonorrhoeae* infection remains under question.

In both males and females, gonococcal infection usually results in shedding of urethral and cervical epithelial cells, respectively. PMNs containing gonococci are also found in exudates from males with gonorrhea. Similar to the urethral epithelial cells, PMNs could also serve as a source of inflammatory cytokines and CMP-NANA and may release viable gonococci to the extracellular environment for disease transmission or reentry into urethral epithelial cells following desialylation by host sialidases. An Opa–CEACAM interaction on PMNs may allow survival of some intracellular gonococci, because engagement of CEACAM initiates a priming signal within these cells that results in activation of adhesion receptors without the release of inflammatory mediators or the induction of a respiratory burst.

The majority of gonococci that are transmitted from males to their female partners have sialylated LOS. However, the presence or absence of sialic acid on gonococcal LOS does not influence the association of the gonococcus with the cervical epithelium (Edwards and Apicella, 2002). In this model of cervical invasion, AP C' components are produced by the cervical epithelium and subsequently released to the extracellular milieu. Upon transmission of gonococci to the lower female genital tract, C3 deposition occurs on the lipid A determinant of gonococcal LOS. Expression of an LOS (as opposed to an LPS) structure may increase fH activity and, consequently, the conversion of C3b to iC3b on the gonococcus surface. Similarly, the resemblance of gonococcal LOS to human paraglobosides and glycosphingolipids (some of which serve as synthetic precursors of cervical mucins) may favor C' inactivation. The interaction of gonococcal pilin with the I-domain may initially serve as an anchor to CR3 that allows this bacterium to overcome repulsive forces that occur with the host cell membrane and, thus, position the gonococcus at the cervical cell membrane where the concentration of C' components may be expected to be sufficient to allow an efficient interaction with CR3.

Porin is very closely associated with LOS on the gonococcal surface, and its role in CR3-mediated endocytosis may be multifactorial; however, the direct association of porin with the I-domain of CR3 is crucial to cervical cell invasion. The juxtaposition of porin to LOS within the outer gonococcal membrane may spatially favor porin and iC3b adherence to the CR3-I-domain. Additionally, pilin- and porin-mediated binding of the gonococcus to CR3 may elicit a biphasic calcium flux in cervical epithelial cells that increases the affinity and avidity of iC3b-, porin-, and pilin-mediated CR3 adherence, as well as the release of additional effector proteins that might augment the gonococcus–CR3 interaction.

Membrane ruffling in professional phagocytic cells is attributed to the activity of the Rho family of GTPases. Although ruffling has not been previously observed in response to CR3 activation, similar responses probably occur within the cervical epithelium in that the inhibition of Rho activation also inhibits invasion of cervical epithelial cells. Jones et al. (1998) recently reported that fMLP- and Fcγ coreceptor-induced stimulation of CR3 results in the activation of p21-activating kinase (PAK1). Activation of PAK1 can elicit lamellipodia and ruffle formation, as well as vinculin accumulation, through a signaling cascade involving the PAK-interacting guanine nucleotide exchange factor (PIX) and Rac. The ability of porin to inhibit fMLP-induced signaling events in neutrophils suggests that these molecules may share effector proteins; however, it is unknown if binding of gonococcal porin to CR3 elicits membrane ruffling in primary cervical cells by augmenting signal transduction of PIX and PAK1.

Binding of CR3 triggers a cascade of events that may involve the activation of PIX and PAK1 and that results in ezrin and vinculin focal complex formation and the cytoskeletal changes required for ruffle formation. Engagement of CR3 by the gonococcus is sufficient to elicit membrane ruffling; however, other fluid phase molecules may facilitate gonococcal internalization. Once membrane ruffles are formed, endocytosis ensues, allowing gonococcal internalization within macropinosomes. These ideas are consistent with the wide variability associated with the incubation period, the clinical manifestations, and the high prevalence of asymptomatic gonococcal cervicitis in women. Clearly, coevolution with its human host has allowed the gonococcus to carefully orchestrate a mechanism of CR3 adherence in which constituents of its outer membrane, that is, porin and pilus, function cooperatively with host cell factors (i.e., AP proteins) to ensure its increased survival.

The deposition of iC3b on the gonococcal lipid A core may confer a survival advantage to the gonococcus by displacing the LOS oligosaccharide chain such that it is unable to bind to the lectin-binding domain of CR3. iC3b-mediated adherence to CR3 occurs independently of a proinflammatory response; however, binding of the lectin domain stimulates a respiratory burst through the generation of a second signal within the eukaryotic (host) cell. Consequently, a CR3 association mediated through the lectin domain of this receptor may result in decreased survival of the invasive organism. Additionally, the presence of iC3b on the lipid A core may obstruct the association of the lipid A core region with CD18. It, therefore, appears that C3b deposition on lipid A facilitates gonococcal infection and may promote bacterial survival in that any or all of these phenomena would favor adherence to the I-domain of CR3 and subsequent bacterial internalization in the absence of a respiratory burst. iC3b-mediated epithelial adherence, in the absence of a respiratory burst, may allow the gonococcus to attain a carrier-like state within the cervical epithelium. Ascent to the upper female reproductive tract may subsequently occur as the result of hormonal changes that alter host cell epithelia, molecules available for gonococcal use, or the expression of gonococcal virulence factors. These ideas are supported by the increased risk to women for PID and DGI that is associated with menses and with the correlation made between the presence or absence of Opa proteins and the site of gonococcus isolation. (Transparent gonococci are predominantly isolated from the cervix at the time of menses and from the fallopian tubes; from asymptomatic men; and from blood, joint, and genital fluids of DGI patients. Opa-expressing gonococci are predominantly isolated from men and from the cervix at the time of ovulation.)

Surface level expression of CR3 decreases progressively from the ectocervix to the upper female genital tract. Conversely, expression of the LHr increases in an ascending manner from the endometrium to the fallopian tubes (where it serves as a receptor for gonococcal adherence; see Gorby et al.,

1991). Although it is currently not known if the expression of CR3 within the cervical epithelia exhibits cyclic variation, synthesis of C3 and surface expression of the LHr is upregulated at the time of menses. In a concerted action, these molecules may contribute to an increased risk of complicated gonococcal infection.

The level of sialic acid found within the cervical milieu exhibits cyclic variation and is inversely related to C3 production. Additionally, although sialidase activity is present within cervical mucus throughout the menses cycle, cyclic variation also exists with the ability of this enzyme to use endogenous or exogenous substrates. With regard to the increased susceptibility to ascending gonococcal infection correlated with menses, the decreased prevalence of sialic acid within the cervical milieu at the time of menses might suggest that sialylated bacteria are impaired in their ability to ascend the female genital tract, although no discernable difference exists in the ability of sialylated or unsialylated gonococci to interact with cervical epithelia. However, removal of sialic acid from the mucosal surfaces of the female genital tract may allow gonococci of a particular isotype to gain access to subepithelial tissues or to ascend the female genital tract and thereby initiate DGI or PID, respectively, and may have important implications with regard to disease transmission.

REFERENCES

Apicella, M.A., Ketterer, M., Lee, F.K.N., Zhou, D., Rice, P.A., and Blake. M.S. (1996). The pathogenesis of gonococcal urethritis in men: confocal and immunoelectron microscopic analysis of urethral exudates from men infected with *Neisseria gonorrhoeae. J. Infect. Dis.* **173**, 636–646.

Apicella, M.A., Mandrell, R.E., Shero, M., Wilson, M.E., Griffiss, J.M., Brooks, G.F., Lammel, C., Breen, J.F., and Rice, P.A. (1990). Modification by sialic acid of *Neisseria gonorrhoeae* lipooligosaccharide epitope expression in human urethral exudates: an immunoelectron microscopic analysis. *J. Infect. Dis.* **162**, 506–512.

Aral, S.O., Mosher, W.D., and Cates, W. (1991). Self-reported pelvic inflammatory disease in the United States. *JAMA* **266**, 2570–2573.

Bjerknes, R., Guttormsen, H.-K., Solberg, C.O., and Wetzler, L.M. (1995). Neisserial porins inhibit human neutrophil actin polymerization, degranulation, opsonin receptor expression, and phagocytosis but prime the neutrophils to increase their oxidative burst. *Infect. Immun.* **63**, 160–167.

Bolan, G., Ehrhardt, A.A., and Wasserheit, J.N. (1999). Gender perspectives and STDs. In *Sexually Transmitted Diseases*, ed. K.K. Holmes, P.-A. Mardh, P.F.

Sparling, S.M. Lemon, W.E. Stamm, P. Piot, and J.N. Wasserheit, 3rd ed., pp. 117–127. New York: McGraw-Hill.

Campagnari, A.A., Spinola, S.M., Lesse, A.J., Kwaik, Y.A., Mandrell, R.E., and Apicella, M.A. (1990). Lipooligosaccharide epitopes shared among gram-negative non-enteric mucosal pathogens. *Microb. Pathog.* **8**, 353–362.

Chen, C.J., Thomas, C.E., McLean, D.S., Rouquette-Loughlin, C., Shafer, W.M., and Sparling, P.F. (2002). PilQ point mutation results in hemoglobin utilization in the absence of HpuA/B. In *Abstracts of the 13th International Pathogenic Neisseria Conference*, ed. D.A. Caugant and E. Wedege, p. 103. Oslo: Nordberg Aksidenstrykkeri AS.

Chen, T.C., Swanson, J., Wilson, J., and Belland, R.J. (1995). Heparin protects Opa+ *Neisseria gonorrhoeae* from the bactericidal action of normal human serum. *Infect. Immun.* **63**, 1790–1795.

Clark, V.L., Campbell, L.A., Palermo, D.A., Evans, T.M., and Klimpel, K.W. (1987). Induction and repression of outer membrane proteins by anaerobic growth of *Neisseria gonorrhoeae*. *Infect. Immun.* **55**, 1359–1364.

Cooke, S.J., Jolley, K., Ison, C.A., Young, H., and Heckels, J.E. (1998). Naturally occurring isolates of *Neisseria gonorrhoeae*, which display anomalous serovar properties, express PIA/PIB hybrid porins, deletions in PIB or novel PIA molecules. *FEMS Microbiol. Lett.* **162**, 75–82.

Cornelissen, C.N., Kelley, M., Hobbs, M.M., Anderson, J.E., Cannon, J.G., Cohen, M.S., and Sparling, P.F. (1998). The transferrin receptor expressed by gonococcal strain FA1090 is required for experimental infection of human male volunteers. *Mol. Microbiol.* **27**, 611–616.

Dehio, C., Gray-Owen, S.D., and Meyer, T.F. (2000). Host cell invasion by pathogenic *Neisseriae*. In *Subcellular Biochemistry, Vol. 33: Bacterial Invasion into Eukaryotic Cells*, ed. T.A. Oelschlaeger and J. Hacker, pp. 61–96. New York: Plenum.

Densen, P., MacKeen, L.A., and Clark, R.A. (1982). Dissemination of gonococcal infection is associated with delayed stimulation of complement-dependent neutrophil chemotaxis in vitro. *Infect. Immun.* **38**, 563–572.

Dilworth, J.A., Hendley, J.O., and Mandell, G.L. (1975). Attachment and ingestion of gonococci by human neutrophils. *Infect. Immun.* **11**, 512–516.

Edwards, J.L. and Apicella, M.A. (2002). The role of lipooligosaccharide in *Neisseria gonorrhoeae* pathogenesis of cervical epithelia: lipid A serves as a C3 acceptor molecule. *Cell. Microbiol.* **4**, 585–598.

Edwards, J.L., Brown, E.J., Ault, K.A., and Apicella, M.A. (2001). The role of complement receptor 3 (CR3) in *Neisseria gonorrhoeae* infection of human cervical epithelia. *Cell. Microbiol.* **3**, 611–622.

Edwards, J.L., Brown, E.J., Uk-Nham, S., Cannon, J.G., Blake, M.S., and Apicella, M.A. (2002). A co-operative interaction between *Neisseria gonorrhoeae* and complement receptor 3 mediates infection of primary cervical epithelial cells. *Cell. Microbiol.* **4**, 571–584.

Edwards, J.L., Shao, J.Q., Ault, K.A., and Apicella, M.A. (2000). *Neisseria gonorrhoeae* elicits membrane ruffling and cytoskeletal rearrangements upon infection of primary human endocervical and ectocervical cells. *Infect. Immun.* **68**, 5354–5363.

Estabrook, M.M., Griffiss, J.M., and Jarvis, G.A. (1997). Sialylation of *Neisseria meningitidis* lipooligosaccharide inhibits serum bactericidal activity by masking lacto-N-neotetraose. *Infect. Immun.* **65**, 4436–4444.

Evans, B.A. (1977). Ultrastructure study of cervical gonorrhea. *J. Infect. Dis.* **136**, 248–255.

Farrell, C.F. and Rest, R.F. (1990). Up-regulation of human neutrophil receptors for *Neisseria gonorrhoeae* expressing PII outer membrane proteins. *Infect Immun.* **58**, 2777–2784.

Forest, K.T., Dunham, S.A., Koomey, M., and Tainer, J.A. (1999). Crystallographic structure reveals phosphorylated pilin from *Neisseria:* phosphoserine sites modify type IV pilus surface chemistry and fiber morphology. *Mol. Microbiol.* **31**, 743–752.

Forest, K.T. and Tainer, J.A. (1997). Type-4 pilus-structure: outside to inside and top to bottom – a minireview. *Gene* **192**, 165–169.

Giardina, P.C., Williams, R., Lubaroff, D., and Apicella, M.A. (1998). *Neisseria gonorrhoeae* induces focal polymerization of actin in primary human urethral epithelium. *Infect. Immun.* **66**, 3416–3419.

Gorby, G.L., Clemens, C.M., Barley, L.R., and McGee, Z.A. (1991). Effect of human chorionic gonadotropin (hCG) on *Neisseria gonorrhoeae* invasion of and IgA secretion by human fallopian tube mucosa. *Microb. Pathogen.* **10**, 373–384.

Haines, K.A., Reibman, J., Tang, X., Blake, M.S., and Weissmann, G. (1991). Effects of protein I of *Neisseria gonorrhoeae* on neutrophil activation: generation of diacylglycerol from phosphatidylcholine via a specific phospholipase C is associated with exocytosis. *J. Cell Biol.* **114**, 433–442.

Haines, K.A., Yeh, L., Blake, M.S., Cristello, P., Korchak, H., and Weissmann, G. (1988). Protein I, a translocatable ion channel from *Neisseria gonorrhoeae*, selectively inhibits exocytosis from human neutrophils without inhibiting O_2^- generation. *J. Biol. Chem.* **263**, 945–951.

Harvey, H.A., Jennings, M.P., Campbell, C.A., Williams, R., and Apicella, M.A. (2001). Receptor-mediated endocytosis of *Neisseria gonorrhoeae* into primary human urethral epithelial cells: the role of the asialoglycoprotein receptor. *Mol. Microbiol.* **42**, 659–672.

Harvey, H.A., Porat, N., Campbell, C.A., Jennings, M.P., Gibson, B.W., Phillips, N.J., Apicella, M.A., and Blake, M.S. (2000). Gonococcal lipooligosaccharide is a ligand for the asialoglycoprotein receptor on human sperm. *Mol. Microbiol.* **36**, 1059–1070.

Harvey, H.A., Ketterer, M.R., Preston, A., Lubaroff, D., Williams, R., and Apicella, M.A. (1997). Ultrastructure analysis of primary human urethral epithelial cell cultures infected with *Neisseria gonorrhoeae*. *Infect. Immun.* **65**, 2420–2427.

Hedges, S.R., Sibley, D.A., Mayo, M.S., Hook, E.W. III, and Russell, M.W. (1998). Cytokine and antibody responses in women infected with *Neisseria gonorrhoeae*: effects of concomitant infections. *J. Infect. Dis.* **178**, 742–751.

Householder, T.C., Belli, W.A., Lissenden, S., Cole, J.A., and Clark, V.L. (1999). *cis-* and *trans-*acting elements involved in regulation of *aniA*, the gene encoding the major anaerobically induced outer membrane protein in *Neisseria gonorrhoeae*. *J. Bacteriol.* **181**, 541–551.

Jarvis, G.A. (1995). Recognition and control of neisserial infection by antibody and complement. *Trends Microbiol.* **3**, 198–201.

Jones, S.L., Knaus, U.G., Bokoch, G.M., and Brown, E.J. (1998). Two signaling mechanisms for activation of $\alpha_M \beta_2$ avidity in polymorphonuclear neutrophils. *J. Biol. Chem.* **273**, 10,556–10,566.

Källström, H., Islam, Md.S., Berggren, P.-O., and Jonsson, A.-B. (1998). Cell signaling by the type IV pili of pathogenic *Neisseria*. *J. Biol. Chem.* **273**, 21,777–21,782.

Källström, H., Liszewski, M.K., Atkinson, J.P., and Jonsson, A.-B. (1997). Membrane cofactor protein (MCP or CD46) is a cellular pilus receptor for pathogenic *Neisseria*. *Mol. Microbiol.* **25**, 639–647.

Lemon, S.M. and Sparling, P.F. (1999). Pathogenesis of sexually transmitted viral and bacterial infections. In *Sexually Transmitted Diseases*, ed. K.K. Holmes, P.-A. Mardh, P.F. Sparling, S.M. Lemon, W.E. Stamm, P. Piot, and J.N. Wasserheit, pp. 433–449. New York: McGraw-Hill.

Mandrell, R.E. and Apicella, M.A. (1993). Lipo-oligosaccharides (LOS) of mucosal pathogens: molecular mimicry and host-modification of LOS. *Immunobiology* **187**, 382–402.

Marceau, M., Forest, K., Béretti, J.-L., Tainer, J., and Nassif, X. (1998). Consequences of the loss of O-linked glycosylation of meningococcal type IV pilin on piliation and pilus-mediated adhesion. *Mol. Microbiol.* **27**, 705–715.

McGee, Z.A., Johnson, A.P., and Taylor-Robinson, D. (1981). Pathogenic mechanisms of *Neisseria gonorrhoeae*: observations on damage to human fallopian tubes in organ culture by gonococci of colony type 1 or type 4. *J. Infect. Dis.* **143**, 413–422.

Ober, W.B. (1970). Boswell's clap. *JAMA* **212**, 91–95.

Parge, H.E., Forest, K.T., Hickey, M.J., Christensen, D., Getzoff, E.D., and Tainer, J.A. (1995). Structure of the fibre-forming protein pilin at 2.6 A resolution. *Nature* **378**, 32–38.

van Putten, J.P.M., Duensing, T.D., and Carlson, J. (1998). Gonococcal invasion of epithelial cells driven by P.IA, a bacterial ion channel with GTP binding properties. *J. Exp. Med.* **188**, 941–952.

van Putten, J.P.M., Duensing, T.D., and Cole, R.L. (1998). Entry of Opa$^+$ gonococci into Hep-2 cells requires concerted action of glycosaminoglycans, fibronectin and integrin receptors. *Mol. Microbiol.* **29**, 369–379.

van Putten, J.P.M. and Robertson, B.D. (1995). Molecular mechanisms and implications for infection of lipopolysaccharide variation in *Neisseria*. *Mol. Microbiol.* **16**, 847–853.

Rahman, M., Källström, H., Normark, S., and Jonsson, A.-B. (1997). PilC of pathogenic *Neisseria* is associated with the bacterial cell surface. *Mol. Microbiol.* **25**, 11–25.

Ram, S., Cullinane, M., Blom, A.M., Gulati, S., McQuillen, D.P., Boden, R., Monks, B.G., O'Connell, C., Elkins, C., Pangburn, M.K., Dahlback, B., and Rice, P.A. (2001). C4bp binding to porin mediates stable serum resistance of *Neisseria gonorrhoeae*. *Intl. Immunopharmacol.* **1**, 423–432.

Ram, S., McQuillen, D.P., Gulati, S., Elkins, C., Pangburn, M.K., and Rice, P.A. (1998). Binding of complement factor H to loop 5 of porin protein 1A: a molecular mechanism of serum resistance of nonsialylated *Neisseria gonorrhoeae*. *J. Exp. Med.* **187**, 743–752.

Ram, S., Sharma, A.K., Simpson, S.D., Gulati, S., McQuillen, D.P., Pangburn, M.K., and Rice, P.A. (1998). A novel sialic acid binding site on factor H mediates serum resistance of sialylated *Neisseria gonorrhoeae*. *J. Exp. Med.* **187**, 743–752.

Schneider, H., Cross, A.S., Kuschner, R.A., Taylor, D.N., Sandoff, J.C., Boslego, J.W., and Deal, C.D. (1995). Experimental human gonococcal urethritis: 250 *Neisseria gonorrhoeae* MS11mkC are infective. *J. Infect. Dis.* **172**, 180–185.

Schneider, H., Schmidt, K.A., Skillman, D.R., Van De Verg, L., Warren, R.L., Wylie, H.J., Sandoff, J.C., Deal, C.D., and Cross, A.S. (1996). Sialylation lessens the infectivity of *Neisseria gonorrhoeae* MS11mkC. *J. Infect. Dis.* **173**, 1422–1427.

Schryvers, A.B. and Stojiljkovic, I. (1999). Iron acquisition systems in the pathogenic *Neisseria*. *Mol. Microbiol.* **32**, 1117–1123.

Snyder, L.A.S., Butcher, S.A., and Saunders, N.J. (2001). Comparative whole-genome analyses reveal over 100 putative phase variable genes in the pathogenic *Neisseria* spp. *Microbiology* **147**, 2321–2332.

Spence, J.M., Chen, J.C.-R., and Clark, V.L. (1997). A proposed role for the lutropin receptor in contact-inducible gonococcal invasion of Hec1B cells. *Infect. Immun.* **65**, 3736–3742.

Stimson, E., Virji, M., Makepeace, K., Dell, A., Morris, H.R., Payne, G., Saunders, J.R., Jennings, M.P., Barker, S., Panico, M., Bcench, I., and Moxon, E.R.

(1995). Meningococcal pilin: a glycoprotein substituted with digalaactosyl 2,4-diacetamido-2,4,6-trideoxyhexose. *Mol. Microbiol.* **17**, 1201–1214.

Sweet, R.L., Blankfort-Doyle, M., Robbie, M.O., and Schacter, J. (1986). The occurrence of chlamydial and gonococcal salpingitis during the menstrual cycle. *JAMA* **255**, 2062–2064.

Tobiason, D.M. and Seifert, H.S. (2001). Inverse relationship between pilus-mediated gonococcal adherence and surface expression of the pilus receptor, CD46. *Microbiology* **147**, 2333–2340.

Vogel, U. and Frosch, M. (1999). Mechanisms of neisserial serum resistance. *Mol. Microbiol.* **32**, 1133–1139.

Weiser, J.N., Goldberg, J.B., Pan, N., Wilson, L., and Virji, M. (1998). The phosphorylcholine epitope undergoes phase variation on a 43-kilodalton protein in *Pseudomonas aeruginosa* and on pili of *Neisseria meningitidis* and *Neisseria gonorrhoeae. Infect. Immun.* **66**, 4263–4267.

Wen, K.-K., Giardina, P.C., Blake, M.S., Edwards, J.L., Apicella, M.A., and Rubenstein, P.A. (2000). Interaction of the gonococcal porin P.IB with G-and F-actin. *Biochemistry* **39**, 8638–8647.

Williams, J.M., Chen, G.-C., Zhu, L., and Rest, R.F. (1998). Using the yeast two-hybrid system to identify human epithelial cell proteins that bind gonococcal Opa proteins: intracellular gonococci bind pyruvate kinase via their Opa proteins and require host pyruvate for growth. *Mol. Microbiol.* **27**, 171–186.

CHAPTER 8

Group A streptococcal invasion of host cells

Harry S. Courtney and Andreas Podbielski

Group A streptococci (GAS), *Streptococcus pyogenes*, are beta-hemolytic, Gram-positive, pyogenic cocci that usually grow in chains. *S. pyogenes* is classified as group A based on serological reactions with its C carbohydrate, which consists of polymers of rhamnose substituted with N-acetylglucosamine (Lancefield, 1933). For epidemiological purposes, GAS are further classified into more than 100 different types based on serological reactions with the variable domains of M proteins, or, more recently, based on 5' *emm* gene sequences (Beall et al., 1996). Other typing schemes based on serological reactions with serum opacity factor, T proteins, and R proteins are also used (Johnson and Kaplan, 1993).

S. *pyogenes* is almost exclusively associated with humans and commonly causes a variety of diseases, including pharyngotonsillitis, impetigo, scarlet fever, and more severe infections, such as puerperal sepsis, myositis, necrotizing fasciitis, and toxic shock syndrome. Among several of the nonsuppurative complications of group A streptococcal infections are acute rheumatic fever and acute glomerulonephritis, which are usually preceded by infections of the throat and skin, respectively. These sequelae are thought to be due to autoimmune T- and B-cell responses induced by streptococcal products. Accumulating evidence also suggests that group A streptococcal infections may lead to other autoimmune diseases, such as obsessive compulsive disorders, or they may exacerbate others such as guttate psoriasis (reviewed by Cunningham, 2000).

ADHESION: PRELUDE TO INVASION?

To establish these infections, the streptococcus must first attach to the epithelium of the host. This attachment is accomplished by specific interactions

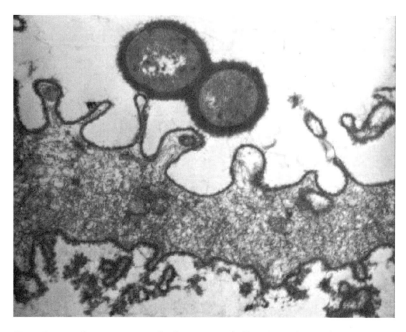

Figure 8.1. An electron micrograph of streptococcal adhesion to a human buccal epithelial cell.

between surface structures of the streptococcus (adhesins) and receptors on host cells (reviewed by Courtney et al., 2002). Attachment of streptococci to a human buccal epithelial cell is shown in Fig. 8.1. Note that the epithelial cell has a ruffled surface that is due to convoluted ridges that cover its surface and that appear as projections in electron micrographs of thin sections. Similar ridges are found on the surfaces of human pharyngeal and tonsillar epithelial cells (Stenfors et al., 2000). Streptococcal attachment to these cells occurs primarily by interactions with receptors on these ridges, and adhesion is often mediated by multiple interactions resulting in high-avidity binding.

Early adhesion experiments demonstrated that streptococci did not attach to all buccal cells but instead attached primarily to those cells coated with fibronectin (Abraham et al., 1983; Simpson and Beachey, 1983). These findings coupled with additional research confirmed that fibronectin was a receptor for GAS (Courtney et al., 1986). Lipoteichoic acid (LTA) was the first streptococcal adhesin to be identified that reacted with fibronectin (Beachey and Ofek, 1976; Courtney et al., 1986). Subsequently, at least 24 other adhesins have been described, many of which react with fibronectin or other

Table 8.1. *Putative streptococcal adhesins–invasins and receptors*

Adhesin–Invasin	Receptor	References
C5a peptidase	Fn	Cheng et al., 2002
C-carbohydrate	?	Botta, 1981
Collagen-binding protein	collagen	Podbielski et al., 1999; Visai et al., 1995
Fba(FbaA)	Fn	Terao et al., 2001
FbaB	Fn	Terao et al., 2002
FBP54	Fn, Fgn	Courtney et al., 1996
Galactose-binding protein	?	Gerlach et al., 1994
G3PDH(or SDH)	Fn, Fgn, 30-kDa protein	Pancholi & Fischetti, 1992, 1997
Hyaluronic acid	CD44	Ashbaugh et al., 2000; Cywes et al., 2000; Schrager et al., 1998; Wessels and Bronze, 1994
Lbp	laminin	Terao et al., 2002
Lsp	laminin	Elsner et al., 2002
Lipoteichoic acid	Fn	Beachey and Ofek, 1976; Courtney et al., 1986
	scavenger receptor, CD14	Dunne et al., 1994
M protein	CD46, galactose, Fn, laminin fucose/fucosylated glycoprotein sialic acid-containing receptors	Courtney et al., 1994; Okada et al., 1994, 1995; Perez-Casal et al., 1995; Berkower et al., 1999; Waostromauo Tylewska, 1982; Wang and Stinson, 1994 Ryan et al., 2001
PFBP	Fn	Rocha & Fischetti, 1999
Protein F1/Sfb1	Fn	Hanski & Caparon, 1992; Sela et al., 1993; Molinari et al., 1997; Talay et al., 1993
	integrins	Ozeri et al., 1998; Okada et al., 1998

(*cont.*)

Table 8.1. (*cont.*)

Adhesin–Invasin	Receptor	References
	caveolin-1	Rohde et al., 2003
	Fn–collagen interactions	Dinkla et al., 2003
Protein F2	Fn	Jaffe et al., 1996
Pullulanase	thyroglobulin, mucin, fetuin	Hytonen et al., 2003
R28	?	Stalhammar-Carlemalm et al., 1999
Scl1, (SclA)	?	Rasmussen et al., 2000;
Scl2 (SclB)	?	Rasmussen & Bjorck, 2001; Whatmore, 2001; Lukomski et al., 2001
SpeB	integrin	Stockbauer et al., 1999;
	laminin, mucin, fetuin, thyrogobulin	Hytonen et al., 2001
Sfbx	Fn	Jeng et al., 2003
SOF	Fn	Kreikemeyer et al., 1995; Rakonjac et al., 1995
Vn-binding protein	Vn	Valentin-Wiegand et al., 1988
28-kDa protein	Fn	Courtney et al., 1992
?	cytokeratin	Tamura & Nittayajarn, 2000
?	heparin sulfate	Duensing et al., 1999

Notes: Fn = fibronectin, Fgn = fibrinogen, Vn = vitronectin, PFBP = *S. pyogenes* Fn-binding protein, FBP54 = Fn-binding protein 54, G3PDH = glyceraldehyde-3-phosphate-dehydrogenase, SOF = serum opacity factor, Scl = streptococcal collagen-like protein, Sfbx = streptococcal Fn-binding protein x, Lbp = laminin-binding protein, Fba = Fn-binding protein a, FbaB = Fn-bindin protein B, and Lsp = lipoprotein of *S. pyogenes*, ? = Unknown

components of the extracellular matrix (Table 8.1). It has been proposed that adhesion of streptococci is a multistep process initiated by LTA, which facilitates the binding of a second adhesin, such as M protein or protein F (Hasty et al., 1992). The secondary adhesin provides host-cell specificity and leads to high-affinity attachment. The list of adhesins in Table 8.1 emphasizes the fact that GAS utilize multiple adhesins to mediate attachment to host cells. There are conflicting reports on some of these adhesins.

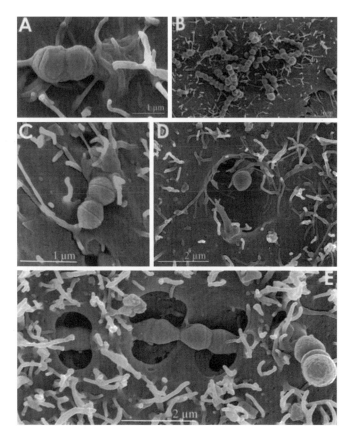

Figure 8.2. Scanning electron micrograph of GAS interacting with HEp-2 cells. M type 49 *S. pyogenes* (A and B) or its *nra* mutant (C–E) were added to monolayers of HEp-2 cells. The attachment and internalization of streptococci are shown after 1 h (A) and 3 h (B) of infection. The microvilli were in close contact with bacteria (A) and no significant morphological changes were apparent in HEp-2 cells incubated with wild-type streptococci (B). The *nra* mutant exhibited a similar interaction pattern after 1 h of infection (C), whereas drastic changes were observed after 3 h of infection (D and E). The *nra* mutant induced large invaginations and was found to enter one such invagination and exit through another. No such alterations were noted with the wild-type parent. (Reproduced with permission from Molinari et al., 2001.)

Some investigators found a role for M protein–CD46 interactions or hyaluronate–CD44 interactions in streptococcal adhesion to keratinocytes (Table 8.1), whereas others reported that inactivation of a variety of virulence factors, including M protein and capsule, did not decrease adhesion or colonization (Darmstadt et al., 2000; Jadoun et al., 2002; Ji et al., 1998).

Although there is, as yet, no entirely satisfactory explanation for all of these discrepancies, it is important to note that the findings for one serotype of GAS cannot necessarily be extrapolated to all other serotypes. Moreover, not all serotypes of GAS have the same array of adhesins and not all adhesins will be expressed simultaneously (McIver et al., 1995). Which adhesin or adhesins that will be used by a particular serotype of GAS will depend on its repertoire of adhesin genes, on environmental signals, and on the receptors expressed by a particular type of host cell.

Some, but not all, of the adhesins listed in Table 8.1 will initiate invasion of the host cells by streptococci. Whether or not a host cell is invaded will depend on the physiological status of the host cell and on the nature of the adhesin–receptor interactions. For example, the adhesion depicted in Fig. 8.1 probably will not lead to invasion because most of the cells in the superficial layer of the buccal epithelium are dead and ready to be desquamated. However, GAS have been found to invade human tonsillar cells *in vivo* (Stenfors et al., 2000) and to invade many types of tissue culture cells (Table 8.1). An example of adhesion that leads to internalization of streptococci is shown in Fig. 8.2.

PATHWAYS FOR INFECTION

There are several possible pathways for infection that streptococci may follow once they have attached to the epithelium (Fig. 8.3). One outcome is the persistence of the streptococci on the host epithelia. Under such circumstances, the bacteria will eventually face the activation of nonspecific, innate, and specific host defense mechanisms. Such effects, as well as a potential induction of bacterial "apoptosis," can result in the complete clearance of the bacteria from the infected host; thus the self-limiting nature of the infection. Alternatively, the individual may become an asymptomatic carrier of the streptococci. Although the phenotype of streptococci within carriers is not known, certain virulence factors are probably downregulated during carriage, resulting in less damage to the host and less activation of host defense mechanisms (Podbielski et al., 2003).

A second pathway is the invasion of tissues and the metastatic spread of the streptococci. This may be accomplished by the release of degradative factors and toxins that can destroy both the intercellular substances and eukaryotic cells, resulting in the spreading of the bacteria to new anatomical sites in the host. At the new sites, the bacteria are exposed to more favorable conditions and can multiply rapidly and continue to spread to adjacent tissues.

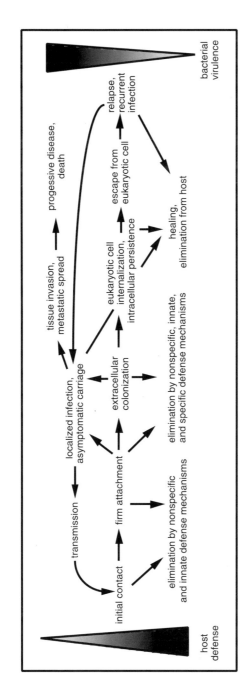

Figure 8.3. Model of the impact of adhesion, invasion, and intracellular persistence on the pathogenesis of streptococcal infections.

A third pathway that can occur during the infectious process is the internalization of the adherent bacteria by the host cells. In the intracellular space, the bacteria will survive provided they can evade or inhibit the activation of phagolysosomes. Thus, the intracellular space can be regarded as a sanctuary where the streptococci can avoid the defenses of the host or antibiotics. Indeed, the internalization of streptococci by host cells has been proposed as an explanation for recurrent infections after antibiotic treatment (Osterlund et al., 1997). Once inside the host cell, the streptococci probably downregulate degradative and lytic factors while upregulating "persistence" factors. In response to a signal, as yet unknown, GAS will escape from their intracellular sanctuary by inducing an apoptotic response in host cells. Another form of invasion has been described for *S. agalactiae*, which can transverse a polarized epithelium without altering its structural integrity (Kallman and Kihlstrom, 1997). However, no such transcytosis has been noted with GAS.

There are varying amounts of supporting data for each of these pathways. That GAS initiate adhesion and persist at a local site for some time before being eliminated has been observed since the days of ancient medicine and is regarded as a textbook standard (Bisno and Stevens, 2000). Similar observations have long been made on the invasion of host tissues and the metastatic spread of streptococci throughout the body of the host.

However, the concept of an intracellular location and the intracellular persistence of GAS is relatively new and does not have extensive, supporting, clinical data. In fact, there are only three studies in which the intracellular presence of *S. pyogenes* was directly demonstrated in tissue specimens from humans (Osterlund and Engstrand, 1997; Osterlund et al., 1997; Podbielski et al., 2003). There are a number of studies in which clinical isolates of GAS were tested for their *in vitro* capability to enter host cells, and even more groups have demonstrated the ability of GAS to be internalized by cultured cell lines (see Cleary and Cue, 2000, for review). The epidemiologic and experimental investigations have focused, so far, on the effect of an intracellular location on the persistence of GAS in upper respiratory tract infections. There is no evidence that persistent and recurrent GAS infections of the skin, especially of the lower legs, are associated with an intracellular location of the bacteria.

EXTRACELLULAR MECHANISMS OF TISSUE INVASION

As just described, streptococci can invade tissues and can be internalized by cells within these tissues (LaPenta et al., 1994). Tissue invasion and internalization are not mutually exclusive and may occur simultaneously. We first discuss the extracellular mechanisms utilized by GAS to invade tissues, and then we discuss the mechanisms required for internalization. One way that streptococci could invade tissues is to blast their way through the epithelial barrier by secreting cytolytic toxins that induce apoptosis of the host cells, thereby clearing a pathway to deeper tissues. Adhesion of streptococci to the host cell would help to effectively deliver the toxin(s) to the desired target (Ofek et al., 1990). Just such a mechanism has been described by Madden et al. (2001). These investigators found that the M-protein-mediated attachment of streptococci to keratinocytes was required for the insertion of streptolysin O (SLO) into cholesterol-rich domains in the cell membranes. The pore formed by SLO facilitated the delivery of streptococcal NAD-glycohydrolase (SPN) into the host cell, resulting in the lysis of host cells. There are several pathways whereby SPN may contribute to virulence. SPN can cleave β-NAD, resulting in a decrease in the intracellular pools of NAD and an increase in nicotinamide, a vasoactive compound. SPN also forms cyclic ADP-ribose and has ADP-ribosylating activities (Stevens et al., 2000). Expression of both SLO and SPN were required for lysis of host cells. Interestingly, *spn* is located adjacent to *slo*, and these genes appear to be cotranscribed from a common promoter (Madden et al., 2001).

Other factors, in addition to SLO and SPN, are involved in the lysis of host cells. Sierig et al. (2003) reported that both SLO and SLS (streptolysin S) contributed to the lysis of keratinocytes. These investigators also found that expression of SLO, but not SLS, impaired the killing of nonencapsulated *S. pyogenes* by neutrophils. The reason why SLO did not impair the killing of encapsulated streptococci is not known.

Another form of extracellular invasion, termed *paracellular translocation*, was recently described by Cywes and Wessels (2001). In this case, adhesion to human keratinocytes was mediated by interactions between the hyaluronic capsule of *S. pyogenes* and CD44, an integrin expressed on several cell types (for an integrin review, see Hynes, 2002). The adhesion of encapsulated streptococci induced membrane ruffling and disruption of intercellular junctions. In some instances, the cell membranes curled back and lifted away from adjacent cells, exposing a passage for streptococci around the cells and into deeper spaces. The soluble form of hyaluronate also induced similar responses in host cells. It was suggested that the binding of hyaluronate to CD44 activated Rac1, a member of the Rho family of GTPases, to phosphorylate a yet unidentified protein that recruited ezrin, an actin linker protein. These interactions triggered the cytoskeletal movements resulting in alterations of cell morphology and intercellular junctions.

In addition to the membrane effects described herein, this activation of the cytoskeleton also caused the formation of lamellipodia. Encapsulated streptococci were found to primarily associate with these lamellipodia, but such interactions did not lead to internalization of the streptococci. Interestingly, an acapsular mutant did not induce such changes, and these streptococci bound primarily to membrane sites devoid of the lamellipodia. The fates of these two different phenotypes of streptococci were also different. Instead of remaining extracellularly, the nonencapsulated streptococci were internalized by the keratinocytes. It was concluded that the hyaluronic acid capsule contributes to translocation by two mechanisms (Cywes and Wessels, 2001). In one, the hyaluronic acid capsule disrupts intercellular junctions and thereby facilitates the passage of streptococci into deeper tissues. In the second, the capsule prevents streptococci from being internalized and trapped within epithelial cells. As a corollary, the acapsular mutant failed to be efficiently translocated, because it became internalized and trapped within epithelial cells. It is interesting to note that the acapsular mutant was more cytolytic than the encapsulated wild type, presumably because the acapsular mutant was internalized more efficiently.

Another intriguing hypothesis concerning the role of ezrin and the streptococcal inhibitor of complement (SIC) in the fate of streptococci that invade human tissues was recently proposed by Hoe et al. (2002). Inactivation of the *sic* gene enhanced adhesion to host cells, suggesting that the SIC protein interfered with adhesion in some manner. SIC was found to rapidly enter epithelial cells and to react with ezrin and moesin. The mode of entry is not known, but perhaps the cytolysin-mediated transfer mechanism described by Madden et al. (2001) may have a role. Because ezrin, radixin, and moesin (ERM) are involved in the formation of lamellipodia, filopodia, and microvilli through linkages with actin filaments (Bretscher et al., 2002), it was suggested that the binding of SIC to ERM alters actin-mediated formation of microvilli at the cell surface. The implication is that such disruptions may also prevent receptor clustering at lamellipodia, which may be required for efficient adhesion.

SIC was also found to impair phagocytosis of the M1 serotype by neutrophils. It was proposed that the ability of SIC to impede adhesion provided a survival advantage within the host (Hoe et al., 2002). It is also possible that SIC could have a role in the dissemination of streptococci. Adhesion may occur while SIC is downregulated, and upregulation of SIC may release the streptococci and enhance their ability to spread to other tissues within the host or to other hosts. In this regard, it would be interesting to determine if the application of SIC to animals that are colonized with *S. pyogenes* would attenuate or enhance the virulence of the streptococci. It should be noted that *sic* has been found only in M1, M12, M55, and M57 serotypes (Akesson et al., 1996; Hartas and Sriprakash, 1999).

Although the effects of hyaluronic acid and SIC on adhesion were different, their effect on internalization of the streptococci was the same, that is, inhibition of uptake. In each case, the ERM proteins were found to be intimately involved. It is not yet clear how hyaluronate recruits actin to help in the formation of lamellipodia yet prevents actin from participating in the uptake of streptococci. The effects of hyaluronic acid and SIC on epithelial cells are similar to those seen in cells in which the gene for moesin was inactivated (Speck et al., 2003), which also resulted in the loss of polarity and intercellular junctions. In addition, these moesin-deficient cells were found to detach and migrate. This finding suggests an intriguing possibility – that the invasion of host cells by GAS could result in the detachment of a portion of these cells and their migration, resulting in the spread of the internalized streptococci to other tissues.

MECHANISMS FOR INTERNALIZATION OF STREPTOCOCCI BY HOST CELLS

Several different mechanisms have been described for the internalization of *S. pyogenes* by host cells. One of the most common mechanisms is that mediated by adhesins that interact with extracellular matrix proteins (Table 8.1). Our discussion is limited primarily to adhesins that are known to be involved in the internalization of the streptococci. Future research will undoubtedly reveal additional adhesins that contribute to the uptake of streptococci by host cells.

Protein F/Sfb1

Protein F1/Sfb1 is a large adhesin that is anchored to the bacterial cell surface by an LPXXG motif. Although the first description of this protein used the designation Sfb1 (Talay et al., 1991), most publications have used protein F1, which is the designation that is used in the following sections. Protein F1 promotes adhesion and invasion of host cells by interactions with fibronectin. The binding of fibronectin to protein F1 involves two binding domains – a repeating peptide domain that is immediately preceded by an upper binding domain (UBD). The effect of these domains on binding is sequentially exerted (Talay et al., 2000). The second binding step appears to trigger a conformational change in fibronectin that exposes its cell attachment domain (RGD), which is cryptic in soluble fibronectin (Tomasini-Johansson et al., 2001). The exposed RGD domain of fibronectin can then interact with integrins on the surface of the host cell and trigger internalization of the bacterium. Protein-F1-mediated invasion of host cells may also require interactions between the

UBD of protein F1 and the collagen-binding domain of fibronectin for the optimal uptake of streptococci (Dinkla et al. 2003). Recently, an alternative internalization mechanism involving protein F1 and caveolin-1 on the membranes of host cells and the formation of omega-shaped cavities was identified (Rohde et al., 2003).

Although the *in vitro* data on the involvement of protein F1 in the internalization process are very convincing, the epidemiological data are not as strong. Protein F1 is not an absolute requirement, because up to 48% of the GAS strains do not carry a *prtF1* gene (Natanson et al., 1995). Some investigators found an association between the expression of protein F1 and internalization (Neeman et al., 1998), whereas others found no such association (Brandt et al., 2001).

M proteins

M proteins are alpha-helical, coiled-coil proteins that are one of the major virulence factors of the organism and contribute to the ability of streptococci to evade phagocytosis and to attach to host cells (reviewed by Fischetti, 1989; Cunningham, 2000). A number of M proteins are adhesins (Table 8.1) and at least two of the M proteins, M1 and M3, bind fibronectin (Schmidt et al., 1993; Cue et al., 1998). An M type 1 *S. pyogenes* bound fibronectin and invaded human lung cell cultures, whereas an M-negative mutant had a reduced capacity to bind fibronectin and to invade these cells (Cue et al., 1998). Laminin was also found to promote invasion by the same M type 1 *S. pyogenes*, indicating that a single strain can utilize multiple pathways that lead to invasion of host cells. It was suggested that invasion is mediated by interactions between M protein and either fibronectin or laminin. The bound fibronectin then reacts with integrins on the cell surface that induce actin polymerization and uptake of streptococci by a zipper-like mechanism (Dombek et al., 1999). The M3 protein also mediates adhesion to HEp-2 cells and HaCat keratinocytes (Berkower et al., 1999). The expression of the M3 protein led to an increase in the invasion of HaCat cells, but the M3-protein-mediated invasion of HEp-2 was dependent on the presence of serum. Although it was not determined what component of serum was responsible for this invasion, it is likely that fibronectin will play a role similar to that found for M1 protein.

It is clear that certain M proteins that bind fibronectin can initiate adhesion and participate in the invasion process. However, most M proteins do not bind fibronectin. Do these M proteins contribute to invasion of host cells? Only a few such types of M proteins have been investigated for their role in the invasion process. The most widely investigated is the type M6 protein, but the results and conclusions of different investigators varied considerably. Fluckiger et al. (1998) found that inactivation of *emm6* had no effect on streptococcal adhesion but dramatically reduced invasion of Detroit 565 pharyngeal cells. However, a more common finding is that protein F is the major component responsible for invasion and that the M6 protein contributes to a lesser degree (Jadoun et al., 1997).

One of the ways that the M6 protein can contribute to invasion is to increase the numbers of adherent streptococci, which can lead to an increase in the numbers of streptococci that invade the host cells by a protein-F-mediated mechanism. This possibility is supported by the findings of Okahashi et al. (2003), who reported that inactivation of *emm6* reduced adhesion to mouse osteoblasts by ~90% but only reduced invasion of these cells by ~ 35%. In contrast, inactivation of *prtf* had no effect on adhesion but reduced invasion by ~90%. In addition, the introduction and expression of *emm6* in *S. gordonii* led to an increase in adhesion but not invasion of HEp-2 tissue culture cells (von Hunolstein et al., 2000). However, the introduction and expression of *emm6* in *Enterococcus faecalis* led to an increase in invasion of host cells (Okada et al., 1998).

From the foregoing paragraphs, it appears that the background in which M proteins are presented can have a dramatic impact on its role in the invasion of host cells. To evaluate the roles of several different M proteins in the same genetic background, Berkower et al. (1999) introduced *emm3*, *emm6*, and *emm18* into an isogenic M-negative and protein-F1-negative mutant and tested the recombinant strains in adhesion and invasion assays. Their results indicated that M3, M18, and M6 proteins do mediate adhesion to HEp-2 cells and to HaCat keratinocytes. These findings are similar to those of Courtney et al. (1997a), who found that introduction of *emm18* and *emm5* into an M-negative strain restored adhesion to HEp-2 cells. However, the role of M proteins in the invasion of host cells appears to depend on the type of host cells and on the serotype of the M protein. Berkower et al. (1999) reported that the expression of the M6 protein, but not M3 and M18 proteins, restored the ability to invade HEp-2 cells. In contrast, expression of both M6 and M3 proteins, but not the M18 protein, led to invasion of HaCat keratinocytes. The effect of serum on invasion was dependent on the M type. M6- and M18-protein-mediated invasion of HEp-2 cells was unaffected by serum, whereas invasion by streptococci expressing the M3 protein was enhanced by serum. These data indicate that the ability of M proteins to act as adhesins and invasins can be dependent on M type; that is, it will depend on the variable sequences in the N-terminus of the molecule (Fig. 8.4). Furthermore, streptococci expressing different M proteins can bind to different receptors on the

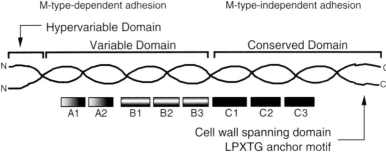

Figure 8.4. Schematic of M proteins and their role in adhesion and invasion of host cells. M protein is the primary virulence factor of *S. pyogenes* and confers the abilities to resist phagocytosis in blood and to attach to host cells. M protein is an α-helical, coiled-coil protein that emanates from the surface of the bacteria. The number and sequence of the A and B repeats will vary depending on the M type. The C repeats are conserved among different M types. Adhesion to host cells can be mediated by either the variable domain or the conserved domain, depending on the receptors that are expressed by host cells. Host cells expressing CD46 can react with the C-repeat domain and mediate adhesion that is independent of the M type. Adhesion and invasion mediated by the variable domain will depend on the type of M protein expressed by the streptococcus and on the kinds of receptors expressed by the targeted tissues. The binding of fibronectin to M proteins (and therefore the internalization of streptococci by host cells) is mediated by the variable domain of M proteins. (Reproduced with permission from Courtney et al., 2002.)

same host cell. Berkower et al. (1999) found that recombinant strains expressing the M6 protein did not inhibit adhesion–invasion of HaCat keratinocytes by a recombinant strain expressing the M18 protein, implying that the M6 and M18 proteins bind to different receptors on these keratinocytes.

Frick et al. (2000) described another property of M proteins that promotes adhesion and invasion. They found a 19 amino acid consensus sequence in certain M and M-like proteins that promotes interbacterial aggregation and the attachment of microcolonies to host cells. Such attachment also enhanced invasion of host cells. A synthetic peptide copying this aggregative sequence not only inhibited adhesion and invasion of host cells but also dramatically increased the survival of mice challenged with M type 1 *S. pyogenes*.

FbaA

FbaA, originally termed OrfX (Podbielski et al., 1996), is a 38-kDa protein with an LPXTG cell-anchoring motif (Terao et al., 2001). FbaA is expressed by M serotypes 1, 2, 4, 22, 28, and 49, but not by other M serotypes or by streptococcal groups B, C, and D. It has three or four proline-rich repeat domains that are highly homologous to the fibronectin-binding repeats found in FbpA. Similar to many other fibronectin-binding proteins, expression of FbaA promoted the streptococcal invasion of HEp-2 tissue culture cells. Although it was not determined how the binding of fibronectin to FbaA led to invasion, it is likely that fibronectin served as a cross-linking agent as described herein. A comparison of invasion of a multigene activator (Mga)-negative mutant and a FbaA-negative mutant with the parental M1 strain suggested that other surface proteins under the control of Mga may also be involved. This finding is not surprising, because others have reported that the M1 protein is under control of Mga and that it mediates adhesion and invasion of host cells (see earlier paragraphs). The FbaA-negative mutant was also less virulent in a mouse model than its wild-type parent, suggesting that FbaA is a virulence factor.

C5a peptidase

The scpA gene encodes the C5a peptidase, which degrades the chemoattractive complement factor C5a and thereby inhibits migration of neutrophils into an infected area. The C5a peptidase also binds fibronectin and is involved in adhesion and invasion of host cells by group B streptococci (Cheng et al., 2002). Whether it functions as an invasin in GAS remains to be demonstrated. The scpa gene is cotranscribed with the fbaA gene. Transcription attenuation in GAS was first demonstrated in the scpa promoter (Pritchard and Cleary, 1996) and later in the downstream fbaA promoter (Podbielski et al., 1996). There are no known stimuli that derepress this transcription attenuation.

FbaB and other RGD-containing proteins

FbaB is a newly described 79-kDa, fibronectin-binding protein found only in M3 and M18 serotypes of S. pyogenes (Terao et al., 2002). FbaB has a repeated peptide that binds fibronectin and has homology to the fibronectin-binding repeats of other streptococcal proteins. In addition, the N-terminal domain of FbaB has ~45% homology with an N-terminal region of protein F1, protein F2, and Cpa. Inactivation of the gene for FbaB resulted in decreased adhesion to and invasion of HEp-2 epithelial cells. The FbaB-negative mutant was less virulent in mice-challenged IP, indicating that FbaB is another virulence factor. Relative to its virulence, the FbaB protein was expressed mainly by type M3 and M18 S. pyogenes isolated from toxic-shock-like syndrome patients. No expression of FbaB was found in pharyngeal isolates of M types 3 and 18.

These findings suggest that FbaB may contribute to the invasive potential of GAS.

One interesting structural feature of this protein is that, in addition to its fibronectin-binding repeat peptide, FbaB also contains the RGD peptide. As already described, many of the adhesins interact with fibronectin, which then interacts, by means of its RGD domain, with integrins on host cell membranes. Thus, a streptococcal surface protein that contains the RGD motif may be able to bind directly to integrins without the need for a cross-linking agent. For example, a variant of SpeB that contained the RGD sequence is expressed by a highly virulent clinical isolate of M type 1 *S. pyogenes* (Stockbauer et al., 1999). This speB variant, called mSpeB2, was found to interact with human integrins $\alpha_v\beta_3$ and $\alpha_{IIb}\beta_3$ and to mediate adhesion to host cells. However, it is not known if the RGD sequence in FbaB or mSpeB2 has a direct role in the invasion of host cells.

Another RGD-containing protein of GAS is the Mac protein (Lei et al., 2002). This protein was so named because of its homology to the human Mac-1 leukocyte integrin, but it has also been termed IdeS for its IgG degrading activity (von Pawel-Rammingen et al., 2002). The streptococcal Mac protein is secreted and is involved in the resistance of streptococci to phagocytosis. Variants of Mac that contained the RGD sequence were found to bind to different integrins and to block uptake of streptococci by neutrophils, but it is not known if the secreted Mac interferes with adhesion and invasion of host cells other than professional phagocytes.

Alternative mechanisms for internalization

Invasion of host cells can also be triggered by interactions that do not involve integrins or fibronectin. Such alternative pathways could be serotype specific and involve the formation of large invaginations (Molinari et al., 2000). For the majority of strains, the responsible mechanisms remain to be elucidated. One potential bacterial receptor candidate is the Lsp (for lipoprotein of *S. pyogenes*) surface protein.

Lsp is a recently described surface molecule involved in adhesion and invasion of host cells by GAS (Elsner et al., 2002). Lsp is a member of the LraI-lipoprotein family that is expressed by a variety of streptococci. Lsp was found to bind directly to laminin, suggesting that laminin–Lsp interactions may have a role in adhesion and invasion. Lsp-negative mutants neither attached as well nor invaded host cells as well as the parental M49 strain. However, the Lsp-negative mutant also demonstrated reduced binding of fibronectin.

Table 8.2. *Responses of host cells to group A streptococci*

M type	Host cells	Response of host cells	References
M6	keratinocytes	IL-1α, IL-1β, IL-6, IL-8	Wang et al., 1997
		prostagland in E$_2$	Ruiz et al., 1998
M24	HEp-2 cells	IL-6	Courtney et al., 1997b
M?	HEp-2 cells	NF-κB	Medina et al., 2002
M49, M52	keratinocytes	β defensin 2	Dinulos et al., 2003
M22, M54, M59	mouse keratinocytes	cathelicidin	Dorschner et al., 2001
M1	A-549, HEp-2 cells	apoptosis	Tsai et al., 1999
M3	human blood cells	CD111b, ROS, $\downarrow$ CD62L	Saetre et al., 2000
M1	human mononuclear cells	IL-1β, IL-6, TNFα, IL-12, IL-18, IFN-γ	Miettinen et al., 1998
M6	Detroit 562, FaDu cells	phosphorylation	Pancholi & Fischetti, 1997
M6	HEp-2 cells	caspase 9, apoptosis	Nakagawa et al., 2001
M49	HEp-2 cells, keratinocytes	annexin V, caspases 2, 3, 7	Kreikemeyer et al., 2003

Because Lsp does not react directly with fibronectin, this suggested that, in addition to Lsp, the expression of other surface proteins may have been altered in the Lsp-negative mutant. Thus, the role of Lsp in streptococcal adhesion and invasion remains to be clarified.

RESPONSES OF THE HOST TO STREPTOCOCCAL ADHESION AND INVASION

The attachment of GAS to host cells is known to stimulate responses from host cells (Table 8.2). For example, the adhesion of an M type 24 *S. pyogenes* to HEp-2 cells resulted in the release of IL-6, whereas an isogenic, M24-protein-negative mutant did not attach to these cells and failed to stimulate the release of IL-6 (Courtney et al., 1997). Similarly, the adhesion of a type M6 *S. pyogenes* to cultures of HaCat keratinocytes caused a dramatic increase in the release of IL-1α, IL-1β, IL-6, IL-8, and prostaglandin E$_2$ (Wang et al., 1997; Ruiz et al., 1998). In contrast, an M6 protein-negative mutant either caused no such increase or did so at later time points.

Wang et al. (1997) reported that adherent streptococci caused injury to keratinocytes that coincided with the release of cytokines. Subsequent research by Ruiz et al. (1998) indicated that both SLO expression and adhesion were required for the stimulation of IL-6, IL-8, and prostaglandin E$_2$ expression in HaCat keratinocytes. The role of SPN in the SLO-mediated stimulation of the cytokines was not determined. Interestingly, expression of SLO had no impact on the stimulation of IL-1 by adherent streptococci (Ruiz et al., 1998), suggesting that other streptococcal products may stimulate the IL-1 response. Indeed, there are a number of streptococcal products including LTA, DNA, and a variety of streptococcal pyrogenic exotoxins that can evoke cytokine responses in the host. These responses are thought to be a central cause of morbidity and mortality associated with toxic shock syndrome and necrotizing fasciitis.

Streptococci can also stimulate phosphorylation of proteins in host cells. SDH (streptococcal dehydrogenase) was identified as one streptococcal protein that can initiate phosphorylation events in host cells (Pancholi and Fischetti, 1992, 1997). SDH is a 35.8-kDa glyceraldhyde-3-phosphate dehydrogenase that is expressed on the surface of streptococci. It also has ADP-ribosylating activity and binds to a variety of mammalian proteins. SDH was found to react with 32-kDa membrane proteins in Detroit 562 and FaDu tissue culture cells derived from human pharyngeal carcinomas. The interaction of SDH and M type 6 *S. pyogenes* with these pharyngeal cells stimulated the phosphorylation of histone H3. SDH did not induce phosphorylation in other tissue culture cells such as Chang cells, Chinese hamster ovarian cells, and human epitheloid carcinoma cells. It was postulated that the induction of tyrosine kinases and protein kinase C may regulate actin polymerization and invasion of host cells by streptococci. The findings that genistein, a tyrosine-kinase inhibitor, and staurosporine, a serine/threonine-kinase inhibitor, had no inhibitory effect on adhesion, yet almost completely blocked invasion, clearly indicate that phosphorylation of host proteins is required for invasion (Pancholi and Fischetti, 1997).

The adhesion of streptococci induces other host cell responses, including the production of IL-1α and TNF-α (Table 8.2). Darmstadt et al. (1999) demonstrated that, in addition to their proinflammatory role, these cytokines inhibited the adhesion of *S. pyogenes* to keratinocytes. The mechanism whereby IL-1α and TNF-α blocked adhesion is not known.

The interaction of streptococci with host cells also enhanced the expression of the transcription factor known as nuclear factor kappa B (NF-κB; Medina et al., 2002; Okahashi et al., 2003). Adhesion induced an initial spike of transcription, whereas internalization of the streptococci resulted in a sustained NF-κB response that was blocked by cytochalisin D. LTA is one of the streptococcal components that can contribute to this response. LTA induced the expression of NF-κB and COX 2 expression in a human pulmonary cell line, and this expression was blocked by genistein (Lin et al., 2002), again indicating a role for tyrosine kinases in host cell responses to streptococci and their products.

Bacterial interactions with host cells can also induce the expression of antibacterial peptides, which is part of the innate immune defenses. There are presently four main groups of antibacterial peptides: α-defensins, β-defensins, θ-defensins, and cathelicidins. GAS stimulated the production of cathelicidin *in vivo* in mouse keratinocytes (Dorschner et al., 2001). Cathelicidin in the 1-μM range inhibited the growth of *S. pyogenes*, indicating that it has the potential to prevent or mitigate streptococcal infections. To evaluate this potential, Nizet et al. (2001) compared the invasiveness of GAS in *Cnlp*-null mice and wild-type mice by using a subcutaneous model of infection. *Cnlp* encodes the antimicrobial peptide CRAMP, the mouse form of cathelicidin. The lesions induced by *S. pyogenes* in wild-type mice were almost completely resolved by day 8 but were still readily apparent in *Cnlp*-null mice. These investigators also showed that a CRAMP-resistant strain of *S. pyogenes* caused greater necrosis than wild-type *S. pyogenes* in their subcutaneous model of infection. These data indicate that cathelicidins are an important part of the innate immune response to streptococcal infections of the skin.

GAS also induced, albeit poorly, the production of β-defensin 2 in human keratinocytes (Dinulos et al., 2003). That *S. pyogenes* is a poor stimulator of β-defensins is somewhat unexpected, because *S. pyogenes* induces the expression of NF-κB (Medina et al., 2002), which can upregulate the expression of defensins. However, it was suggested that this lack of stimulation of β-defensins could contribute to the virulence of *S. pyogenes* in skin infections, because β-defensin 2 is a potent killer of *S. pyogenes*.

RESPONSES OF GAS TO INTERACTIONS WITH HOST CELLS

The interaction of streptococci with host cells not only induces a response in the host cell but also induces a response in the streptococci. Broudy et al. (2001, 2002) reported that coculturing a type M6 *S. pyogenes* with Detroit 562 pharyngeal cells induced the expression of a phage-encoded streptococcal pyrogenic exotoxin C (SpeC) and DNase (called Spd1). SpeC and Spd1 are coexpressed from the same transcript. The coculturing of *S. pyogenes* strain CS112 with pharyngeal cells also resulted in the induction of lysogenic bacteriophage particles. These investigators went on to show that a factor, released by the pharyngeal cells, was responsible for these responses. The soluble factor was <10 kDa and resistant to heat and proteolytic degradation.

The expression of a variety of virulence factors is upregulated when GAS invade the host. The bacterial factors regulated in response to invasion versus those regulated in response to internalization with host cells have to be distinguished from each other. Investigators have long known that expression of M protein increases when the bacteria enter the bloodstream. Gryllos et al. (2001) found that introduction of GAS into the pharynx of baboons induced expression of the hyaluronic acid capsule. A similar increase in capsule production was found when the streptococci were introduced into the blood and the peritoneal cavity of mice. However, it is not entirely clear if this induction of virulence factors streptococci is in response to signals from the host or is due to the growth phase of the streptococci. The expression of the capsule and a variety of other virulence factors are upregulated during the exponential growth phase and downregulated during the stationary phase (McIver and Scott, 1997; Gryllos et al., 2001).

Only recently, some data on bacterial virulence gene expression were collected from GAS during their internalization or intracellular persistence. During such behavior, it seems favorable for the bacteria to suppress the expression of secreted lytic factors. This suppression is at least partially controlled by the negative global regulator Nra (Podbielski et al., 1999), a member of the RALP (RofA-like proteins) regulator family (Fogg et al., 1997; Granok et al., 2000). Inactivation of Nra in an M49 strain greatly increased the cytolysis of HEp-2 cells. Utilizing whole genome array hybridization, Nra was shown to regulate between 120 and 290 genes depending on the growth phase and filtering levels used (Podbielski et al., unpublished results). Approximately two thirds of the genes are negatively regulated by Nra and include the capsule synthesis operon *has*, the genes for surface factors such as protein F2, Cpa, ScpA, SOF/SfbII, SfbX, SclA, and SclB; and genes for secreted proteins such as the CAMP factor, SLS, SpeB, and SpeA. Consequently, inactivation of *nra* resulted in an increase in the expression of these genes. However, the expression of SLO was unaffected. It is interesting to note that the Nra-negative mutant induced large invaginations in the membranes of HEp-2 cells, and chains of streptococci were found to enter one such invagination and resurface through another (Fig. 8.2). No such phenomenon was seen with the parent.

Another predominantly negative regulator is encoded by the Fas-regulon. It comprises two histidine kinases, FasB/FasC, the response regulator, FasA, and an effector RNA, FasX. Judged by its sequence, its temporal expression pattern with a maximum at the end of the exponential growth phase, and the type of dependent genes (positively regulated secreted factors such as SLS, streptokinase, and SpeB; negatively regulated surface proteins–adhesins such as the M-related protein, SCPA, and FBP54), the Fas-regulon resembles the *Staphylococcus aureus* Agr-regulon. However, unlike the Agr-regulon, the Fas-regulon has not been demonstrated to be involved in a cell-density-dependent type of regulation (Kreikemeyer et al., 2001). A recent whole genome array hybridization analysis revealed that 18 to 90 genes are influenced by a functional Fas-regulon. This fact, as well as the potential convergence of several signaling pathways in the two histidine kinase sensors of the Fas-regulon, indicate that there is probably more than growth-phase-dependent regulation that is exerted by this regulon.

By using various Fas-regulon mutants and complemented strains, researchers recently showed that the Fas-regulon exerts a negative control of bacterial fibronectin binding and adhesion to keratinocytes and HEp-2 epithelial cells – albeit at different time points of the growth curve (keratinocytes, exponential phase; HEp-2 cells, stationary phase). Inactivation of the Fas-regulon not only enhanced adhesion but also increased internalization of the streptococci. Once internalized, the mutants exhibited a strongly reduced induction of apoptotic pathways as measured by annexin V, and caspase-2, -3, and -7 expression (Kreikemeyer et al., unpublished data). The GAS isolate used for this analysis was a serotype M49 strain. Utilizing a serotype M6 strain, Nagakawa et al. (2001) found a somewhat different pattern of apoptosis induction that relied on caspase-9 activation and subsequent selective degradation of mitochondrial functions. Because the M6 strain also carried a typical Fas-regulon, there could more than one way that GAS induce apoptosis.

GAS exhibit the highest tendency for internalization once they enter the stationary phase. The Fas-regulon and the *nra* gene are maximally transcribed during the transition from exponential to stationary growth phase. This temporal expression profile fits well into the concept of FasBCAX and Nra (and other RALPs, Beckert et al., 2001) as central regulators that keep the delicate balance between extracellular aggressiveness and intracellular "tameness" that would be necessary for the bacteria to best survive in these different environments.

Voyich et al. (2003) used a gene array encompassing the whole genome of an M1 serotype of GAS to probe gene expression during phagocytosis by human neutrophils. A two-component regulatory system, Ihb–Irr regulon, was identified that was expressed during phagocytosis. The Ihb–Irr system played a fundamental role in the survival of GAS within phagocytes. In addition, the *irr* gene was highly expressed by multiple serotypes of GAS during infections of the human pharynx, indicating that it regulates gene expression in response to interactions with the host *in vivo*.

INTRASPECIES AND INTERSPECIES SIGNALING BETWEEN STREPTOCOCCI

Cell-to-cell signaling among bacteria is a relatively new discovery and has been referred to as quorum sensing to indicate that thresholds of signals are achieved only when a sufficient density of bacteria is reached. These signaling peptides have also been called pheromones because of their role in regulating competence and conjugation. For many Gram-positive bacteria, this signal will consist of small, linear, or cyclic peptides (reviewed by Dunny and Leonard, 1997). Upton et al. (2001) described a lantibiotic peptide, salivaricin A (SalA), that is secreted by streptococci and whose expression is autoregulated. The secreted form of this peptide, called SalA1 in GAS, not only upregulated expression of *salA1* in *S. pyogenes* but also upregulated the expression of *salA* in *S. salivarius*. Similarly, SalA upregulates expression of *salA1* in *S. pyogenes*. In addition to autoregulation, SalA/SalA1 also induces the expression of an immunity factor that is required for resistance to the antibiotic effect of SalA. Thus, strains of streptococci that do not produce SalA/SalA1 are sensitive to its antibiotic effect. Hence, strains of streptococci that express SalA/SalA1 could modulate the bacterial population in the host. Although the effect of adhesion on secretion of SalA has not been determined, adhesion may enhance secretion as a result of an increase in the density of the bacterial population. In addition, adherent bacteria can multiply and form microcolonies that are very dense.

Another factor that seems to be involved in interbacterial signaling with effects on the GAS virulence in animals, as well as on a putative natural competence of the bacteria, is the Sil/Blp-like system (Hidalgo-Grass et al., 2002; Smoot et al., 2002). The *sil*A-E genes encode a regulatory secreted peptide, a potential processing machinery, a two-component regulatory system, and another protein with an unknown function. The *sil* locus resembles the *S. aureus agr* and *S. pneumoniae com* quorum-sensing regulons. Inactivation of *silC* reduced the invasiveness of an *emm14* strain of *S. pyogenes* in a mouse model. Coinjection of the *silC* mutant with an avirulent *emm14*-negative mutant restored virulence and invasiveness. It was suggested that the *emm14*-negative mutant complemented the defect of the *silC* mutant by secreting a signaling peptide that activated *sil*A/B, which then activated other genes required for invasion. So far, *sil* has been identified in only two serotype strains, M14 and M18. It apparently resides on a 13.8-kb genomic island that is not contained in two other genomes of *S. pyogenes*. Exactly how the various components of the *sil* locus interact with each other, and with other genes involved in the infectious process, remains to be elucidated. Sil may be a target for novel therapies because of its involvement in the regulation of factors required for invasion.

ASSAY FOR INTERNALIZATION

The most widely used assay for internalization of bacteria by host cells is the antibiotic protection assay. This assay is based on the assumptions that internalized bacteria will not be killed by the addition of antibiotics to the medium, and that only those bacteria on the surface will be killed. A measure of both adhesion and internalization can be obtained by determining the total number of CFU before and after treatment with antibiotics. However, investigators utilizing this assay should be aware of a potential problem that has been described by Molinari et al. (2001). These investigators found a low rate of internalization of streptococci using the antibiotic protection assay, but double-staining, immunofluorescent, microscopic analysis indicated that the internalization rate was actually high. It was suggested that the host cell membrane is damaged during invasion, allowing antibiotics to enter host cells and kill the bacteria. Thus, investigators need to ensure that the integrity of the host cell membrane is maintained during the assay to obtain valid results. The antibiotic protection assay also cannot distinguish between the rate of uptake of bacteria and the rate of intracellular multiplication, or between adherent bacteria and those that have been internalized and released by apoptosis.

FUTURE CONSIDERATIONS

It is clear that many types of eukaryotic cells can internalize streptococci and that GAS can invade host cells *in vivo*. Some of the molecular mechanisms involved in these processes have been identified. Additional research has to be done to sort out the adhesins and mechanisms of invasion that are truly involved in streptococcal infections from those that have a role only under *in vitro* conditions. Much progress has been made in these directions, but there are many remaining questions. How long can GAS persist intracellularly in humans? Do intracellular GAS multiply or are they dormant? Do L-forms of GAS have a role in intracellular persistence? Do intracellular GAS have the means to initiate their efflux from the cell without apoptosis? If so, are there specific stimuli for the bacterial switch between persistence and efflux? What are the genes involved in persistence versus those involved in tissue invasion?

A hint to the answers of some of these questions has come from research on bacterial regulation, which has identified a number of genes that are regulated in response to internalization. Whole genome gene arrays in concert with real-time reverse transcription polymerase chain reaction (RT-PCR) analyses hold great promise in identifying genes required for invasion of tissues, metastatic spreading, internalization, persistence, and resistance to innate defense mechanisms. These approaches are only possible as a result of the genome sequencing work of Ferretti et al. (2001), Smoot et al. (2002), and Beres et al. (2002). Currently being conducted are gene array analyses of the genomes of M1, M3, M18, and M49, and real-time RT-PCR analyses of *fas* and *nra* at three different phases of growth (Podbielski et al, unpublished work). Undoubtedly, these types of investigative approaches will generate a wealth of information, a better understanding of streptococcal virulence, and novel directions to pursue for preventive therapies.

REFERENCES

Abraham, S.N., Beachey, E.H., and Simpson, W.A. (1983). Adherence of *Streptococcus pyogenes*, *Escherichia coli*, and *Pseudomonas aeruginosa* to fibronectin-coated and uncoated epithelial cells. *Infect. Immun.* **41**, 1261–1268.

Akesson, P., Sjoholm, A.G., and Bjorck, L. (1996). Protein SIC, a novel extracellular protein of *Streptococcus pyogenes* interfering with complement function. *J. Biol. Chem.* **271**, 1081–1088.

Ashbaugh, C., Moser, T., Shearer, M., White, G., Kennedy, R., and Wessels, M. (2000). Bacterial determinants of persistent throat colonization and the associated immune response in a primate model of human group A streptococcal pharyngeal infection. *Cell. Microbiol.* **2**, 283–292.

Beachey, E.H. and Ofek, I. (1976). Epithelial cell binding of group A streptococci by lipoteichoic acid on fimbriae denuded of M protein. *J. Exp. Med.* **143**, 759–771.

Beall, B., Facklam, R., and Thompson, T. (1996). Sequencing *emm*-specific PCR products for routine and accurate typing of group A streptococci. *J. Clin. Microbiol.* **34**, 953–958.

Beckert, S., Kreikmeyer, B., and Podbielski, A. (2001). Group A streptococcal *rofA* gene is involved in the control of several virulence genes and eucaryotic cell attachment and internalization. *Infect. Immun.* **69**, 534–537.

Beres, S., Sylva, G., Barbian, K., Lei, B., Hoff, J., Mammarella, N., Liu, M., Smoot, J., Porcella, S., Parkins, L., Campbell, D., Smith, T., McCormick, J., Leung, D., Schlievert, P., and Musser, J.M. (2002). Genome sequence of a serotype M3 strain of group A streptococcus: phage-encoded toxins, the high-virulence phenotype, and clone emergence. *Proc. Natl Acad. Sci. USA* **99**, 10,078–10,083.

Berkower, C., Ravins, M., Moses, A., and Hanski, E. (1999). Expression of different group A streptococcal M proteins in an isogenic background demonstrates diversity in adherence to and invasion of eukaryotic cells. *Mol. Microbiol.* **31**, 1463–1475.

Bisno, A. and Stevens, D. (2000). *Streptococcus pyogenes* (including streptococcal shock and necrotizing fasciitis). In *Principles and Practice of Infectious Diseases*, ed, Mandell, G., Bennet, J., and Dolin, R. pp. 2102–2117. Philadelphia, Churchill Livingstone.

Botta, G. (1981). Surface components in adhesion of group A streptococci to pharyngeal epithelial cells. *Curr. Microbiol.* **6**, 101–104.

Brandt, C., Allerberger, F., Spellerberg, B., Holland, R., Lutticken, R., and Haase, G. (2001). Characteristics of consecutive *Streptococcus pyogenes* isolates from patients with pharyngitis and bacteriological treatment failure: special reference to *prt*F1 and *sic/drs. J. Infect. Dis.* **183**, 670–674.

Bretscher, A., Edwards, K., and Fehon, R.G. (2002). ERM proteins and merlin: integrators at the cell cortex. *Nat. Rev. Mol. Cell. Biol.* **3**, 586–599.

Broudy, T., Pancholi, V., and Fischetti, V. (2001). Induction of lysogenic bacteriophage and phage-associated toxin from group A streptococci during co-culture with human pharyngeal cells. *Infect. Immun.* **69**, 1440–1443.

Broudy, T., Pancholi, V., and Fischetti, V. (2002). The *in vitro* interaction of *Streptococcus pyogenes* with human pharyngeal cells induces a phage-encoded extracellular DNase. *Infect. Immun.* **70**, 2805–2811.

Cheng, Q., Stafslien, D., Purushothaman, S., and Cleary, P. (2002). The group B streptococcal C5a peptidase is both a specific protease and invasin. *Infect. Immun.* **70**, 2408–2413.

Cleary, P. and Cue, D. (2000). High frequency invasion of mammalian cells by β hemolytic streptococci. *Subcell. Biochem.* **33**, 137–166.

Courtney, H.S., Bronze, M.S., Dale, J.B., and Hasty, D.L. (1994). Analysis of the role of M24 protein in group A streptococcal adhesion and colonization by use of Ω-interposon mutagenesis. *Infect. Immun.* **62**, 4868–4873.

Courtney, H.S., Dale, J.B., and Hasty, D.L. (1996). Differential effects of the streptococcal fibronectin-binding protein, FBP54, on adhesion of group A streptococci to human buccal cells and HEp-2 tissue culture cells. *Infect. Immun.* **64**, 2415–2419.

Courtney, H.S., Hasty, D.L., and Dale, J.B. (2002). Molecular mechanisms of adhesion, colonization, and invasion of group A streptococci. *Ann. Med.* **34**, 77–87.

Courtney, H.S., Hasty, D.L., Dale, J.B., and Poirier, T.P. (1992). A 28 kilodalton fibronectin-binding protein of group A streptococci. *Curr. Microbiol.* **25**, 245–250.

Courtney, H.S., Liu, S., Dale, J.B., and Hasty, D.L. (1997a). Conversion of M serotype 24 of *Streptococcus pyogenes* to M serotypes 5 and 18: effect on resistance to phagocytosis and adhesion to host cells. *Infect. Immun.* **65**, 2472–2474.

Courtney, H.S., Ofek, I., Hasty, and D.L. (1997b). M protein mediated adhesion of M type 24 *Streptococcus pyogenes* stimulates release of interleukin-6 by HEp-2 tissue culture cells. *FEMS Microbiol. Lett.* **151**, 65–70.

Courtney, H.S., Ofek, I., Simpson, W.A., Hasty, D.L., and Beachey, E.H. (1986). Binding of *Streptococcus pyogenes* to soluble and insoluble fibronectin. *Infect. Immun.* **53**, 454–459.

Cue, D., Dombek, P., Lam, H., and Cleary, P. (1998). *Streptococcus pyogenes* serotype M1 encodes multiple pathways for entry into human epithelial cells. *Infect. Immun.* **66**, 4593–4601.

Cunningham, M. (2000). Pathogenesis of group A streptococcal infections. *Clin. Microbiol. Rev.* **13**, 470–511.

Cywes, C., Stamenkovic, I., and Wessels, M.R. (2000). CD44 as a receptor for colonization of the pharynx by group A streptococci. *J. Clin. Invest.* **106**, 995–1002.

Cywes, C. and Wessels, M. (2001). Group A streptococcus tissue invasion by CD-44-mediated cell signalling. *Nature* **414**, 648–652.

Darmstadt, G.L., Fleckman, P., and Rubens, C.E. (1999). Tumor necrosis factor-alpha and interleukin-1alpha decrease the adherence of *Streptococcus pyogenes* to cultured keratinocytes. *J. Infect. Dis.* **180**, 1718–1721.

Darmstadt, G.L., Mentele, L., Podbielski, A., and Rubens, C.E. (2000). Role of group A streptococcal virulence factors in adherence to keratinocytes. *Infect. Immun.* **68**, 1215–1221.

Dinkla, K., Rohde, M., Jansen, W., Carapetis, J., Chhatwal, G., and Talay, S. (2003). *Streptococcus pyogenes* recruits collagen via surface-bound fibronectin: a novel colonization and immune evasion mechanism. *Mol. Microbiol.* **47**, 861–869.

Dinulos, J., Mentele, L., Fredericks, L., Dale, B., and Darmstadt, G. (2003). Keratinocyte expression of human β defensin 2 following bacterial infection: role in cutaneous host defense. *Clin. Diagnost. Lab. Immunol.* **10**, 161–166.

Dombek, P., Cue, D., Sedgewick, J., Ruschkowski, S., Finlay, B., and Cleary, P. (1999). High frequency intracellular invasion of epithelial cells by serotype M1 group A streptococci: M1 protein-mediated invasion and cytoskeletal rearrangements. *Mol. Microbiol.* **31**, 859–70.

Dorschner, R., Pestonjamasp, V., Tamakuwala, S., Ohtake, T., Rudisill, J., Nizet, V., Agerberth, B., Gudmundsson, G., and Gallo, R. (2001). Cutaneous injury induces the release of cathelicidin anti-microbial peptides active against group A streptococcus. *J. Invest. Dermatol.* **117**, 91–97.

Duensing, T., Wing, J., and van Putten, J. (1999). Sulfated polysaccharide-directed recruitment of mammalian host proteins: a novel strategy in microbial pathogenesis. *Infect. Immun.* **67**, 4463–4468.

Dunne, D.W., Resnick, D., Greenberg, D.J., Kreiger, J.M., and Joiner, K.A. (1994). The type 1 macrophage scavenger receptor binds to Gram-positive bacteria and recognizes lipoteichoic acid. *Proc. Natl Acad. Sci. USA* **91**, 1863–1867.

Dunny, G. and Leonard, B. (1997). Cell-cell communication in gram-positive bacteria. *Annu. Rev. Microbiol.* **51**, 527–564.

Elsner, A., Kreikmeyer, B., Braukn-Kiewnick, A., Spellerberg, B., Buttaro, B., and Podbielski, A. (2002). Involvement of Lsp, a member of the LraI-lipoprotein family in *Streptococcus pyogenes*, in eukaryotic cell adhesion and internalization. *Infect. Immun.* **70**, 4859–4869.

Ferretti, J., McShan, W., Ajdic, D., Savic, D., Savic, G., Lyon, K., Primeaux, C., Sezate, S. Suvorov, A., Kenton, S., Lai, H.S., Lin, S.P., Qian, Y., Jia, H.G., Najar, F.Z., Ren, Q., Zhu, N., Song, L., White, J., Yuan, X., Clifton, S.W., Roe, B.A., and McLaughlin, R. (2001). Complete genome sequence of an M1 strain of *Streptococcus pyogenes*. *Proc. Natl Acad. Sci. USA* **98**, 4658–4663.

Fischetti, V.A. (1989). Streptococcal M protein: molecular design and biological behaviour. *Clin. Microbiol.* **2**, 285–314.

Fluckiger, U., Jones, K., and Fischetti, V. (1998). Immunoglobulins to group A streptococcal surface molecules decrease adherence to and invasion of human pharyngeal cells. *Infect. Immun.* **66**, 974–979.

Fogg, G.C. and Caparon, M.G. (1997). Constitutive expression of fibronectin binding in *Streptococcus pyogenes* as a result of anaerobic activation of *rofA*. *J. Bacteriol.* **179**, 6172–6180.

Frick, I., Morgelin, M., and Bjorck, L. (2000). Virulent aggregates of *Streptococcus pyogenes* are generated by homophillic protein-protein interactions. *Mol. Microbiol.* **37**, 1232–1247.

Gerlach, D., Schalen, C., Tigyi, Z., Nilsson, B., Forsgren, A., and Naidu, A.S. (1994). Identification of a novel lectin in *Streptococcus pyogenes* and its possible role in bacterial adherence to pharyngeal cells. *Curr. Microbiol.* **28**, 331–338.

Granok, A., Parsonage, D., Ross, R., and Caparon, M. (2000). The RofA binding site in *Streptococcus pyogenes* is utilized in multiple transcriptional pathways. *J. Bacteriol.* **182**, 1529–1540.

Gryllos, I., Cywes, C., Shearer, M., Cary, M., Kennedy, R., and Wessels, M. (2001). Regulation of capsule gene expression by group A streptococcus during pharyngeal colonization and invasive infection. *Mol. Microbiol.* **42**, 61–74.

Hanski, E. and Caparon, M. (1992). Protein F, a fibronectin-binding protein, is an adhesin of the group A streptococcus, *Streptococcus pyogenes. Proc. Natl Acad. Sci. USA* **89**, 6172–6176.

Hartas, J. and Sriprakash, K. (1999). *Streptococcus pyogenes* strains containing *emm12* and *emm55* possess a novel gene coding for distantly related SIC protein. *Microb. Pathogen.* **26**, 25–33.

Hasty, D.L., Ofek, I., Courtney, H.S., and Doyle, R. (1992). Multiple adhesins of streptococci. *Infect. Immun.* **60**, 2147–2152.

Hidalgo-Grass, C., Ravins, M., Dan-Goor, M., Jaffe, J., Moses, A., and Hanski, E. (2002). A locus of group A streptococci involved in invasive disease and DNA transfer. *Mol. Microbiol.* **46**, 87–99.

Hoe, N., Ireland, R., DeLeo, F., Gowen, B., Dorward, D., Voyich, J., Liu, M., Burns, E. Jr., Culnan, D., Bretscher, A., and Musser, J.M. (2002). Insight into the molecular basis of pathogen abundance: group A streptococcus inhibitor of complement inhibits bacterial adherence and internalization into human cells. *Proc. Natl Acad. Sci. USA* **99**, 7646–7651.

Hynes, R.O. (2002). Integrins: bidirectional, allosteric signaling machines. *Cell* **110**, 673–687.

Hytonen, J., Haataja, S., and Finne, J. (2003). *Streptococcus pyogenes* glycoprotein-binding strepadhesin activity is mediated by a surface-associated carbohydrate-degrading enzyme, pullanase. *Infect. Immun.* **71**, 784–793.

Hytonen, J., Haataja, S., Gerlach, D., Podbielski, A., and Finne, J. (2001). The speB virulence factor of *Streptococcus pyogenes*, a multifunctional secreted and cell surface molecule with strepadhesin, laminin-binding and cysteine protease activity. *Mol. Microbiol.* **39**, 512–519.

Jadoun, J., Burstein, E., Hanski, E., and Sela, S. (1997). Proteins M6 and F1 are required for efficient invasion of group A streptococci into cultured epithelial cells. *Adv. Exp. Med.* **418**, 511–515.

Jadoun, J., Eyal, O., and Sela, S. (2002). Role of CsrR, hyaluronic acid, and SpeB in the internalization of *Streptococcus pyogenes* M type 3 strain by epithelial cells. *Infect. Immun.* **70**, 462–469.

Jaffe, J., Natanson-Yaron, S., Caparon, M., and Hanski, E. (1996). Protein F2, a novel fibronectin-binding protein from *Streptococcus pyogenes*, possesses two binding domains. *Mol. Microbiol.* **21**, 373–384.

Jeng, A., Sakota, V., Li, Z., Datta, V., Beall, B., and Nizet, V. (2003). Molecular genetic analysis of a group A streptococcus operon encoding serum opacity factor and a novel fibronectin-binding protein, SfbX. *J. Bacteriol.* **185**, 1208–1217.

Ji, Y., Schnitzler, N., DeMaster, E., and Cleary, P. (1998). Impact of M49, Mrp, Enn, and C5a peptidase proteins of colonization of the mouse oral mucosa by *Streptococcus pyogenes*. *Infect. Immun.* **66**, 5399–5405.

Johnson, D. and Kaplan, E. (1993). A review of the correlation of T-agglutination patterns and M-protein typing and opacity factor production in the identification of group A streptococci. *J. Med. Microbiol.* **38**, 311–315.

Kallman, J. and Kihlstrolm, E. (1997). Penetration of group B streptococci through polarized Madin-Darby canine kidney cells. *Ped. Res.* **42**, 799–804.

Kreikemeyer, B., Boyle, M., Buttaro, B., Heinemann, M., and Podbielski, A. (2001). Group A streptococcal growth-phase associated virulence factor regulation by a novel operon (Fas) with homologies to two-component-type regulators requires a small RNA molecule. *Mol. Microbiol.* **39**, 392–406.

Kreikemeyer, B., McIver, K., and Podbielski, A. (2003). Virulence factor regulation and regulatory networks in *Streptococcus pyogenes* and their impact on pathogen-host interactions. *Trends Microbiol.* **11**, 224–232.

Kreikemeyer, B., Talay, S.R., and Chhatwal, G.S. (1995). Characterization of a novel fibronectin-binding surface protein in group A streptococci. *Mol. Microbiol.* **17**, 137–145.

Lancefield, R.C. (1933). A serological differentiation in human and other groups of hemolytic streptococci. *J. Exp. Med.* **57**, 571–595.

LaPenta, D., Rubens, C., Chi, E., and Cleary, P. (1994). Group A streptococci efficiently invade human respiratory epithelial cells. *Proc. Natl Acad. Sci. USA* **91**, 12,115–12,119.

Lei, B., Deleo, F., Reid, S., Voyich, J., Magoun, L., Liu, M., Braughton, K., Ricklefs, S., Hoe, N., Cole, R.L., Leong, J.M., Musser, J.M. (2002). Opsonophagocytosis-inhibiting Mac protein of group A streptococcus: identification and characteristics of two genetic complexes. *Infect. Immun.* **70**, 6880–6890.

Lin, C., Kuan, I., Wang, C., Lee, H., Lee, W., Sheu, J., Hsiao, G., Wu, C., and Kuo, H. (2002). Lipoteichoic acid-induced cyclooxygenase-2 expression requires activation of p44/42 and p38 mitogen-activated protein kinase signal pathways. *Eur. J. Pharmacol.* **450**, 1–9.

Lukomski, S., Nakashima, K., Abdi, I., Cipriano, V., Shelvin, B., Graviss, E., and Musser, J. (2001). Identification of characterization of a second extracellular collagen-like protein made by group A streptococcus: control of production at the level of transcription. *Infect. Immun.* **69**, 1729–1738.

Madden, J., Ruiz, N., and Caparon, M. (2001). Cytolysin-mediated translocation (CMT): a functional equivalent of type III secretion in gram-positive bacteria. *Cell* **104**, 143–152.

McIver, K.S., Heath, A.S., and Scott, J.R. (1995). Regulation of virulence by environmental signals in group A streptococci: influence of osmolarity, temperature, gas exchange, and iron limitation on *emm* transcription. *Infect. Immun.* **63**, 4540–4542.

McIver, K.S. and Scott, J.R. (1997). Role of *mga* in growth phase regulation of virulence genes of the group A streptococcus. *J. Bacteriol.* **179**, 5178–5187.

Medina, E., Anders, D., and Chhatwal, G.S. (2002). Induction of NF-kappaB nuclear translocation in human respiratory epithelial cells by group A streptococci. *Microbiol. Pathog.* **33**, 307–313.

Miettinen, M., Matikainen, S., Vuopio-Varjkila, J., Pirhonen, J., Varkila, K., Kurimoto, M., and Julkunen, I. (1998). Lactobacilli and streptococci induce interleukin-12 (IL-12), IL-18, and gamma interferon production in human periphereal blood mononuclear cells. *Infect. Immun.* **66**, 6058–6062.

Molinari, G., Rohde, M., Guzman, C., and Chhatwal, G. (2000). Two distinct pathways for the invasion of streptococci in non-phagocytic cells. *Cell. Microbiol.* **2**, 145–154.

Molinari, G., Rohde, M., Talay, S.R., Chhatwal, G.S., Beckert, S., and Podbielski, A. (2001). The role played by the group A streptococcal negative regulator Nra on bacterial interactions with epithelial cells. *Mol. Microbiol.* **40**, 99–114.

Molinari, G., Talay, S.R., Valentin-Weigand, P., Rohde, M., and Chhatwal, G. (1997). The fibronectin-binding protein of *Streptococcus pyogenes*, Sfbl, is involved in internalization of group A streptococci by epithelial cells. *Infect. Immun.* **65**, 1357–1363.

Nakagawa, I., Nakata, M., Kawabata, S., and Hamada, S. (2001). Cytochrome C-mediated caspase-9 activation triggers apoptosis in *Streptococcus pyogenes* infected epithelial cells. *Cell. Microbiol.* **3**, 359–405.

Natanson, S., Sela, S., Moses, A., Musser, J., Caparon, M., and Hanski, E. (1995). Distribution of fibronectin-binding proteins among group a streptococci of different M types. *J. Infect. Dis.* **171**, 871–878.

Neeman, R., Keller, N., Barzilai, A., Korenman, Z., and Sela, S. (1998). Prevalence of internalization-associated gene, prtF1, among persisting group-A streptococcus strains isolated from asymptomatic carriers. *Lancet* **352**, 1974–1977.

Nizet, V., Ohtake, T., Lauth, X., Trowbridge, J., Rudisill, J., Dorschner R., Pestonjamasp, V., Piraino, J., Huttner, K., and Gallo, R.L. (2001). Innate antimicrobial peptide protects the skin from invasive bacterial infections. *Nature* **414**, 454–457.

Ofek, I., Zafiri, I., Goldhar, J., and Eisenstein, B. (1990). Inability of toxin inhibitors to neutralize enhanced toxicity caused by bacteria adherent to tissue culture cells. *Infect. Immun.* **58**, 3737–3742.

Okada, N., Liszewski, M., Atkinson, J., and Caparon, M. (1995). Membrane cofactor protein (CD46) is a keratinocyte receptor for the M protein of the group A streptococcus. *Proc. Natl Acad. Sci. USA* **92**, 2489–2493.

Okada, N., Pentland, A., Falk, P., and Caparon, M. (1994). M protein and protein F act as important determinants of cell-specific tropism of *Streptococcus pyogenes* in skin tissue. *J. Clin. Invest.* **94**, 965–977.

Okada, N., Tatsuno, I., Hanski, E., Caparon, M., and Sasakawa, C. (1998). *Streptococcus pyogenes* protein F promotes invasion of HeLa cells. *Microbiology* **144**, 3079–3086.

Okahashi, N., Sakurai, A., Nakagawa, I., Fujiwara, T., Kawabata, S., Amano, A., and Hamada, S. (2003). Infection by *Streptococcus pyogenes* induces the receptor activator of NF-kB ligand expression in mouse osteoblastic cells. *Infect. Immun.* **71**, 948–955.

Osterlund, A. and Engstrand, L. (1997). An intracellular sanctuary for *Streptococcus pyogenes* in human tonsillar epithelium – studies of asymptomatic carriers and in vitro cultured biopsies. *Acta Otolaryngol.* **117**, 883–888.

Osterlund, A., Popa, R., Nikkila, T., Scheynius, A., and Engstrand, L. (1997). Intracellular reservoir of *Streptococcus pyogenes* in vivo: a possible explanation for recurrent pharyngotonsillitis. *Laryngoscope* **107**, 640–646.

Ozeri, V., Rosenshine, I., Mosher, D., Fassler, R., and Hanski, E. (1998). Roles of integrins and fibronectin in the entry of *Streptococcus pyogenes* into cells via protein F1. *Mol. Microbiol.* **30**, 625–637.

Pancholi, V. and Fischetti, V. (1992). A major surface protein on group A streptococci glyceraldehyde-3-phosphate-dehydrogenase with multiple binding activity. *J. Exp. Med.* **176**, 415–426.

Pancholi, V. and Fischetti, V. (1997). Regulation of the phosphorylation of human pharyngeal cell proteins by group A streptococcal surface dehydrogenase: signal transduction between streptococci and pharyngeal cells. *J. Exp. Med.* **186**, 1633–1643.

Perez-Casal, J., Okada, N., Caparon, M., and Scott, J. (1995). Role of the conserved C-repeat region of the M protein of *Streptococcus pyogenes*. *Mol. Microbiol.* **15**, 907–916.

Podbielski, A., Beckert, S., Schattke, R., Leithauser, F., Lestin, F., Gobler, B., and Kreikemeyer, B. (2003). Epidemiology and virulence gene expression of intracellular group A streptococci in tonsils of recurrently infected adults. *Int. J. Med. Microbiol.* **293**, 179–190.

Podbielski, A., Woischnik, M., Leonard, B., and Schmidt, K. (1999). Characterization of *nra*, a global negative regulator gene in group A streptococci. *Mol. Microbiol.* **31**, 1051–1064.

Podbielski, A, Woischnik, M., Pohl, B., and Schmidt, K.H. (1996). What is the size of the group A streptococcal vir regulon? The Mga regulator affects expression of secreted and surface virulence factors. *Med. Microbiol. Immunol.* **185**, 171–181.

Pritchard, K. and Cleary, P. (1996). Differential expression of genes in the vir regulon of *Streptococcus pyogenes* is controlled by transcription termination. *Mol. Gen. Genet.* **250**, 207–213.

Rakonjac, J.V., Robbins, J.C., and Fischetti, V. (1995). DNA sequence of the serum opacity factor of group A streptococci: identification of a fibronectin-binding repeat domain. *Infect. Immun.* **63**, 622–631.

Rasmussen, M. and Bjorck, L. (2001). Unique regulation of SclB – a novel collagen-like surface protein of *Streptococcus pyogenes*. *Mol. Microbiol.* **40**, 1427–1438.

Rasmussen, M., Eden, A., and Bjorck, L. (2000). SclA, a novel collagen-like surface protein of *Streptococcus pyogenes*. *Infect. Immun.* **68**. 6370–6377.

Rocha, C. and Fischetti, V. (1999). Identification and characterization of a novel fibronectin-binding protein on the surface of group A streptococci. *Infect. Immun.* **67**, 2720–2728.

Rodhe, M., Muller, E., Chhatwal, G., and Talay, S. (2003). Host cell caveolae act as an entry port for group A streptococci. *Cell. Microbiol.* **5**, 323–342.

Ruiz, N., Wang, B., Pentland, A., and Caparon, M. (1998). Streptolysin O and adherence synergistically modulate proinflammatory responses of keratinocytes to group A streptococci. *Mol. Microbiol.* **27**, 337–346.

Ryan, P., Pancholi, V., and Fischetti, V. (2001). Group A streptococci bind to mucin and human pharyngeal cells through sialic acid-containing receptors. *Infect. Immun.* **69**, 7402–7412.

Saetre, T., Hoiby, E., Kahler, H., and Lyberg, T. (2000). Changed expression of leukocyte adhesion molecules and increased production of reactive oxygen species caused by *Streptococcus pyogenes* in human whole blood. *APMIS* **108**, 573–580.

Schmidt, K.H., Mann, K., Cooney, J., and Kohler, W. (1993). Multiple binding of type 3 streptococcal M protein to fibrinogen, albumin, and fibronectin. *FEMS Immun. Med. Microbiol.* **7**, 135–144.

Schrager, H.M., Alberti, S., Cywes, S., Dougherty, G.J., and Wessels, M.R. (1998). Hyaluronate acid capsule modulates M protein mediated adherence and acts as a ligand for attachment of group A streptococci to CD44 on human keratinocytes. *J. Clin. Invest.* **101**, 1708–1716.

Sela, S., Aviv, A., Tovi, A., Burstein, I., Caparon, M., and Hanski, E. (1993). Protein F: an adhesin of *Streptococcus pyogenes* binds fibronectin via two distinct domains. *Mol. Microbiol.* **10**, 1049–1055.

Sierig, G., Cywes, C., Wessels, M., and Ashbaugh, C. (2003). Cytotoxic effects of streptolysin O and streptolysin S enhance the virulence of poorly encapsulated group A streptococci. *Infect. Immun.* **71**, 446–455.

Simpson, W.A. and Beachey, E.H. (1983). Adherence of group A streptococci to fibronectin on oral epithelial cells. *Infect. Immun.* **39**, 275–279.

Smoot, J., Barbian, K., Van Gompel, J., Smoot, L., Chaussee, M., Sylva, G., Sturdevant D., Ricklefs, S., Porcella, S.F., Parkins, C.D., Beres, S.B., Campbell, D.S., Smith, T.M., Zhang, Q., Kapur, V., Daly, J.A., Veasy, L.G., and Musser, J.M. (2002). Genome sequences and comparative microarray analysis of serotype M18 group A streptococcus strains associated with acute rheumatic fever outbreaks. *Proc. Natl Acad. Sci. USA* **99**, 4668–4673.

Speck, O., Hughes, S., Noren, N., Kulikauskas, R., and Fehon, R. (2003). Moesin functions antagonistically to the Rho pathway to maintain epithelial integrity. *Nature* **421**, 83–87.

Stalhammar-Carlemalm, M., Areschoug, T., Larsson, C., and Lindahl, G. (1999). The R28 protein of *Streptococcus pyogenes* is related to several group B streptococcal surface proteins, confers protective immunity and promotes binding to human epithelial cells. *Mol. Microbiol.* **33**, 208–219.

Stenfors, L., Bye, H., Raisanen, S., and Myklebust, R. (2000). Bacterial penetration into tonsillar surface epithelium during infectious mononucleosis. *J. Laryngol. Otol.* **114**, 848–852.

Stevens, D., Salmi, D., McIndoo, E., and Bryant, A. (2000). Molecular epidemiology of nga and NAD glycohydralase/ADP-ribosyltransferase activity among *Streptococcus pyogenes* causing toxic shock syndrome. *J. Infect. Dis.* **182**, 1117–1128.

Stockbauer, K.E., Magoun, L., Liu, M., Burns, E.H. Jr., Gubbam S., Renish, S., Pan, X., Bodary, S.C., Baker, E., Coburn, J., Leong, J.M., and Musser, J.M. (1999). A natural variant of the cysteine protease virulence factor of group A streptococcus with an arginine-glycine-aspartic acid (RGD) motif preferentially binds human integrins $\alpha v\beta 3$ and $\alpha IIb\beta 3$. *Proc. Natl Acad. Sci. USA* **96**, 242–247.

Talay, S.R., Valentin-Weigand, P., Jerlstrom, P.G., Timmis, K.N., and Chhatwal, G.S. (1993). Fibronectin-binding protein of *Streptococcus pyogenes*: sequence of the binding domain involved in adherence of streptococci to epithelial cells. *Infect. Immun.* **60**, 3837–3844.

Talay, S.R., Zock, A., Rohde, M., Molinari, G., Oggioni, M., Pozzi, G., Guzman, C.A., and Chhatwal, G.S. (2000). Co-operative binding of human fibronectin to Sfb1 protein triggers streptococcal invasion into respiratory epithelial cells. *Cell. Microbiol.* **2**, 521–535.

Talay, S., Ehrenfeld, E., Chhatwal, G., and Timmis, K. (1991). Expression of the fibronectin-binding components of *Streptococcus pyogenes* in *Escherichia coli* demonstrates that they are proteins. *Mol. Microbiol.* **5**, 1727–1734.

Tamura, G. and Nittayajarn, A. (2000). Group B streptococci and other gram-positive cocci bind to cytokeratin 8. *Infect. Immun.* **68**, 2129–2134.

Terao, Y., Kawabata, S., Kunitomo, E., Murakami, J., Nakagawa, I., and Hamada, S. (2001). Fba, a novel fibronectin-binding protein from *Streptococcus pyogenes*, promotes bacterial entry into epithelial cells, and the *fba* gene is positively transcribed under the Mga regulator. *Mol. Microbiol.* **42**, 75–86.

Terao, Y., Kawabata, S., Nakata, M., Nakagawa, I., and Hamada, S. (2002). Molecular characterization of a novel fibronectin-binding protein of *Streptococcus pyogenes* strains isolated from toxic shock-like syndrome patients. *J. Biol. Chem.* **277**, 47,428–47,435.

Terao, Y., Kawabata, S., Kunitomo, E., Nakagawa, I., and Hamada, S. (2002). Novel laminin-binding protein of *Streptococcus pyogenes*, Lbp, is involved in adhesion to epithelial cells. *Infect. Immun.* **70**, 993–997.

Tomasini-Johansson, B., Kaufman, N., Ensenberger, M., Ozeri, V., Hanski, E., and Mosher, D. (2001). A 49-residue peptide from adhesin F1 of *Streptococcus pyogenes* inhibits fibronectin-matrix assembly. *J. Biol. Chem.* **276**, 23,430–23,439.

Tsai, P., Lin, Y., Kuo C., Lei, H., and Wu, J. (1999). Group A streptococcus induces apoptosis in human epithelial cells. *Infect. Immun.* **67**, 4334–4339.

Upton, M., Tagg, J.R., Wescombe, P., and Jenkinson, H.F. (2001). Intra- and inter-species signalling between *Streptococcus salivarius* and *Streptococcus pyogenes* mediated by SalA and SalA1 lantibiotic peptides. *J. Bacteriol.* **183**, 3931–3938.

Valentin-Wiegand, P., Grulich-Henn, J., Chhatwal, G., Muller-Berhasus, G., Blobel, H., and Preissner, K. (1988). Mediation of adherence of streptococci to human endothelial cells by complement S protein (vitronectin). *Infect. Immun.* **56**, 2851–2855.

Visai, L., Bozzini, S., Raucci, G., Toniolo, A., and Speziale, P. (1995). Isolation and characterization of a novel collagen-binding protein from *Streptococcus pyogenes* strain 6414. *J. Biol. Chem.* **270**, 347–353.

von Hunolstein, C., Greco, R., Ajello, M., Orefici, G., and Valenti, P. (2000). *Streptococcus pyogenes* internalization by Hela cells is not mediated by M6 protein. In *Streptococci and Streptococcal Diseases – Entering the New Millennium*, ed. D. Martin and J. Tagg, pp. 681–683, Wellington, New Zealand: Securacopy.

Von Pawel-Rammingen, U., Johansson, B., and Bjorck, L. (2002). IdeS, a novel streptococcal cysteine proteinase with unique specificity for immunoglobulin G. *EMBO J.* **21**, 1607–1615.

Voyich, J., Sturdevant, D., Braughton, K., Kobayashi, S., Lei, B., Virtaneva, K., Dorward, D., Musser, J., and Deleo, F. (2003). Genome-wide protective response used by group A streptococcus to evade destruction by human polymorphonuclear leukocytes. *Proc. Natl Acad. Sci. USA* **100**, 1996–2001.

Wadstrom, T. and Tylewska, S. (1982). Glycoconjugates as possible receptors for *Streptococcus pyogenes. Curr. Microbiol.* **7**, 343–346.

Wang, B., Ruiz, N., Pentland, A., and Caparon, M. (1997). Keratinocyte proinflammatory responses to adherent and nonadherent group A streptococci. *Infect. Immun.* **65**, 2119–2126.

Wang, J. and Stinson, M. (1994). Streptococcal M6 protein binds to fucose-containing glycoproteins on cultured human epithelial cells. *Infect. Immun.* **62**, 1268–1274.

Wessels, M.R. and Bronze, M.S. (1994). Critical role of the group A streptococcal capsule in pharyngeal colonization and infection in mice. *Proc. Natl Acad. Sci. USA* **91**, 12,238–12,242.

Whatmore, A.M. (2001). *Streptococcus pyogenes* sclB encodes a putative hypervariable surface protein with a collagen-like repetitive structure. *Microbiology* **147**, 419–429.

CHAPTER 9

Invasion of oral epithelial cells by *Actinobacillus actinomycetemcomitans*

Diane Hutchins Meyer, Joan E. Lippmann, and Paula Fives-Taylor

ACTINOBACILLUS ACTINOMYCETEMCOMITANS: ADHERENCE MECHANISMS REQUIRED FOR INVASION

Colony phase variation

A. actinomycetemcomitans produces three distinct colonial morphologies on solid medium. A rough colony phenotype is typically generated by organisms upon isolation from the gingiva. These are small (~0.5–1 mm in diameter), translucent circular colonies with rough surfaces and irregular edges (Fig. 9.1). An internal star-shaped or crossed cigar morphology that embeds the agar is a distinguishing characteristic that gives *A. actinomycetemcomitans* its name (Zambon, 1985). In liquid culture, the rough colony phenotype cells form aggregates on the vessel walls, resulting in a clear medium (Fig. 9.1). Repeated subculture on agar of rough phenotypic isolates yields two distinct colonial variants; one is smooth surfaced and transparent, and the other is smooth surfaced and opaque (Slots, 1982; Scannapieco et al., 1987; Rosan et al., 1988; Inouye et al., 1990). The transparent smooth-surfaced variants appear to be an intermediate between the transparent rough-surfaced and opaque smooth-surfaced types (Inouye et al., 1990). In broth, the smooth-surfaced opaque type grows as a turbid homogeneous suspension, whereas the smooth-surfaced transparent type aggregates and adheres to the vessel walls (Inouye et al., 1990). In general, isolates undergo a rough-to-smooth variant transition soon after culture *in vitro*. In contrast, a smooth-to-rough variant transition that appears to be associated with nutritional requirements occurs only rarely during *in vitro* culture (Inouye et al., 1990; Meyer et al., 1991; Meyer, unpublished observation).

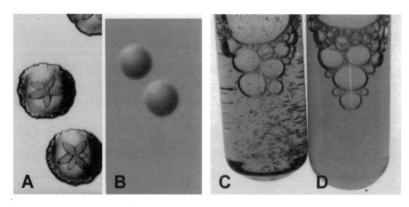

Figure 9.1. Micrographs of *A. actinomycetemcomitans* phenotypes. Rough (A and C) and smooth (B and D) colonial variants cultured in agar (top panel) and broth (lower panel). A unique characteristic of the rough colony morphology is the lovely star form that is embedded in the agar (A). In broth culture, the rough phenotype forms tight aggregates (C).

Surface-associated molecules and organelles

Bacterial colonial variation is indicative of the differential expression of cell surface components (Braun, 1965). *A. actinomycetemcomitans* rough colony variants express 43- and 20-kDa outer membrane proteins, rough colony protein A and B, respectively, that are not expressed in smooth colony variants (Haase et al., 1999). These proteins are encoded by genes that have homology to genes known to encode fimbriae-associated proteins. In that regard, *A. actinomycetemcomitans* rough colony variants are heavily fimbriated, whereas smooth colony variants have few or no fimbriae (Scannapieco et al., 1987; Rosan et al., 1988; Inouye et al., 1990; Meyer and Fives-Taylor, 1994). In accordance, rough colony variants adhere better than smooth colony variants *in vitro* to epithelial cells (Meyer and Fives-Taylor, 1994). In contrast, smooth colony variants invade epithelial cells *in vitro* significantly better than do rough variant colonies (Meyer et al., 1991). *A. actinomycetemcomitans* strain SUNY 465, the invasion prototype, exhibits a smooth-to-rough variant shift after anaerobic growth on agar that is accompanied by cell fimbriation and increased adherence to KB oral epithelial cells (Meyer and Fives-Taylor, 1994). Although the role of the phenotypic variation is not known, it has been proposed that it may play some role in the episodic nature of periodontitis (Meyer et al., 1991).

A. actinomycetemcomitans fimbriae are peritrichous, ~2 μm in length and 5 nm in diameter, and frequently occur in bundles (Holt et al., 1980;

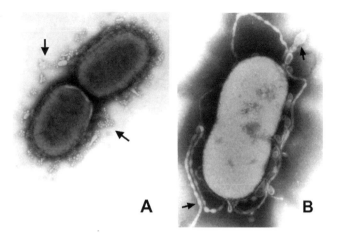

Figure 9.2. Scanning electron micrographs of vesicles (blebs) associated with the *A. actinomycetemcomitans* surface. Vesicles may vary in number and occur in different forms or shapes and sizes. The vesicles associated with the organism in panel A (35,000×) are relatively short and peritrichous (arrows), whereas those in panel B (70,000×) are extremely long and essentially wrap around the organism (arrows).

Scannapieco et al., 1983, 1987; Preus et al., 1988; Rosan et al., 1988). They are composed of a 54-kDa fimbrial subunit (Inouye et al., 1990). Adhesion of *A. actinomycetemcomitans*-fimbriated strains to both buccal epithelial cells and the Gin-1 fibroblast cell line is inhibited by antiserum to the 54-kDa protein, indicating a role for it in adhesion. Fimbrial-associated protein (fap), an *A. actinomycetemcomitans* attachment factor, is expressed in fimbriated, but not in nonfimbriated, strains (Ishihara et al., 1997). Flp, a 6.5-kDa protein which is a component of *A. actinomycetemcomitans* fimbriae, exhibits some amino acid sequence similarity to type IV pilin (Inoue et al., 1998). Clearly, there is a correlation between *A. actinomycetemcomitans* fimbriation and adhesion. However, *A. actinomycetemcomitans* cells devoid of fimbriae exhibit adhesiveness as well, indicating that nonfimbrial components also function in adhesion (Inouye et al., 1990; Meyer and Fives-Taylor, 1994).

Membraneous vesicles (blebs) are a prominent feature of the surface of *A. actinomycetemcomitans*. These structures, which are thought to be predominantly lipopolysaccharide in nature, are extensions of the outer membrane that either remain attached to or bud off from the cell surface (Fig. 9.2). Large numbers of vesicles are released into the external environment during culture (Holt et al., 1980). The formation and morphology of vesicles are altered by growth conditions (Meyer and Fives-Taylor, 1993). Cells grown on agar have vesicles characterized by thick fibrils with knob-like ends. *A.*

actinomycetemcomitans vesicles contain endotoxin, bone resorption activity, actinobacillin (a bacteriocin), and leukotoxin (Hammond et al., 1981, 1987; Nowotny et al., 1982; Stevens et al., 1987; Lucia et al., 2002). Highly leukotoxic *A. actinomycetemcomitans* strains have an abundance of vesicles, whereas low-leukotoxic or nonleukotoxic strains have few or no vesicles (Lai et al., 1981). The leukotoxicity of vesicles associated with highly leukotoxic strains is 5 to 10 times greater than that of minimally leukotoxic strains (Kato et al., 2001).

A. *actinomycetemcomitans* vesicles also demonstrate adhesiveness. The addition of vesicles to weakly adherent or nonadherent strains significantly increases the ability of those strains to attach to epithelial cells (Meyer and Fives-Taylor, 1993). The adhesive nature of the vesicles prompted the hypothesis that these organelles function in *A. actinomycetemcomitans* as delivery vehicles for the toxic materials that they harbor (Meyer and Fives-Taylor, 1993). Scanning electron microscopy of the process of invasion of *A. actinomycetemcomitans* into epithelial cells revealed that bacteria in contact with, and in the process of being internalized by, the cells have surface-associated vesicles. In contrast, bacteria not in contact with epithelial cells do not possess vesicles, suggesting that the internalization process may be a trigger for vesicle formation (Meyer et al., 1996). The role of vesicles in the *A. actinomycetemcomitans* internalization process is not known, but clearly further investigation is warranted.

The surface of *A. actinomycetemcomitans* may also be associated with an extracellular amorphous material (ExAmMat) that can embed groups of cells in a matrix (Holt et al., 1980). Whereas an early study reported that cells grown in liquid medium lacked amorphous material, other reports indicate that ExAmMat can occur on cells grown in liquid culture (Wilson et al., 1985; Meyer and Fives-Taylor, 1994). Expression of ExAmMat has some association with growth in tryptone-based medium (Wilson et al., 1985). Therefore, similar to fimbriae and vesicles, culture conditions can modify expression of ExAmMat. It has been determined that ExAmMat is proteinacious, most likely a glycoprotein, and has adhesive properties (Meyer and Fives-Taylor, 1993). ExAmMat is not affixed firmly to the cell surface, as the washing of cells with phosphate-buffered saline removes ExAmMat and results in reduced adhesion to epithelial cells (Meyer and Fives-Taylor, 1993). Moreover, weakly adherent *A. actinomycetemcomitans* strains adhere strongly to epithelial cells after suspension in ExAmMat, a process termed conveyed adhesion. The surfaces of *A. actinomycetemcomitans* suspended in ExAmMat are associated with large amounts of material, indicating that the conveyed adhesion is the result of a direct transfer of ExAmMat onto the bacterial surface (Fives-Taylor et al., 1995).

Specific adhesion to epithelial cells

Most *A. actinomycetemcomitans* strains that have been tested to date adhere to epithelial cells strongly; however, the adhesiveness of strains does vary (Meyer and Fives-Taylor, 1994). The rapid process reaches saturation levels within 1 h of infection (Mintz and Fives-Taylor, 1994). Adhesion is affected by growth conditions (Meyer and Fives-Taylor, 1994), which likely determine the expression of specific adhesins. Cell surface entities that mediate adherence include fimbriae (Rosan et al., 1988; Meyer and Fives-Taylor, 1994), ExAmMat (Meyer and Fives-Taylor, 1994), and vesicles (Meyer and Fives-Taylor, 1993). Trypsin and protease treatment of smooth, nonfimbriated strains reduces adhesion of *A. actinomycetemcomitans* to epithelial cells, indicating that these nonfimbriae types of adhesins are proteinacious (Mintz and Fives-Taylor, 1994).

Recently, a surface-associated protein determined to be an autotransporter was shown to be involved in *A. actinomycetemcomitans* adhesion to epithelial cells (Rose et al., 2003). The adhesin, called Aae for adhesion to epithelial cells, is encoded by a gene (*aae*) that is homologous to autotransporter genes of *Haemophilus influenzae* and *Neisseria* species (St. Geme et al., 1994; St. Geme and Cutter, 2000). A unique feature of *aae* is a 135-base repeat sequence that varies in number from strain to strain (Rose et al., 2003). Four alleles have been identified to date. Lactoferrin in human milk whey was shown to decrease epithelial cell binding of *A. actinomycetemcomitans* strain 29523, but not of SUNY 465, a strain having an allele with fewer repeats (Rose et al., 2003). On the basis of these results it was suggested that the repeats may play a role in the binding of lactoferricin, the peptide on the N-terminus of lactoferrin that interacts with and causes damage to the cells' outer membrane (Yamauchi et al., 1993). Fewer copies of the repeats would effectively reduce the opportunity for binding.

In summary, the adhesion of *A. actinomycetemcomitans* to epithelial cells is multifactorial, with several adhesins and mechanisms playing a role.

HOST CELL SIGNALING PATHWAYS MODULATED IN RESPONSE TO THE BACTERIAL CHALLENGE

Several lines of evidence suggest that *A. actinomycetemcomitans* infection and subsequent internalization is associated with activation of host cell signaling pathways. An active metabolic state and novel protein synthesis by both *A. actinomycetemcomitans* and the epithelial cell are required for invasion (Sreenivasan et al., 1993). In addition, entry is associated with an elevation of intracellular Ca^{2+} levels (Fives-Taylor et al., 1996), a process associated with

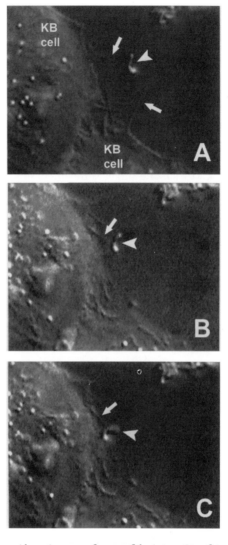

Figure 9.3. Real-time video microscopy frames of the interaction of *A. actinomycetemcomitans* strain SUNY 465 with KB epithelial cells. KB microvilli (arrow) become highly active and seek out bacteria (arrowhead) soon after infection (A). A microvillus makes contact with a bacterium (B). The microvillus seizes the *A. actinomycetemcomitans* and draws it toward and eventually into the KB cell (C).

signaling events (Berridge, 1995) and characteristic of invasion by some bacteria (Baldwin et al., 1991; Pace et al., 1993; Izutsu et al., 1996). It has also been shown that staurosporin, a wide-spectrum protein kinase inhibitor, decreases *A. actinomycetemcomitans* internalization, whereas genistein, a specific inhibitor of tyrosine protein kinase, increases internalization (Fives-Taylor and Meyer, 1998). These results led to the hypothesis that staurosporin may be modulating a signaling pathway involved in the *A. actinomycetemcomitans* internalization process, whereas genistein is modulating one involved in its egression from the host cell.

In related studies it was shown that an infection of KB cells produces changes in both its SDS-PAGE tyrosine-phosphorylated protein profile and immunofluorescence microscopy phosphotyrosine-labeling pattern (Fives-Taylor and Meyer, 1998). The fact that cross talk or signaling is an absolute requirement for *A. actinomycetemcomitans* invasion was ascertained by use of real-time video immunofluorescence microscopy (Fig. 9.3). Uninfected KB cells and those subjected to nonviable *A. actinomycetemcomitans* are absolutely quiescent. In contrast, infection with viable *A. actinomycetemcomitans* arouses the KB cells; their microvilli become highly active — seeking out, sequestering, drawing in, and eventually engulfing the bacterium (Lippmann and Fives-Taylor, 1999).

PHYSICAL MECHANISM OF INTERNALIZATION AND INTRACELLULAR LOCATION, FATE OF BACTERIA, AND CELL-TO-CELL SPREAD

A. actinomycetemcomitans penetration of the gingival epithelium was demonstrated in early clinical studies (Saglie et al., 1986; Christersson et al., 1987). Those *in vivo* studies revealed that *A. actinomycetemcomitans* occurs in very specific intracellular locations and exhibits a very distinctive penetration pattern (Saglie et al., 1982, 1986, 1988). It has also been established that intracellular *A. actinomycetemcomitans* is present *in vivo* in human buccal epithelial cells, where it occurs in clusters (Rudney et al., 2001). *A. actinomycetemcomitans* invasion of epithelial cells, both in an oral cell line (Meyer et al., 1991) and in human primary gingival cells (Fives-Taylor et al., 1995), has been demonstrated by use of *in vitro* studies. The overall *A. actinomycetemcomitans* invasion process is dynamic and complex: it involves attachment to the host cell, entry in a vacuole, escape from the vacuole into the cytoplasm, intracellular spread, and cell-to-cell spread (Meyer at al., 1996). Entry of the epithelial cell is initiated when interaction of *A. actinomycetemcomitans* with the epithelial cell triggers effacement of microvilli and formation of craters on the epithelial cell surface (Sreenivasan et al., 1993; Meyer et al., 1996). The majority of *A. actinomycetemcomitans* strains use an actin-dependent mechanism for invasion, but the mode of entry of a few strains is actin independent (Brissette and Fives-Taylor, 1998). The invasion process in strains that utilize the actin-independent mechanism is not well studied. Thus, a role for the two different mechanisms is not known; nor is it known whether actin dependence or independence is the only characteristic that separates these two invasion processes.

The process of invasion described here is that of the *A. actinomycetemcomitans* invasion prototype, strain SUNY 465, a strain that utilizes the actin-dependent mechanism. Attachment of *A. actinomycetemcomitans* to the epithelial cell promotes rearrangement of actin from the periphery of the epithelial cell to a focal point beneath the organism at the point of entry (Fives-Taylor et al., 1995). *A. actinomycetemcomitans* enters the epithelial cell through ruffled, lip-rimmed apertures (Fives-Taylor et al., 1996; Meyer et al., 1996). To date, two pathways believed to be associated with the entry of *A. actinomycetemcomitans* into epithelial cells have been identified; the transferrin receptor is implicated in one pathway, and integrins are implicated in the other (Meyer et al., 1997a).

Subsequent to the initial interaction and attachment, *A. actinomycetemcomitans* enters the host cell in a membrane-bound vacuole by receptor-mediated endocytosis. In the classical endocytic pathway, macromolecules are taken into early endosomes and delivered to lysosomes. Internalized organisms are known to use a variety of means to avoid lysosomal degradative enzymes; these include blockage of delivery to lysosomes, inhibition of endosome acidification, and acid activation of virulence factors that modify the lysosome. *A. actinomycetemcomitans* trafficking within the vacuole, a process that does not require endosomal acidification, is as follows. Within 30 min of infection, 40% of internalized *A. actinomycetemcomitans* are in the early endosome. By 60 min, the number in the early endosome is greatly reduced and *A. actinomycetemcomitans* are also present in the late endosome. By 2 h, essentially all *A. actinomycetemcomitans* are associated with the late endosome, and by 3–4 h the vacuoles are devoid of bacteria and *A. actinomycetemcomitans* are present in both the cytoplasm and cell culture medium. Thus, the *A. actinomycetemcomitans* organisms avoid degradation by lysosomal enzymes by escaping from these organelles 3–4 h postinfection (Lippmann and Fives-Taylor, 2000).

The mechanism(s) used by *A. actinomycetemcomitans* for lysis of the host vacuole is not known. Preliminary studies suggest a role for leukotoxin in lysis of the vacuole (Meyer, unpublished results). *A. actinomycetemcomitans* possesses phospholipase C (PLC; see Meyer et al., 1997b), a molecule used by certain enteric pathogens for vacuole lysis (Camilli et al., 1991; Smith et al., 1995); thus, PLC is another possibility.

Clearly, intracellular *A. actinomycetemcomitans* are not quiescent. A short time after escape from the vacuole into the cytoplasm, *A. actinomycetemcomitans* interact in a highly specific manner with host cell microtubules and utilize a microtubule-mediated mechanism for intracellular spread, as well as for spread to neighboring epithelial cells (Meyer et al., 1999). The spread to neighboring cells is by means of *A. actinomycetemcomitans*-induced intercellular protrusions, which are extensions of the host cell membrane that extend from one epithelial cell to another (Meyer et al., 1996). Bacteria can be seen within these protrusions by scanning, transmission, and fluorescent microscopy (Meyer et al., 1996). The cell-to-cell spread involves movement and transfer through protrusions (Meyer et al., 1996), not engulfment of protrusions as is the case with both *Shigella* and *Listeria* (Tilney and Portnoy, 1989; Kadurugamuwa et al., 1991). In addition to microtubules, the protrusions contain microfilaments that may be involved in the structural framework of protrusions rather than mechanistically in the movement process (Meyer et al., 1999).

Whereas the precise means by which *A. actinomycetemcomitans* usurps host cell microtubules for movement is not known, *in vitro* studies show that *A. actinomycetemcomitans* localizes exclusively with the plus ends of microtubules of taxol-induced microtubule asters (Fig. 9.4), indicating a specific *A. actinomycetemcomitans*–microtubule interaction (Rose et al., 1998, 1999). Furthermore, both immunofluorescence microscopy and bactELISA indicate the presence of a kinesin-like entity on the surface of *A. actinomycetemcomitans*, thus implicating motor proteins in the movement along the microtubules (Meyer et al., 2000). It is hypothesized that the kinesin entity on the *A. actinomycetemcomitans* surface mediates the interaction with microtubules and that the bacterium is transported in a manner similar to that of organelles and vesicles, the usual cargo of microtubules. Whereas certain bacteria, such as *Shigella* and *Listeria*, usurp host cell actin to move within cells and to spread to adjacent cells (Tilney and Portnoy, 1989; Kadurugamuwa et al., 1991), this provides the first evidence that host cell dispersion of an intracellular pathogen involves usurpation of the host cell microtubule transport system.

Two genes have been implicated in *A. actinomycetemcomitans* invasion. One is homologous to *apaH*, a gene that encodes RGD, a sequence known to bind integrins (Saarela et al., 1999). *A. actinomycetemcomitans* DNA that contains the *apaH* gene confers on noninvasive *Escherichia coli* the ability to invade epithelial cells (Meyer et al., 1995; Saarela et al., 1999). It was also determined that insertional inactivation of *apaH* in *A. actinomycetemcomitans* substantially reduced its invasion (Lippmann and Fives-Taylor, unpublished

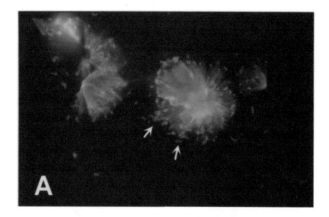

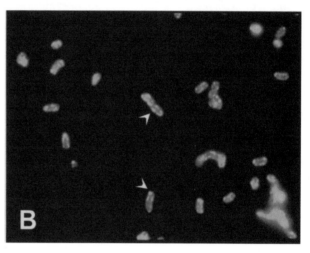

Figure 9.4. Microtubule asters and the *A. actinomycetemcomitans* kinesin-like entity. Immunofluorescent microscopy of the interaction of *A. actinomycetemcomitans* strain SUNY 465 (arrows) with taxol-induced asters (panel A) and the kinesin-like motor protein (arrowheads) associated with the *A. actinomycetemcomitans* surface (panel B). The *A. actinomycetemcomitans* (green) bind specifically to the plus ends (periphery) of the asters (orange) (panel A). Binding of *A. actinomycetemcomitans* to the microtubules is believed to be mediated by the kinesin-like entities (yellow) on the *A. actinomycetemcomitans* surface (orange) (panel B). See color section.

observation). Furthermore, RGD peptides inhibit *A. actinomycetemcomitans* invasion, whereas RAD peptides have no effect on invasion (Saarela et al., 1999). The *apaH* gene is a homolog of *invA*, *ialA*, and *ygdP*, genes that are associated with invasion by *Rickettsia prowazekii* (Gaywee et al., 2002), *Bartonella bacilliformis* (Mitchell and Minnick, 1995), and *E. coli* K1 (Bessman et al., 2001), respectively. These genes produce proteins that are members

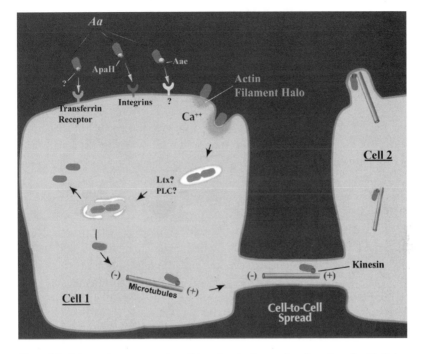

Figure 9.5. Schematic representation of *A. actinomycetemcomitans* invasion of epithelial cells. Aae and ApaH have been identified as molecules on the surface of *A. actinomycetemcomitans* that can mediate its contact with epithelial cells. ApaH interacts with host cell integrins; the receptor to which Aae binds is not known. The transferrin receptor is also implicated as a receptor for *A. actinomycetemcomitans*. Entry associated with actin rearrangement and a Ca^{++} flux occurs in a membrane-bound vacuole. The mechanism of escape from the vacuole is unclear, but leukotoxin (Ltx) and phospholipase C (PLC) are implicated. Once in the cytoplasm, *A. actinomycetemcomitans* can move within the host cell and spread to adjacent cells by usurping host cell microtubules. Kinesin-like motor proteins on the *A. actinomycetemcomitans* surface are implicated in mediating its interaction with microtubules. See color section.

of the Nudix family of hydrolases, which catalyze the dinucleoside polyphosphates, a class of signaling nucleotides (Conyers and Bessman, 1999; Saarela et al., 1999; Bessman et al., 2001). It has also been reported that *A. actinomycetemcomitans* invasion involves genes that share some homology to *spa* genes, genes that are involved in the export of proteins from cells (Laing-Gibbard et al., 1998).

In summary, it is clear that the process by which *A. actinomycetemcomitans* enters and escapes from host cells is very complex (Fig. 9.5). Future studies should reveal even more dynamic interactions and lead to targeting of mechanisms suitable for therapeutic intervention.

CONSEQUENCES OF INVASION WITH REGARD TO INNATE HOST RESPONSE

The first line of defense of the host against intrusive bacteria is phagocyte recruitment (chemotaxis) to the region. A number of steps are involved in this process: binding of chemotactic signaling factors, upregulation of adhesion receptors, binding to the endothelium, and movement of phagocytes to the underlying tissues. Whereas invasion per se is a means by which bacteria can ultimately escape this response, it initially represents a significant challenge to intrusive organisms. Thus, the ability to disrupt chemotaxis promotes survival of pathogenic organisms. *A. actinomycetemcomitans* secretes a low-molecular-weight protein that can inhibit polymorphonuclear leukocyte chemotaxis (Van Dyke et al., 1982; Ashkenazi et al., 1992). It is also known that *A. actinomycetemcomitans* capsular-like serotype-specific polysaccharide antigen (SPA) plays an important role in its ability to resist phagocytosis and killing by polymorphonuclear lymphocytes (Yamaguchi et al., 1995). Expression of chemoattractant protein 1 (MCP-1) and neutrophil chemotactic factor IL-8 mRNA is increased in monocytes after stimulation with SPA (Yamaguchi et al., 1996).

Polymorphonuclear leukocytes can also kill bacteria by fusing with lysosomes from which they acquire potent antibacterial agents. Bacteria able to inhibit the fusion or ward off the antibactericidal action are protected. *A. actinomycetemcomitans* can inhibit the production by polymorphonuclear leukocytes of some of these compounds, and it is resistant to others. A heat-stable protein in *A. actinomycetemcomitans* inhibits the production of hydrogen peroxide by polymorphonuclear leukocytes (Ashkenazi et al., 1992), and many strains are intrinsically resistant to high concentrations of hydrogen peroxide (Miyasaki et al., 1984). In addition, *A. actinomycetemcomitans* is resistant to a number of defensins, which are cationic peptides that occur in neutrophils (Miyasaki et al., 1990).

A. actinomycetemcomitans induces apoptotic cell death in murine macrophages (J774.1 cells) *in vitro*. Entry of *A. actinomycetemcomitans* into the macrophages is an absolute requirement for the cytotoxicity, that is, nuclear morphology changes and an increase in the proportion of fragmented DNA (Kato et al., 1995). Studies suggest that protein kinase C signaling regulates the apoptosis (Nonaka et al., 1997). With the use of LR-9 cells, CD14-defective mutants of J774.1 cells, it was determined that CD14 molecules likely participate in the phagocytosis of *A. actinomycetemcomitans*, as well as in the regulation of the apoptotic events (Muro et al., 1997). Both caspase-1 and caspase-3 appear to play a role in the *A. actinomycetemcomitans*-induced apoptosis in macrophages (Nonaka et al., 2001). In general, infected macrophages kill bacteria within phagosomes by means of nitric oxide (NO). In this regard, it has been reported that NO affords *A. actinomycetemcomitans*-infected murine macrophages partial protection from apoptosis by decreasing caspase activity (Nakashima et al., 2002). An interesting caveat that requires further investigation is the finding that *A. actinomycetemcomitans* LPS stimulates the production of NO, a process that involves activation of protein kinase C and protein tyrosine kinase, as well as the regulatory control of cytokines (Sosroseno et al., 2002).

CORRELATION AMONG INVASION, PATHOGENICITY, AND CLINICAL PRESENTATION

A lack of an adequate animal model in which to study invasion precludes, at this juncture, the ability to establish a correlation among *A. actinomycetemcomitans* invasion, pathogenicity, and clinical presentation. On the basis of the collective *in vivo* and *in vitro* observations made to date, the hypothesis is put forth that invasion of epithelial cells per se and the dynamic process of intercellular and intracellular spread are the means by which *A. actinomycetemcomitans* migrates to gingival and connective tissue and initiates the destruction associated with periodontal disease. The demonstration of *A. actinomycetemcomitans* in buccal epithelial cells prompted the hypothesis that bacteria inside exfoliated cells would be afforded a protected environment for exchange between various oral niches both within and between human subjects (Rudney et al., 2001). The rough-to-smooth and smooth-to-rough variant shifts, with concomitant shifts in adhesive-to-invasive and invasive-to-adhesive capabilities, respectively, may account in part for the episodic nature of periodontal disease.

A total understanding of the *A. actinomycetemcomitans* invasion process and its role in pathogenicity (Fives-Taylor et al., 1999) is all the more important today in light of increasing evidence that indicates that there is a link between periodontal disease and systemic disorders (DeStefano et al., 1993; Beck et al., 1996; Meyer and Fives-Taylor, 1998; Teng et al., 2002).

REFERENCES

Ashkenazi, M., White, R.R., and Dennison, D.K. (1992). Neutrophil modulation by *Actinobacillus actinomycetemcomitans*. I. Chemotaxis, surface receptor expression and F-actin polymerization. *J. Periodontal Res.* **27**, 264–273.

Baldwin, T.J., Ward, W., Aitken, A., Knutton, S., and Williams, P.H. (1991). Elevation of intracellular free calcium levels in HEp-2 cells infected with enteropathogenic *Escherichia coli*. *Infect. Immun.* **59**, 1599–1604.

Beck, J., Garcia, R., Heiss, G., Vokonas, P.S., and Offenbacher, S. (1996). Periodontal disease and cardiovascular disease. *J. Periodontol.* **67**, 1123–1137.

Berridge, M.J. (1995). Calcium signaling and cell proliferation. *Bioessays* **17**, 491–500.

Bessman, M.J., Walsh, J.D., Dunn, C.A., Swaminathan, J., Weldon, J.E., and Shen, J. (2001). The gene *ygdP*, associated with the invasiveness of *Escherichia coli* K1, designates a nudix hydrolase, Orf176, active on adenosine (5′)-pentaphospho-(5′)-adenosine (Ap$_5$A). *J. Biol. Chem.* **276**, 37,834–37,838.

Braun, W. (1965). *Bacterial genetics*. Philadelphia: W.B. Saunders.

Brissette, C.A. and Fives-Taylor, P.M. (1998). *Actinobacillus actinomycetemcomitans* may utilize either actin-dependent or actin-independent mechanisms of invasion. *Oral Microbiol. Immunol.* **13**, 137–142.

Camilli, A., Goldfine, H., and Portnoy, D.A. (1991). *Listeria monocytogenes* mutants lacking phosphatidylinositol-specific phospholipase C are avirulent. *J. Exp. Med.* **173**, 751–754.

Christersson, L.A., Albini, B., Zambon, J.J., Wikesjo, U.M., and Genco, R.J. (1987). Tissue localization of *Actinobacillus actinomycetemcomitans* in human periodontitis. I. Light, immunofluorescence and electron microscopic studies. *J. Periodontol.* **58**, 529–539.

Conyers, G.B. and Bessman, M.J. (1999). The gene, *ialA*, associated with invasion of human erythrocytes by *Bartonella bacilliformis*, designates a nudix hydrolase active on dinucleoside 5′-polyphosphate. *J. Biol. Chem.* **274**, 1203–1206.

DeStefano, F., Anda, R.F., Kahn, S., Williamson, D.F., and Russell, C.M. (1993). Dental disease and risk of coronary heart disease and mortality. *BMJ* **306**, 688–691.

Fives-Taylor, P.M. and Meyer, D.H. (1998). The complex, multistep process of invasion of epithelial cells by the periodontopathogen, *Actinobacillus actinomycetemcomitans*, In *The 2nd Indiana Conference: Microbial Pathogenesis, Current and Emerging Issues*, ed. D.J. LeBlanc, M.S. Lantz, and L.M. Switalski, pp. 3–16 Indianapolis: Indiana University.

Fives-Taylor, P.M., Hutchins Meyer, D., Mintz, K.P., and Brissette, C. (1999). *Actinobacillus actinomycetemcomitans*: a causative agent of destructive periodontal disease. *Periodontology* **20**, 136–167.

Fives-Taylor, P., Meyer, D., and Mintz, K. (1995). Characteristics of *Actinobacillus actinomycetemcomitans* invasion of and adhesion to cultured epithelial cells. *Adv. Dent. Res.* **9**, 55–62.

Fives-Taylor, P., Meyer, D., and Mintz, K. (1996). Virulence factors of the periodontopathogen *Actinobacillus actinomycetemcomitans*. *J. Periodontol.* **67**, 291–297.

Gaywee, J., Xu, W., Radulovic, S., Bessman, M.J., and Azad, A.F. (2002). The *Rickettsia prowazekii* invasion gene homolog (*invA*) encodes a nudix hydrolase active on adenosine (5′)-pentaphospho-(5′)-adenosine. *Mol. Cell Proteomics* **1**, 179–183.

Haase, E.M., Zmuda, J.L., and Scannapieco, F.A. (1999). Identification and molecular analysis of rough-colony-specific outer membrane proteins of *Actinobacillus actinomycetemcomitans*. *Infect. Immun.* **67**, 2901–2908.

Hammond, B.F., Darkes, M., Lai, C., and Tsai, C.C. (1981). Isolation and characterization of membrane vesicles of *Actinobacillus actinomycetemcomitans*. *J. Dent. Res.* **60**, 333.

Hammond B.F., Lillard, S.E., and Stevens, R.H. (1987). A bacteriocin of *Actinobacillus actinomycetemcomitans*. *Infect. Immun.* **55**, 686–691.

Holt, S.C., Tanner, A.C., and Socranzky, S.S. (1980). Morphology and ultra structure of oral strains of *Actinobacillus actinomycetemcomitans* and *Haemophilus aphrophilus*. *Infect. Immun.* **30**, 588–600.

Inoue, T., Tanimoto, I., Ohta, H., Kato, K., Murayama, Y., and Fukui, K. (1998). Molecular characterization of low-molecular-weight component protein, Flp, in *Actinobacillus actinomycetemcomitans* fimbriae. *Microbiol. Immunol.* **42**, 253–258.

Inouye, T., Ohta, H., Kokeguchi, S., Fukui, K., and Kato, K. (1990). Colonial variation and fimbriation of *Actinobacillus actinomycetemcomitans*. *FEMS Microbiol. Lett.* **57**, 13–17.

Ishihara, K., Honma, K., Miura, T., Kato, T., and Okuda, K. (1997). Cloning and sequence analysis of the fimbriae associated protein (*fap*) gene from *Actinobacillus actinomycetemcomitans*. *Microb. Pathog.* **23**, 63–69.

Izutsu, K.T., Belton, C.M., Chan, A., Fatherazi, S., Kanter, J.P., Park, Y., and Lamont, R.J. (1996). Involvement of calcium in interactions between gingival epithelial cells and *Porphyromonas gingivalis*. *FEMS Microbiol. Lett.* **144**, 145–150.

Kadurugamuwa, J.L., Rohde, M., Wehland, J., and Timmis, K.N. (1991). Intracellular spread of *Shigella flexneri* through a monolayer mediated by membranous protrusions and associated with reorganization of the cytoskeletal protein vinculin. *Infect. Immun.* **59**, 3463–3471.

Kato S., Muro, M., Akifusa, S., Hanada, N., Semba, I., Fujii, T., Kowashi, Y., and Nishihara, T. (1995). Evidence for apoptosis of murine macrophages by *Actinobacillus actinomycetemcomitans* infection. *Infect. Immun.* **63**, 3914–3919.

Kato, S., Kowashi, Y., and Demuth, D.R. (2001). Outer membrane-like vesicles secreted by *Actinobacillus actinomycetemcomitans*. *Microb. Pathog.* **32**, 1–13.

Lai, C.H., Listgarten, M.A., and Hammond, B.F. (1981). Comparative ultrastructure of leukotoxic and non-leukotoxic strains of *Actinobacillus actinomycetemcomitans*. *J. Periodontal Res.* **16**, 379–389.

Laing-Gibbard, L.P., Lepine, G., and Ellen, R.P. (1998). DNA fragments of *Actinobacillus actinomycetemcomitans* involved in invasion of KB cells. *J. Dent. Res.* **77SI-B**, 770.

Lippmann, J.E. and Fives-Taylor, P.M. (2000). Co-localization of intracellular *A. actinomycetemcomitans* with vacuoles containing early endosomal and late endosomal proteins. *J. Dent. Res.* **79SI**, 255.

Lucia, L.F., Farias, F.F., Eustaquio, C.J., Auxiliadora, M., Carvalho, R., Alviano, C.S., and Farias L.M. (2002). Bacteriocin production by *Actinobacillus actinomycetemcomitans* isolated from the oral cavity of humans with periodontal disease, periodontally healthy subjects and marmosets. *Res. Microbiol.* **153**, 45–52.

Meyer, D.H. and Fives-Taylor, P.M. (1993). Evidence that extracellular components function in adherence of *Actinobacillus actinomycetemcomitans* to epithelial cells. *Infect. Immun.* **61**, 4933–4936.

Meyer, D.H. and Fives-Taylor, P.M. (1994). Characteristics of adherence of *Actinobacillus actinomycetemcomitans* to epithelial cells. *Infect. Immun.* **62**, 928–935.

Meyer, D.H. and Fives-Taylor, P.M. (1998). Oral pathogens: from dental plaque to cardiac disease. *Curr. Opin. Microbiol.* **1**, 88–95.

Meyer, D.H., Lippmann, J.E., and Fives-Taylor, P.M. (1996). Invasion of epithelial cells by *Actinobacillus actinomycetemcomitans*: a dynamic, multistep process. *Infect. Immun.* **64**, 2988–2997.

Meyer, D.H., Rose, J.E., Lippmann, J.E., and Fives-Taylor, P.M. (1999). Microtubules are associated with intracellular movement and spread of the periodontopathogen *Actinobacillus actinomycetemcomitans*. *Infect. Immun.* **67**, 6518–6525.

Meyer, D.H., Wei, J., and Fives-Taylor, P.M. (1995). Cloning of a DNA fragment associated with *Actinobacillus actinomycetemcomitans* invasion. *J. Dent. Res.* **74SI**, 200.

Meyer, D.H., Mintz, K.P., and Fives-Taylor, P.M. (1997a). Models of invasion of enteric and periodontal pathogens into epithelial cells: a comparative analysis. *Crit. Rev. Oral Biol. Med.* **8**, 389–409.

Meyer, D.H., Sreenivasan, P.K., and Fives-Taylor, P.M. (1991). Evidence for invasion of a human oral cell line by *Actinobacillus actinomycetemcomitans*. *Infect. Immun.* **59**, 2719–2726.

Meyer, D.H., Fives-Taylor, P.M., and Rose, J.E. (2000). *Actinobacillus actinomycetemcomitans* displays an entity that binds antibody to a kinesin-like microtubule motor protein. *J. Dent. Res.* **79SI**, 256.

Meyer, D.H., Mackie, T.N., and Fives-Taylor, P.M. (1997b). *Actinobacillus actinomycetemcomitans* exhibits phospholipase C-B (PC-PLC) activity. *J. Dent. Res.* **76SI**, 26.

Mintz, K.P. and Fives-Taylor, P.M. (1994). Adhesion of *Actinobacillus actinomycetemcomitans* to a human oral cell line. *Infect. Immun.* **62**, 3672–3678.

Mitchell, S.J. and Minnick, M.F. (1995). Characterization of a two-gene locus from *Bartonella bacilliformis* associated with the ability to invade human erythrocytes. *Infect. Immun.* **63**, 1552–1562.

Miyasaki, K.T., Bodeau, A.L., Ganz, T., Selsted, M.E., and Lehrer, R.I. (1990). *In vitro* sensitivity of oral, gram-negative, facultative bacteria to the bactericidal activity of human neutrophil defensins. *Infect. Immun.* **58**, 3934–3940.

Miyasaki, K.T., Wilson, M.E., Reynolds, H.S., and Genco, R.J. (1984). Resistance of *Actinobacillus actinomycetemcomitans* and differential susceptibility of oral *Haemophilus* species to the bactericidal effects of hydrogen peroxide. *Infect. Immun.* **46**, 644–648.

Muro, M., Koseki, T., Akifusa, S., Kato, S., Kowashi, Y., Ohsaki, Y., Yamato, Y., Nishijima, M., and Nishihara, T. (1997). Role of CD14 molecules in internalization of *Actinobacillus actinomycetemcomitans* by macrophages and subsequent induction of apoptosis. *Infect. Immun.* **65**, 1147–1151.

Nakashima K., Tomioka, J., Kato, S., Nishihara, T., and Kowashi, Y. (2002). Nitric oxide-mediated protection of *A. actinomycetemcomitans*-infected murine macrophages against apoptosis. *Nitric Oxide* **6**, 61–68.

Nonaka K., Ishisaki, A., Muro, M., Kato, S., Oido, M., Nakashima, K., Kowashi, Y., and Nishihara, T. (1997). Possible involvement of protein kinase C in apoptotic cell death of macrophages infected with *Actinobacillus actinomycetemcomitans*. *FEMS Microbiol. Lett.* **159**, 247–254.

Nonaka K., Ishisaki, A., Okahashi, N., Koseki, T., Kato, S., Muro, M., Nakashima, K., Nishihara, T., and Kowashi, Y. (2001). Involvement of caspases in apoptotic cell death of murine macrophages infected with *Actinobacillus actinomycetemcomitans*. *J. Periodontal Res.* **36**, 40–47.

Nowotny, A., Behling, U.H., Hammond, B., Lai, C.H., Listgarten, M., Pham, P.H., and Sanavi, F. (1982). Release of toxic microvesicles by *Actinobacillus actinomycetemcomitans*. *Infect. Immun.* **37**, 151–154.

Pace, J., Hayman, M.J., and Galan, J.E. 1993. Signal transduction and invasion of epithelial cells by *Salmonella typhimurium*. *Cell* **72**, 505–514.

Preus, H.R., Namork, E., and Olsen, I. (1988). Fimbriation of *Actinobacillus actinomycetemcomitans*. *Oral Microbiol. Immunol.* **3**, 93–94.

Rosan, B., Slots, J., Lamont, R.J., Listgarten, M.A., and Nelson, G.M. (1988). *Actinobacillus actinomycetemcomitans* fimbriae. *Oral Microbiol. Immunol.* **3**, 58–63.

Rose, J.E., Meyer, D.H., and Fives-Taylor, P.M. (1998). Detection of bacteria-microtubule interactions in a cell-free extract. *Meth. Cell Sci.* **19**, 325–330.

Rose, J.E., Meyer, D.H., and Fives-Taylor, P.M. (1999). *Actinobacillus actinomycetemcomitans* binds specifically to the plus ends of microtubules. *J. Dent. Res.* **78SI**, 133.

Rose, J.E., Meyer, D.H., and Fives-Taylor, P.M. (2003). Aae, an autotransporter involved in adhesion of *Actinobacillus actinomycetemcomitans* to epithelial cells. *Infect. Immun.* **71**, 2386–2393.

Rudney, J.D., Chen, R., and Sedgewick, G.J. (2001). Intracellular *Actinobacillus actinomycetemcomitans* and *Porphyromonas gingivalis* in buccal epithelial cells collected from human subjects. *Infect. Immun.* **69**, 2700–2707.

Saarela, M., Lippmann, J.E., Meyer, D.H., and Fives-Taylor, P.M. (1999). *Actinobacillus actinomycetemcomitans apaH* is implicated in invasion of epithelial cells. *J. Dent. Res.* **78SI**, 259.

Saglie, F.R., Marfany, A., and Camargo, P. (1988). Intragingival occurrence of *Actinobacillus actinomycetemcomitans* and *Bacteroides gingivalis* in active destructive periodontal lesions. *J. Periodontol.* **59**, 259–265.

Saglie, F.R., Smith, C.T., Newman, M.G., Carranza, F.A. Jr., Pertuiset, J.H., Cheng, L., Auil, E., and Nisengard, R.J. (1986). The presence of bacteria in the oral epithelium in periodontal disease. II. Immunohistochemical identification of bacteria. *J. Periodontol.* **57**, 492–500.

Saglie, F.R., Carranza, F.A. Jr., Newman, M.G., Cheng, L., and Lewin, K.J. (1982). Identification of tissue-invading bacteria in human periodontal disease. *J. Periodontal Res.* **17**, 452–455.

Scannapieco, F.A., Kornman, K.S., and Coykendall, A.L. (1983). Observation of fimbriae and flagella in dispersed subgingival dental plaque and fresh bacterial isolates from periodontal disease. *J. Periodontal Res.* **18**, 620–633.

Scannapieco, F.A., Millar, S.J., Reynolds, H.S., Zambon, J.J., and Levine, M.J. (1987). Effect of anaerobiosis on the surface ultrastructure and surface proteins of *Actinobacillus actinomycetemcomitans (Haemophilus actinomycetemcomitans)*. *Infect. Immun.* **55**, 2320–2323.

Slots, J. (1982). Selective medium for isolation of *Actinobacillus actinomycetemcomitans*. *J. Clin. Microbiol.* **15**, 606–609.

Smith, G.A., Marquis, H., Jones, S., Johnston, N.C., Portnoy, D.A., and Goldfine, H. (1995). The two distinct phospholipases C of *Listeria monocytogenes* have overlapping roles in escape from a vacuole and cell-to-cell spread. *Infect. Immun.* **63**, 4231–4237.

Sosroseno, W., Barid, I., Herminajeng, E., and Susilowati, H. (2002). Nitric oxide production by a murine macrophage cell line (RAW 264.7) stimulated with lipopolysaccharide from *Actinobacillus actinomycetemcomitans*. *Oral Microbiol. Immunol.* **17**, 72–78.

Sreenivasan, P.K., Meyer, D.H., and Fives-Taylor, P.M. (1993). Requirements for invasion of epithelial cells by *Actinobacillus actinomycetemcomitans*. *Infect. Immun.* **61**, 1239–1245.

Stevens, R.H., Lillard, S.E., and Hammond, B.F. (1987). Purification and biochemical properties of a bacteriocin from *Actinobacillus actinomycetemcomitans*. *Infect. Immun.* **55**, 692–697.

St. Geme, J.W. III and Cutter, D. (2000). The *Haemophilus influenzae* Hia adhesin is an autotransporter protein that remains uncleaved at the C terminus and fully cell associated. *J. Bacteriol.* **182**, 6005–6013.

St. Geme, J.W. III, de la Morena, M.L., and Falkow, S. (1994). A *Haemophilus influenzae* IgA protease-like protein promotes intimate interaction with human epithelial cells. *Mol. Microbiol.* **14**, 217–233.

Teng, Y.T., Taylor, G.W., Scannapieco, F., Kinane, D.F., Curtis, M., Beck, J.D., and Kogon, S. (2002). Periodontal health and systemic disorders. *J. Can. Dent. Assoc.* **68**, 188–192.

Tilney, L.G. and Portnoy, D.A. (1989). Actin filaments and the growth, movement, and spread of the intracellular bacterial parasite, *Listeria monocytogenes*. *J. Cell Biol.* **109**, 1597–1608.

Van Dyke, T.E., Bartholomew, E., Genco, R.J., Slots, J., and Levine, M.J. (1982). Inhibition of neutrophil chemotaxis by soluble bacterial products. *J. Periodontol.* **53**, 502–508.

Wilson, M., Kamin, S., and Harvey, W. (1985). Bone resorbing activity of purified capsular material from *Actinobacillus actinomycetemcomitans*. *J. Periodontal Res.* **20**, 484–491.

Yamaguchi, N., Kawasaki, M., Yamashita, Y., Nakashima, K., and Koga, T. (1995). Role of capsular polysaccharide-like serotype-specific antigen in resistance of *Actinobacillus actinomycetemcomitans* to phagocytosis by human polymorphonuclear leukocytes. *Infect. Immun.* **63**, 4589–4594.

Yamaguchi, N., Yamashita, Y., Ikeda, D., and Koga, T. (1996). *Actinobacillus actinomycetemcomitans* serotype b-specific polysaccharide antigen stimulates production of chemotactic factors and inflammatory cytokines by human monocytes. *Infect. Immun.* **64**, 2563–2570.

Yamauchi, K., Tomita, M., Giehl, T.J., and Ellison, R.T. III. (1993). Antibacterial activity of lactoferrin and a pepsin-derived lactoferrin peptide fragment. *Infect. Immun.* **61**, 719–728.

Zambon, J.J. (1985). *Actinobacillus actinomycetemcomitans* in human periodontal disease. *J. Clin. Periodontol.* **12**, 1–20.

CHAPTER 10

Invasion by *Porphyromonas gingivalis*

Özlem Yilmaz and Richard J. Lamont

Porphyromonas gingivalis cells are Gram-negative, anaerobic, nonmotile short rods that produce black pigmented colonies on blood agar. The taxonomy of the species dates back to 1921 when Oliver and Wherry isolated an organism from a variety of oral and nonoral sites that they were to designate *Bacterium melaninogenicum*. This heterogeneous grouping was later subdivided into nonfermenters, weak fermenters, and strong fermenters. After a number of status changes within the genus *Bacteroides*, asaccharolytic oral isolates were assigned to the taxon *P. gingivalis*. The primary ecological niche of *P. gingivalis* is in the subgingival crevice, the gap between the surfaces of the tooth and the gingiva (gum); however, the organism can be found elsewhere in the mouth, including supragingival (above the gum) tooth surfaces, the tongue, tonsils, and buccal (cheek) mucosa. Although the species has been associated with odontogenic abscesses and nonoral infections (discussed later), the primary pathogenic potential of *P. gingivalis* is in periodontal disease. The periodontal tissues include the gingiva, periodontal ligament, and alveolar bone, and they constitute the supporting tissues of the teeth. Chronic destruction of the periodontium, such as occurs in periodontal diseases, can eventually lead to exfoliation of teeth and is the most common cause of tooth loss in adults. Periodontal diseases vary in severity and age of onset, and *P. gingivalis* is associated, either alone or in combination with other bacteria, with the most severe manifestations. However, the frequent occurrence of the organism in healthy adults and in young children indicates that a complex interplay between host and pathogen exists, and that disruption of this ecological balance is required for disease to ensue.

In the gingival crevice the area of contact between the gingiva and the tooth is known as the junctional epithelium, and it is characterized by a lack of keratinization, limited differentiation, and a relatively permeable structure. In destructive periodontal disease there is migration of the junctional epithelium, resulting in enlargement of the crevice into a deeper periodontal pocket that contains inflammatory cells such as neutrophils and T cells. The gingiva itself also contains immune cells, including B cells, T cells, and dendritic cells. The microbiota of the gingival area in both health and disease is complex, with at least 500 species of bacteria present in the gingival crevice. Colonization and persistence by periodontal pathogens require, therefore, successful encounters with antecedent bacterial species and with host eukaryotic cells. Consistent with these constraints, *P. gingivalis* can bind to, invade, and survive inside a variety of host cells, including epithelial and endothelial cells.

P. GINGIVALIS ADHESION

An early event in the process of bacterial internalization is adhesion to the host cell surface. Attachment of bacteria is often required to trigger the signaling pathways within the host cells that ultimately induce the membrane and cytoskeletal rearrangements that bring the bacteria into the cell. *P. gingivalis* is endowed with a multiplicity of adhesins that mediate adhesion to epithelial cells, endothelial cells, fibroblasts, and erythrocytes, and to components of the extracellular matrix, namely laminin, elastin, fibronectin, type I collagen, thrombospondin, and vitronectin (Sojar et al., 1995, 1999; Kontani et al., 1997; Nakamura et al., 1999; Dorn et al., 2000; Lamont and Jenkinson, 2000). Adhesive activity has been demonstrated for fimbriae, outer membrane proteins, and proteases. These molecules may function collectively, and their activities are integrated and controlled at the transcriptional and posttranslational levels (reviewed in Lamont and Jenkinson, 1998, 2000).

The major fimbriae of *P. gingivalis* are distinct from other Gram-negative fimbriae and do not appear to belong to any of the existing classifications. The fimbriae are composed of an ~43-kDa fimbrillin (FimA) monomer that possesses a number of binding domains for individual substrate recognition. The functional domain of FimA for epithelial cells has been localized to a region spanning amino acid residues 49–90, although the boundaries of the interactive site are not known precisely (Sojar et al., 1999). Fimbriae-mediated binding is important for subsequent invasion, as FimA-deficient mutants are attenuated in their ability to internalize, and both purified fimbrillin and antibodies to fimbriae can block invasion (Njoroge et al., 1997; Weinberg et al., 1997). Furthermore, fimbrillin-coated microspheres are efficiently taken up by epithelial cells (Nakagawa et al., 2002). A number of epithelial cell molecules have been found to function as cognate receptors for fimbrillin, including a 48-kDa surface protein, cytokeratins, and integrins (Weinberg et al., 1997; Sojar et al., 2002; Yilmaz et al., 2002). With regard to stimulation of invasion, fimbriae–integrin binding may be the most significant interaction, as integrin antibodies can inhibit *P. gingivalis* invasion of epithelial cells (Yilmaz et al., 2002). Moreover, as integrins are initiators of signal transduction pathways, the engagement of an integrin receptor by *P. gingivalis* fimbriae may be a means by which the organism begins to seize control of host cell signaling machinery. *P. gingivalis* fimbriae are also involved in the adhesion-dependent invasion of endothelial cells (Deshpande et al., 1998; Khlgatian et al., 2002), although other adhesins may work in concert (Dorn et al., 2000).

The *fimA* gene is monocistronic, although immediately downstream are four genes whose products may be associated with the mature fimbriae (Watanabe et al., 1996). The *fimA* upstream region contains a functionally active sigma-70-like promoter consensus sequence along with a potential UP element (Xie and Lamont, 1999). AT-rich sequences upstream of the RNA polymerase binding sites are involved in positive regulation of transcriptional activity. Environmental cues to which the *fimA* promoter responds include temperature, hemin concentration, and salivary molecules (Xie et al., 1997), which are parameters with relevance to conditions in the oral cavity. The *fimA* gene can also be positively autoregulated by the FimA protein (Xie et al., 2000). In addition, expression of FimA decreases following association of *P. gingivalis* with epithelial cells (Wang et al., 2002). Thus, after completing the task of inducing invasion, fimbriae may no longer be required by the intracellular *P. gingivalis* cells. As fimbrillin has a number of immunostimulatory properties, such as induction of cytokines and chemokines (Ogawa et al., 1994), reduced FimA expression may aid in immune avoidance by the organism.

Although the primary function of proteinases secreted by the asaccharolytic *P. gingivalis* is the provision of nutrients, proteinases are also involved both directly and indirectly in adhesion. Several distinct proteinases are produced by *P. gingivalis* (reviewed in Potempa et al., 1995; Kuramitsu, 1998; Curtis et al., 1999), and direct enzyme–substrate interactions may be able to effectuate adhesion, although such binding is unlikely to persist for extended periods. Of greater importance, the C-terminal coding regions of the Arg-X and Lys-X specific proteases RgpA and Kgp contain extensive regions with hemagglutinin activity (Barkocy-Gallagher et al., 1996; Lamont and Jenkinson, 1998; Curtis et al., 1999). These regions, therefore, can be predicted to bind directly to human cell surface receptors. Proteinases can also contribute to adherence through the partial degradation of substrates resulting in the subsequent exposure of epitopes for adhesin recognition. For example, hydrolysis of fibronectin or other matrix proteins by the Arg-X specific proteases RgpA and RgpB displays C-terminal Arg residues that mediate fimbriae-dependent binding (Kontani et al., 1996). Other means by which RgpA and RgpB can contribute to adhesion are through processing the leader peptide from the fimbrillin precursor (Nakayama et al., 1996), and by upregulating transcription of the *fimA* gene (Tokuda et al., 1996; Xie et al., 2000).

In addition to the hemagglutinin-associated activities of the RgpA and Kgp proteinases, several additional hemagglutinin (*hag*) genes are present in the *P. gingivalis* genome. The *hagA* gene encodes a large protein of over 230 kDa containing four contiguous direct 440–456 aa residue repeat blocks (Han et al., 1996). Each repeat block may represent a functional hemagglutinin domain, and similar *hagA*-like sequences are found at multiple sites in the chromosome. The *hagB* and *hagC* genes are at distinct chromosomal loci, although the HagB and C proteins (~40 kDa) are very similar (Progulske-Fox et al., 1995). A minimal peptide motif PVQNLT has been shown to be associated with hemagglutinating activity and is found within the proteinase–hemagglutinin sequences (Shibita et al., 1999) and at multiple chromosomal sites.

Thus, *P. gingivalis* possesses a variety of adhesins with differing receptor specificities and affinities that can potentially impinge to varying degrees upon diverse receptor-dependent host cell biochemical pathways. This may allow the organism to utilize more than one pathway for internalization. For example, although fimbriae-deficient mutants show reduced uptake into epithelial cells, a low level of invasion remains (Weinberg et al., 1997). Nonfimbrial dependent uptake, although less efficient than FimA-mediated uptake, may result from attachment by other surface adhesins.

UPTAKE OF *P. GINGIVALIS* BY HOST CELLS

P. gingivalis can invade epithelial cells, endothelial cells, and dendritic cells (reviewed in Lamont and Jenkinson, 1998; Lamont and Yilmaz, 2002). Interestingly, although the overall mechanistic basis is similar in these cell systems, the signal transduction pathways activated by the organism and the intracellular locations and trafficking of the bacteria differ according to cell type. However, in all cases invasion is an active, bacterially driven process that has a significant impact on the phenotypic properties and fate of the host cell.

Primary gingival epithelial cells

The study of primary cultures of gingival epithelial cells (GEC) provided the first evidence for intracellular invasion by *P. gingivalis* in 1992 (Lamont et al., 1992). GEC are cultured from basal epithelial cells extracted from gingival explants, and they can be maintained in culture for several generations. Immunohistochemical staining has shown that the cells are nondifferentiated and noncornified, which are characteristics that are in common with the junctional epithelium. Thus, although not derived from the junctional epithelium, GEC demonstrate similar properties and thus provide a relevant *ex vivo* model for the events that occur at the base of the gingival crevice.

   P. gingivalis lacks the components and effectors of the type III secretion machinery that are important in the invasive processes of other organisms and that are discussed elsewhere in this volume. However, when in contact with GEC, *P. gingivalis* is induced to secrete a novel set of extracellular proteins (Park and Lamont, 1998). Some functional equivalence to type III effectors is implied by the finding that one of the secreted proteins of *P. gingivalis* is a homologue of a phosphoserine phosphatase (Lamont and Yilmaz, 2002), although the functionality of this molecule, either intracellularly or extracellularly, remains to be established. Nevertheless, despite the absence of a classical type III secretion apparatus, *P. gingivalis* invasion of GEC is swift and profuse (Belton et al., 1999).

   Fluorescent microscopic imaging has shown that invasion is complete within 15 min, and that all GEC exposed to *P. gingivalis* take up large numbers of organisms: this is invasion on a scale unsurpassed by any of the enteropathogens. Once inside the cells, the bacteria are not confined to a membrane-bound vacuole and congregate in the perinuclear region (Belton et al., 1999). *P. gingivalis* cells remain viable and capable of intracellular replication (Lamont et al., 1995). Interestingly, despite the burden of large numbers of intracellular bacteria, GEC do not undergo necrotic or apoptotic cell death (Nakhjiri et al., 2001); however, the cells contract and there is condensation of the actin cytoskeleton after prolonged cohabitation with *P. gingivalis* (Belton et al., 1999).

   The signaling events required for uptake into GEC are thought to ensue primarily from fimbriae–integrin interactions, although other signal transduction pathways may be operational (Yilmaz et al., 2002). Integrins have both a structural role, in linking extracellular matrix proteins with the cellular actin cytoskeleton in order to regulate cell shape and tissue architecture, and a signaling role, through intracellular signals generated by integrin–receptor coupling that regulate cell migration, gene expression, growth, survival, and inflammatory responses. Integrins are heterodimeric transmembrane receptors with no catalytic activity; therefore, the signals initiated by integrin–ligand interactions are transduced into cells through the activation of a number of specialized cytoplasmic proteins. Enhanced tyrosine phosphorylation

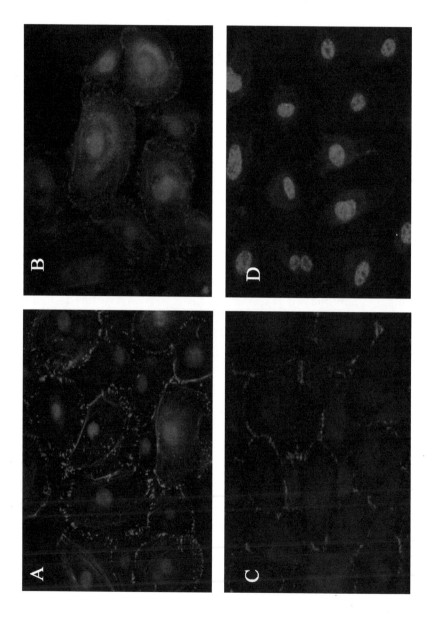

of these proteins plays an essential role in the perpetuation of the intracellular signals created by integrins into specific targets in cells. Among those proteins, FAK (focal adhesion kinase) and paxillin have emerged as key signal-transducing components (Turner, 2000). Phosphorylation of paxillin is associated with coordinate formation of focal adhesions and stress fibers. Paxillin thereby provides a platform for efficient propagation of signals from one component to the next. Invasion by fimbriated *P. gingivalis* promotes a significant amount of tyrosine phosphorylation of paxillin in GEC during early infection (between 5 and 20 min; Yilmaz et al., 2002).

Recent immunofluorescent studies (Yilmaz et al., 2003) have visualized the subcellular distribution of paxillin during *P. gingivalis* invasion (Fig. 10.1). Immediately after infection by *P. gingivalis*, paxillin aggregates at the plasma membrane and forms microspikes or lamellipodial-like extensions at the edges of cells. These results suggest that *P. gingivalis* recruits paxillin to the plasma membrane and promotes formation of focal adhesion complexes that may potentiate efficient bacterial uptake in GEC. In addition to the early redistribution of paxillin, after 24 h there is colocalization of paxillin with *P. gingivalis* in the perinuclear space. Immunofluorescence also indicates that FAK is activated and recruited to the membrane of GEC during early periods of *P. gingivalis* exposure (Fig. 10.1), and later it relocates to the perinuclear area with *P. gingivalis*. Paxillin and FAK phosphorylations, therefore, appear to be important in both early and late events of *P. gingivalis* invasion and intracellular trafficking.

Integrin signaling also modulates actin cytoskeleton and microtubule dynamics. Sensitivity of the invasion process to the inhibitors cytochalasin D and nocodazole provides indirect evidence that both actin microfilament and microtubule rearrangements are required for *P. gingivalis* entry into the host cell (Lamont et al., 1995). Indeed, immunofluorescent microscopy has shown that actin is assembled into filament-rich microspikes at the periphery of the GEC during *P. gingivalis* invasion. Later, both the actin microfilament and microtubules are dramatically depolymerized and nucleated.

Signaling pathways downstream of integrin focal adhesions often funnel through the MAP kinase family of signal transduction mediators. Consistent

INVASION BY PORPHYROMONAS GINGIVALIS

Figure 10.1. (*facing page*). Immunofluorescence microscopy (400×) of *P. gingivalis* invasion of GEC. *P. gingivalis* (green) induces recruitment of paxillin (red), A, and FAK (red), B, to the cell peripheries. GEC in C and D are stained with antibodies to paxillin or FAK, respectively, in the absence of *P. gingivalis*. See color section.

with this, invasion by *P. gingivalis* results in activation of JNK, a stress-activated protein kinase of the MAP kinase family (Watanabe et al., 2001). Also consonant with integrin activation, *P. gingivalis* invasion is associated with a transient increase in the intracellular Ca^{2+} concentration. The calcium ion increase results, at least in part, from release of calcium from a thapsigargin-sensitive intracellular store (Izutsu et al., 1996). However, *P. gingivalis* is capable of multiple independent interventions on GEC signal transduction pathways as, in contrast to JNK, internalized *P. gingivalis* cause dephosphorylation of ERK1/2 (Watanabe et al., 2001). In addition, *P. gingivalis* does not induce activation and nuclear translocation of the eukaryotic transcriptional activator NF-κB. Such an ability to selectively activate and suppress different components of related signaling pathways signifies a degree of versatility and sophistication of *P. gingivalis* in its dealings with GEC, pointing toward a long evolutionary association between the two cell types. The interactions between *P. gingivalis* and GEC are represented schematically in Fig. 10.2.

KB (HeLa) cells

Another cell type used to study of *P. gingivalis* invasion is the KB cell. Long thought to be derived from the oral epithelium, KB cells are now recognized to be in fact HeLa cells that contaminated the original cell culture. Invasion of these transformed epithelial cells by *P. gingivalis* is somewhat less efficient than GEC, with values less than 0.1% of the initial inoculum generally reported (Duncan et al., 1993; Sandros et al., 1994; Njoroge et al., 1997). This may be a consequence of the alterations in signal transduction pathways and surface protein expression that accompany transformation. Initial adherence to KB cells can be mediated by both cysteine proteases (Chen et al., 2001) and FimA (Njoroge et al., 1997). In contrast to GEC, engulfment of bacteria then occurs by classic receptor-mediated endocytosis (Sandros et al., 1996), and bacteria can be found both free in the cytoplasm and contained within membrane-bound vacuoles (Njoroge et al., 1997). Features in common with GEC include the accumulation of *P. gingivalis* cells in the perinuclear region (Houalet-Jeanne et al., 2001) and subsequent bacterial replication (Madianos et al., 1996).

Endothelial cells

Invasion of bovine and human heart and aortic endothelial cells by *P. gingivalis* has been established (Deshpande et al., 1998; Dorn et al., 1999).

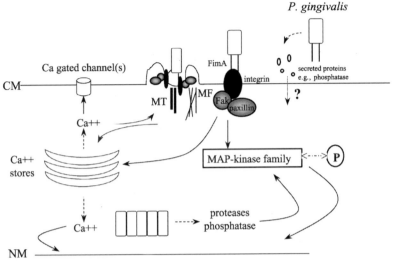

P. gingivalis

Ca gated channel(s)

FimA

integrin

secreted proteins
e.g., phosphatase

CM

MT

MF

Fak
paxillin

?

Ca++

Ca++
stores

MAP-kinase family

P

303

proteases
phosphatase

Ca++

NM

Disruption of nuclear transcription factor activity and
modulation of gene expression (e.g., IL-8, Bcl-2)

Figure 10.2. Model of currently understood *P. gingivalis* interactions with primary gingival epithelial cells. *P. gingivalis* cells bind through adhesins such as fimbriae to integrins on gingival cells. FAK and paxillin are recruited, and microtubules and microfilaments are rearranged to facilitate invagination of the membrane that results in the engulfment of bacterial cells. *P. gingivalis* rapidly locate in the perinuclear area where they replicate. Calcium ions are released from intracellular stores, which may regulate calcium-gated pores in the cytoplasmic membrane. Other signaling molecules such as the MAP-kinase family can be phosphorylated/dephosphorylated or degraded. Gene expression in the epithelial cells is ultimately affected. Abbreviations: Ca = calcium; CM = cytoplasmic membrane; IL-8 = interleukin-8; MF = actin microfilaments; MT = tubulin microtubules; NM = nuclear membrane; P = phosphate; ⟶ is for a pathway with potential intermediate steps; ┄┄➤ is for translocation; ┄┄➤ is for release; ←- - -› is for reversible association.

Invasion requires remodeling of the microfilament and microtubule cytoskeleton through protein phosphorylation-dependent signaling (Deshpande et al., 1998). Once inside the cells the bacteria are present in multimembranous vacuoles that resemble autophagosomes (Dorn et al., 2001). These vacuoles are positive for the early endosomal marker Rab5, and they rapidly acquire HsGsa7p, required for the formation of the autophagosome. The bacteria then traffic to a late autophagosome that contains both the rough endoplasmic reticulum protein BiP and the lysosomal protein LGP120, but that lacks cathepsin L and late endosomal markers (Dorn et al., 2001). In endothelial cells, therefore, *P. gingivalis* evades the endocytic pathway leading to

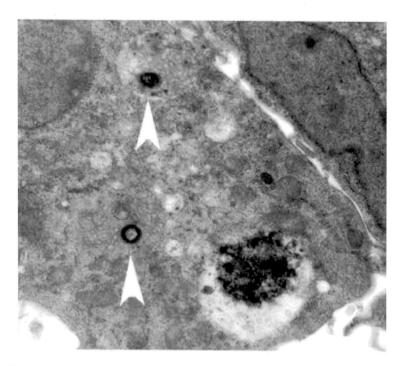

Figure 10.3. Transmission electron microcopy (12,500×) of *P. gingivalis* captured by a dendritic cell and located within multivesiculated compartments (arrowheads). (Image provided by C. W. Cutler.)

lysosomes, and instead it traffics to the autophagosome, where development of this organelle is impeded. Invasion of endothelial cells by *P. gingivalis* could allow access to cells of vascular walls, where the induction of autophagocytic pathways may alter the properties of the cells. The persistence of *P. gingivalis* could then exacerbate the immune response along the vasculature. Invasion of vascular endothelial cells may also provide a portal for bacterial entry into the bloodstream with subsequent systemic spread.

Dendritic cells

Dendritic cells are antigen-presenting cells that can activate lymphocytes, including, distinctively naïve T cells (Banchereau and Steinman, 1998). Dendritic cells increase in number at sites of diseased oral epithelium that contain intragingival bacteria. *P. gingivalis* can internalize within cultured dendritic cells (Fig. 10.3), and invasion is associated with sensitization and activation of the dendritic cells (Cutler et al., 1999). This process has parallels with contact hypersensitivity responses, and hence such reactions may play a role in periodontal diseases. The mechanistic basis for invasion in these cells has yet to be investigated.

CONSEQUENCES OF *P. GINGIVALIS* INVASION

Epithelial cells

The presence of large numbers of intracellular *P. gingivalis* would, at first glance, appear to represent an insult that could be terminal for the epithelial cells. Indeed, after several hours of infection, GEC begin to shrink and round up, features indicative of apoptotic cell death. However, the cells do not detach from the substratum and remain capable of trypan blue exclusion, calcein hydrolysis, and maintenance of physiologic intracellular calcium ion concentrations (Belton et al., 1999). An examination of molecular markers of apoptosis showed that at early time points of *P. gingivalis* invasion there is an increase in proapoptotic molecules such as Bax. However, after extended incubation the Bax levels decline, and there is an increase in the expression of the antiapoptotic molecule Bcl-2 (Nakhjiri et al., 2001). Furthermore, *P. gingivalis* is even capable of blocking apoptosis induced by the human topoisomerase I inhibitor, camptothecin. These results can be interpreted in terms in which the initial response of GEC to the *P. gingivalis* onslaught is one of activation of programmed cell death. *P. gingivalis*, however, having located in a nutritionally rich, immune-privileged site, blocks the cell death pathways in order to maintain its intracellular lifestyle. Furthermore, the proteases of *P. gingivalis* may protect the organism from β-defensins, which are intracellular antimicrobial peptides produced by epithelial cells (Devine et al., 1999). Unconstrained replication of *P. gingivalis* may be prevented, however, by calprotectin, an S100 calcium-binding protein with broad-spectrum antimicrobial activity produced inside epithelial cells (Nisapakultorn et al., 2001).

The interaction between *P. gingivalis* and GEC is more than a life or death struggle for supremacy and can result in more subtle phenotypic changes in the host. Transcription and secretion of interleukin (IL)-8 (a potent neutrophil chemokine) by GEC is inhibited following *P. gingivalis* invasion. Moreover, *P. gingivalis* can antagonize IL-8 secretion following stimulation of epithelial cells by common plaque commensals (Darveau et al., 1998). Reduced expression of epithelial cell intercellular adhesion molecule (ICAM)-1 may also contribute to downregulation of the innate host response (Madianos et al., 1997). Regulation of matrix metalloproteinase (MMP) production by gingival epithelial cells is disrupted following contact with *P. gingivalis* (Fravalo et al., 1996), thus interfering with extracellular matrix repair and reorganization.

Endothelial cells

Although the ultimate metabolic fate of endothelial cells and their internalized organisms remains to be determined, it is becoming apparent that many important properties of endothelial cells are modulated by *P. gingivalis* infection. Invasion by *P. gingivalis* leads to upregulation of ICAM-1 and vascular cell adhesion molecule (VCAM)-1, along with P- and E-selectins (Khlgatian et al., 2002). The effect appears to be mediated through the major fimbriae, as it can be blocked with fimbrial antibodies and does not occur with a fimbriae-deficient mutant. Increased expression of these molecules can be predicted to elevate the levels of leukocytes recruited to sites of *P. gingivalis* infection. Levels of IL-8 and MCP-1 are also modulated in response to *P. gingivalis*. In these cases, however, whole cells of *P. gingivalis* abolish normal IL-8 and MCP-1 responses, whereas isolated outer membrane components and fimbrillin can stimulate production (Nassar et al., 2002). The inhibitory effect of whole cells is not invasion dependent, as it occurs in the absence of detectable internalized *P. gingivalis*. Indeed, under certain conditions, endothelial cells can be induced to secrete MCP-1 by internal *P. gingivalis* (Kang and Kuramitsu, 2002). Thus endothelial cell responses to *P. gingivalis* are complex and multithreaded, possibly a means for facilitating the long-term association of the organism with the host cell.

RELEVANCE TO HEALTH AND DISEASE

Periodontal diseases

The pathogenesis of periodontal diseases involves multiple bacteria with a range of virulence factors interfacing with a variety of host cells and immune effector molecules. Within this framework, *P. gingivalis* invasion could play a number of important roles in the disease process. An intracellular location will shelter the bacteria from the ravages of the immune system, and it may allow the organisms to increase in number to exceed a threshold required to initiate disease. The phenotypic changes in epithelial cells that are related to *P. gingivalis* invasion also have the potential to impinge on several aspects of pathogenesis (Lamont and Yilmaz, 2002).

The gene encoding the neutrophil chemokine IL-8 is transcriptionally downregulated by *P. gingivalis* even in the presence of otherwise stimulatory organisms such as *Fusobacterium nucleatum* (Darveau et al., 1998; Huang et al., 1998, 2001). In clinically healthy tissue, IL-8 forms a concentration gradient that increases from the gingival tissue toward the surface (Tonetti et al., 1994) and will thus direct neutrophils to sites of bacterial accumulation. Hence, low level expression of IL-8 is considered to be important in ensuring gingival health by controlling bacteria and preventing neutrophil-mediated damage in the tissues. Inhibition of IL-8 accumulation by *P. gingivalis* at sites of bacterial invasion could impede innate host defense at the bacteria–epithelia interface, as the host would no longer be able to detect the presence of bacteria and direct neutrophils for their removal. The ensuing overgrowth of bacteria would then contribute to a burst of disease activity.

Nonetheless, host polymorphonuclear neutrophils and other defense mechanisms do eventually become mobilized, as evidenced by the inflammatory nature of *P. gingivalis*-associated periodontal diseases. The overgrowth of subgingival plaque bacteria, or of *P. gingivalis* itself, that ensues after initial immune suppression may therefore reactivate the immune response. Indeed, epithelial cells can be induced to secrete IL-8 by *P. gingivalis*, depending on the conditions of stimulation (Huang et al., 2001). Moreover, the encounter with different host cells as the infection progresses may also stimulate the immune response. For example, *P. gingivalis* invasion of dendritic cells results in maturation, increased costimulatory molecule expression, and stimulatory activity for T cells (Cutler et al., 1999). The migration and proliferation of *P. gingivalis*-specific effector T cells could be one means by which the immune system gears up in periodontal disease (Saglie et al., 1987; Cutler et al., 1999).

In addition to effects on immune modulators, invasion by *P. gingivalis* can impinge on the activity of MMP enzymes. MMPs are members of a family of zinc-dependent endopeptidases with a broad spectrum of proteolytic activity that collectively can degrade all of the components of the extracellular matrix (DeCarlo et al., 1998). MMPs are required for epithelial cell migration and to detach the cells from the underlying matrix, and they are also involved in tissue remodeling by facilitating the removal of damaged tissue. These activities are important in maintenance of the integrity of the epithelial layer. However, destruction of the extracellular matrix, which is a feature of periodontal lesions, may be caused by elevated MMP activity (Tonetti et al., 1994; DeCarlo et al., 1997, 1998), and a higher percentage of tissues from periodontitis sites have mRNA for MMPs as compared with tissue from healthy sites (Tonetti et al., 1994). Control of MMP activity is, therefore, important in sustaining gingival health. Invasion of *P. gingivalis* disrupts the expression of MMPs by gingival epithelial cells (Fravalo et al., 1996), which could contribute both to tissue destruction and to failure to repair a periodontal lesion. These activities are distinct from, but probably complementary to, the direct action of proteolytic enzymes that will be delivered in close proximity to their substrates during the adhesion and entry process. Indeed, *P. gingivalis* proteinases can activate and upregulate the transcription of MMP enzymes (DeCarlo et al., 1997, 1998). Furthermore, *P. gingivalis* proteinases can degrade IL-8 and other cytokines along with occludin, cadherins, catenins, and integrins (Fletcher et al., 1997; Darveau et al., 1998; Yun et al., 1999; Zhang et al., 1999; Katz et al., 2000), which are proteins that are important in maintaining the barrier function of the epithelium.

Systemic diseases

Although tissue destruction in periodontal diseases is limited to the supporting structures of the teeth, epidemiological evidence is emerging for an association between periodontal infections and serious systemic diseases, including coronary artery disease (Scannapieco and Genco, 1999). Several observations provide a credible, though still preliminary, basis for a causal link between infections with periodontal organisms such as *P. gingivalis* and heart disease. *P. gingivalis* has been detected in carotid and coronary atheromas (Chiu, 1999; Haraszthy et al., 2000), and the organism can induce platelet aggregation, which is associated with thrombus formation (Herzberg et al., 1994). Furthermore, *P. gingivalis* infection accelerates the progression of atherosclerosis in a heterozygous apolipoprotein E-deficient murine model (Li et al., 2002). Although common dental procedures, even vigorous tooth brushing, can lead to the presence of oral bacteria in the bloodstream, it is also possible that tissue and cell invasion by *P. gingivalis* in the highly vascularized gingiva may be a means by which these bacteria can gain access to the circulating blood and establish infections at remote sites. Once located at sites such as the heart vessel walls, the invasion of endothelial cells (Deshpande et al., 1998; Dorn et al., 1999) could constitute a chronic insult to arterial walls that may increase susceptibility to tissue damage. In addition, modulation of factors that regulate recruitment of leukocytes (ICAM-1, VCAM-1, MCP-1, and selectins) could enhance arthrosclerosis progression (Kuramitsu et al., 2001; Khlgatian et al., 2002; Nassar et al., 2002). For example, monocytes migrating through the endothelial layer into the subendothelial spaces may produce foam cells (Kang and Kuramitsu, 2002), a feature of both early and late artherosclerotic lesions. *P. gingivalis* can also directly induce foam cell production in a murine macrophage cell line (Kuramitsu et al., 2001).

It is likely that we are only beginning to uncover the full range of consequences of the interactions between invasive oral bacterial and host cells. It is also important to consider that *P. gingivalis* is present in the mouths of healthy individuals and can be found in high numbers inside epithelial cells (Rudney et al., 2001). An intracellular location may initially serve as an integral part of the interactions that maintain a balance between host and pathogen. Disruption of this balance, either by the action of the internal bacteria or by other pathogen or host factors, may be required to activate the disease process.

REFERENCES

Banchereau, J. and Steinman, R.M. (1998). Dendritic cells and the control of immunity. *Nature* **392**, 245–252.

Barkocy-Gallagher, G.A., Han, N., Patti, J.M., Whitlock, J., Progulske-Fox, A., and Lantz, M.S. (1996). Analysis of the *prtP* gene encoding porphypain, a cysteine proteinase of *Porphyromonas gingivalis. J. Bacteriol.* **178**, 2734–2741.

Belton, C.M., Izutsu, K.T., Goodwin, P.C., Park, Y., and Lamont, R.J. (1999). Fluorescence image analysis of the association between *Porphyromonas gingivalis* and gingival epithelial cells. *Cell. Microbiol.* **1**, 215–224.

Chen, T., Nakayama, K., Belliveau, L., and Duncan, M.J. (2001). *Porphyromonas gingivalis* gingipains and adhesion to epithelial cells. *Infect. Immun.* **69**, 3048–3056.

Chiu, B. (1999). Multiple infections in carotid atherosclerotic plaques. *Am. Heart J.* **138**, S534–536.

Curtis, M.A., Kuramitsu, H.K., Lantz, M., Macrina, F.L., Nakayama, K., Potempa, J., Reynolds, E.C., and Aduse-Opoku, J. (1999). Molecular genetics and nomenclature of proteases of *Porphyromonas gingivalis. J. Periodont. Res.* **34**, 464–472.

Cutler, C.W., Jotwani, R., Palucka, K.A., Davoust, J., Bell, D., and Banchereau, J. (1999). Evidence and a novel hypothesis for the role of dendritic cells and *Porphyromonas gingivalis* in adult periodontitis. *J. Periodont. Res.* **34**, 406–412.

Darveau, R.P., Belton, C.M., Reife, R.A., and Lamont, R.J. (1998). Local chemokine paralysis: a novel pathogenic mechanism of *Porphyromonas gingivalis. Infect. Immun.* **66**, 1660–1665.

DeCarlo, A.A., Grenett, H.E., Harber, G.J., Windsor, L.J., Bodden, M.K., Birkedal-Hansen, B., and Birkedal-Hansen, H. (1998). Induction of matrix metalloproteinases and a collagen-degrading phenotype in fibroblasts and epithelial cells by secreted *Porphyromonas gingivalis* proteinase. *J. Periodont. Res.* **33**, 408–420.

DeCarlo, A.A., Windsor, L.J., Bodden, M.K., Harber, G.J., Birkedal-Hansen, B., and Birkedal-Hansen, H. (1997). Activation and novel processing of matrix metalloproteinases by thiol-proteinase from the oral anaerobe *Porphyromonas gingivalis. J. Dent. Res.* **76**, 1260–1270.

Deshpande, R.G., Khan, M.B., and Genco, C.A. (1998). Invasion of aortic and heart endothelial cells by *Porphyromonas gingivalis. Infect. Immun.* **66**, 5337–5343.

Devine, D.A., Marsh, P.D., Percival, R.S., Rangarajan, M., and Curtis, M.A. (1999). Modulation of antibacterial peptide activity by products of *Porphyromonas gingivalis* and *Prevotella* spp. *Microbiology* **145**, 965–971.

Dorn, B.R., Burks, J.N., Seifert, K.N., and Progulske-Fox, A. (2000). Invasion of endothelial and epithelial cells by strains of *Porphyromonas gingivalis. FEMS Microbiol. Lett.* **187**, 139–144.

Dorn, B.R., Dunn, W.A. Jr., and Progulske-Fox, A. (1999). Invasion of human coronary artery cells by periodontal pathogens. *Infect. Immun.* **67**, 5792–5798.

Dorn, B.R., Dunn, W.A. Jr., and Progulske-Fox, A. (2001). *Porphyromonas gingivalis* traffics to autophagosomes in human coronary artery endothelial cells. *Infect. Immun.* **69**, 5698–5708.

Duncan, M.J., Nakao, S., Skobe, Z., and Xie, H. (1993). Interactions of *Porphyromonas gingivalis* with epithelial cells. *Infect. Immun.* **61**, 2260–2265.

Fletcher, J., Reddi, K., Poole, S., Nair, S., Henderson, B., Tabona, P., and Wilson, M. (1997). Interactions between periodontopathogenic bacteria and cytokines. *J. Periodont. Res.* **32**, 200–205.

Fravalo, P., Menard, C., and Bonnaure-Mallet, M. (1996). Effect of *Porphyromonas gingivalis* on epithelial cell MMP-9 type IV collagenase production. *Infect. Immun.* **64**, 4940–4945.

Han, N., Whitlock, J., and Progulske-Fox, A. (1996). The hemagglutinin gene A (*hagA*) of *Porphyromonas gingivalis* 381 contains four large, contiguous, direct repeats. *Infect. Immun.* **64**, 4000–4007.

Haraszthy, V.I., Zambon, J.J., Trevisan, M., Zeid, M., and Genco, R.J. (2000). Identification of periodontal pathogens in atheromatous plaques. *J. Periodontol.* **71**, 1554–1560.

Herzberg, M.C., MacFarlane, G.D., Liu, P., and Erickson, P.R. (1994). The platelet as an inflammatory cell in periodontal diseases: the interactions with *Porphyromonas gingivalis*. In *Molecular Pathogenesis of Periodontal Disease*, ed. R. Genco, S. Hamada, T. Lehner, J. McGhee, and S. Mergenhagen, pp. 247–256. Washington, DC: ASM Press.

Houalet-Jeanne, S., Pellen-Mussi, P., Tricot-Doleux, S., Apiou, J., and Bonnaure-Mallet, M. (2001). Assessment of internalization and viability of *Porphyromonas gingivalis* in KB epithelial cells by confocal microscopy. *Infect. Immun.* **69**, 7146–7151.

Huang, G.T., Haake, S.K., Kim, J.W., and Park, N. (1998). Differential expression of interleukin-8 and intercellular adhesion molecule-1 by human gingival epithelial cells in response to *Actinobacillus actinomycetemcomitans* or *Porphyromonas gingivalis* infection. *Oral Microbiol. Immunol.* 13, 301–309.

Huang, G.T., Kim, D., Lee, J.K., Kuramitsu, H.K., and Haake, S.K. (2001). Interleukin-8 and intercellular adhesion molecule 1 regulation in oral epithelial cells by selected periodontal bacteria: multiple effects of *Porphyromonas gingivalis* via antagonistic mechanisms. *Infect. Immun.* 69, 1364–1372.

Izutsu, K.T., Belton, C.M., Chan, A., Fatherazi, S., Kanter, J.P., Park, Y., and Lamont, R.J. (1996). Involvement of calcium in interactions between gingival epithelial cells and *Porphyromonas gingivalis*. *FEMS Microbiol. Lett.* 144, 145–150.

Kang, I.C. and Kuramitsu, H.K. (2002). Induction of monocyte chemoattractant protein-1 by *Porphyromonas gingivalis* in human endothelial cells. *FEMS Immunol. Med. Microbiol.* 34, 311–317.

Katz, J., Sambandam, V., Wu, J.H., Michalek, S.M., and Balkovetz, D.F. (2000). Characterization of *Porphyromonas gingivalis*-induced degradation of epithelial cell junctional complexes. *Infect. Immun.* 68, 1441–1449.

Khlgatian, M., Nassar, H., Chou, H.H., Gibson, F.C., and Genco, C.A. (2002). Fimbria-dependent activation of cell adhesion molecule expression in *Porphyromonas gingivalis*-infected endothelial cells. *Infect. Immun.* 70, 257–267.

Kontani, M., Kimura, S., Nakagawa, I., and Hamada, S. (1997). Adherence of *Porphyromonas gingivalis* to matrix proteins via a fimbrial cryptic receptor exposed by its own arginine-specific protease. *Mol. Microbiol.* 24, 1179–1187.

Kontani, M., Ono, H., Shibata, H., Okamura, Y., Tanaka, T., Fujiwara, T., Kimura, S., and Hamada, S. (1996). Cysteine protease of *Porphyromonas gingivalis* 381 enhances binding of fimbriae to cultured human fibroblasts and matrix proteins. *Infect. Immun.* 64, 756–762.

Kuramitsu, H.K. (1998). Proteases of *Porphyromonas gingivalis*: what don't they do? *Oral Microbiol. Immunol.* 13, 263–270.

Kuramitsu, H.K., Qi, M., Kang, I.C., and Chen, W. (2001). Role for periodontal bacteria in cardiovascular diseases. *Ann. Periodontol.* 6, 41–47.

Lamont, R.J., Chan, A., Belton, C.M., Izutsu, K.T., Vasel, D., and Weinberg, A. (1995). *Porphyromonas gingivalis* invasion of gingival epithelial cells. *Infect. Immun.* 63, 3878–3885.

Lamont, R.J. and Jenkinson, H.F. (1998). Life below the gum line: pathogenic mechanisms of *Porphyromonas gingivalis*. *Microbiol. Mol. Biol. Rev.* 62, 1244–1263.

Lamont, R.J. and Jenkinson, H.F. (2000). Subgingival colonization by *Porphyromonas gingivalis*. *Oral Microbiol. Immunol.* 15, 341–349.

Lamont, R.J. and Yilmaz, O. (2002). In or out: the invasiveness of oral bacteria. *Periodontology 2000* **30**, 61–69.

Lamont, R.J., Oda, D., Persson, R.E., and Persson, G.R. (1992). Interaction of *Porphyromonas gingivalis* with gingival epithelial cells maintained in culture. *Oral Microbiol. Immunol.* **7**, 364–367.

Li, L., Messas, E., Batista, E.L. Jr., Levine, R.A., and Amar, S. (2002). *Porphyromonas gingivalis* infection accelerates the progression of atherosclerosis in a heterozygous apolipoprotein E-deficient murine model. *Circulation* **105**, 861–867.

Madianos, P.N., Papapanou, P.N., Nannmark, U., Dahlen, G., and Sandros, J. (1996). *Porphyromonas gingivalis* FDC381 multiplies and persists within human oral epithelial cells in vitro. *Infect. Immun.* **64**, 660–664.

Madianos, P.N., Papapanou, P.N., and Sandros, J. (1997). *Porphyromonas gingivalis* infection of oral epithelium inhibits neutrophil transepithelial migration. *Infect. Immun.* **65**, 3983–3990.

Nakagawa, I., Amano, A., Kuboniwa, M., Nakamura, T., Kawabata, S., and Hamada, S. (2002). Functional differences among FimA variants of *Porphyromonas gingivalis* and their effects on adhesion to and invasion of human epithelial cells. *Infect. Immun.* **70**, 277–285.

Nakamura, T., Amano, A., Nakagawa, I., and Hamada, S. (1999). Specific interactions between *Porphyromonas gingivalis* fimbriae and human extracellular matrix proteins. *FEMS Microbiol. Lett.* **175**, 267–272.

Nakayama, K., Yoshimura, F., Kadowaki, T., and Yamamoto, K. (1996). Involvement of arginine-specific cysteine proteinase (Arg-gingipain) in fimbriation of *Porphyromonas gingivalis*. *J. Bacteriol.* **178**, 2818–2824.

Nakhjiri, S.F., Park, Y., Yilmaz, O., Chung, W.O., Watanabe, K., El-Sabaeny, A., Park, K., and Lamont, R.J. (2001). Inhibition of epithelial cell apoptosis by *Porphyromonas gingivalis*. *FEMS Microbiol. Lett.* **200**, 145–149.

Nassar H, Chou, H.H., Khlgatian, M., Gibson, F.C., Van Dyke, T.E., and Genco, C.A. (2002). Role for fimbriae and lysine-specific cysteine proteinase gingipain K in expression of interleukin-8 and monocyte chemoattractant protein in *Porphyromonas gingivalis*-infected endothelial cells. *Infect. Immun.* **70**, 268–276.

Nisapakultorn, K., Ross, K.F., and Herzberg, M.C. (2001). Calprotectin expression in vitro by oral epithelial cells confers resistance to infection by *Porphyromonas gingivalis*. *Infect. Immun.* **69**, 4242–4247.

Njoroge, T., Genco, R.J., Sojar, H.T., and Genco, C.A. (1997). A role for fimbriae in *Porphyromanas gingivalis* invasion of oral epithelial cells. *Infect. Immun.* **65**, 1980–1984.

Ogawa, T., Ogo, H., Uchida, H., and Hamada, S. (1994). Humoral and cellular immune responses to the fimbriae of *Porphyromonas gingivalis* and their synthetic peptides. *J. Med. Microbiol.* **40**, 397–402.

Park, Y. and Lamont, R.J. (1998). Contact-dependent protein secretion in *Porphyromonas gingivalis*. *Infect. Immun.* **66**, 4777–4782.

Potempa, J., Pavloff, N., and Travis, J. (1995). *Porphyromonas gingivalis*: a proteinase/gene accounting audit. *Trends Microbiol.* **3**, 430–434.

Progulske-Fox, A., Tumwasorn, S., Lepine, G., Whitlock, J., Savett, D., Ferretti, J.J., and Banas, J.A. (1995). The cloning, expression and sequence analysis of second *Porphyromonas gingivalis* gene that encodes for a protein involved in hemagglutination. *Oral Microbiol. Immunol.* **10**, 311–318.

Rudney, J.D., Chen, R., and Sedgewick, G.J. (2001). Intracellular *Actinobacillus actinomycetemcomitans* and *Porphyromonas gingivalis* in buccal epithelial cells collected from human subjects. *Infect. Immun.* **69**, 2700–2707.

Saglie, F.R., Pertuiset, J.H., Smith, C.T., Nestor, M.G., Carranza, F.A., Newman, M.G., Rezende, M.T., and Nisengard, R. (1987). The presence of bacteria in the oral epithelium in periodontal disease. III. Correlation with Langerhans cells. *J. Periodontol.* **58**, 417–422.

Sandros, J., Madianos, P.N., and Papapanou, P.N. (1996). Cellular events concurrent with *Porphyromonas gingivalis* invasion of oral epithelium in vitro. *Eur. J. Oral Sci.* **104**, 363–371.

Sandros, J., Papapanou, P.N., Nannmark, U., and Dahlen, G. (1994). *Porphyromonas gingivalis* invades human pocket epithelium *in vitro*. *J. Periodont. Res.* **29**, 62–69.

Scannapieco, F.A. and Genco, R.J. (1999). Association of periodontal infections with atherosclerotic and pulmonary diseases. *J. Periodont. Res.* **34**, 340–345.

Shibita, Y., Hayakawa, M., Takiguchi, H., Shiroza, T., and Abiko, Y. (1999). Determination and characterization of the hemagglutinin-associated short motifs found in *Porphyromonas gingivalis* multiple gene products. *J. Biol. Chem.* **274**, 5012–5020.

Sojar, H.T., Han, Y., Hamada, N., Sharma, A., and Genco, R.J. (1999). Role of the amino-terminal region of *Porphyromonas gingivalis* fimbriae in adherence to epithelial cells. *Infect. Immun.* **67**, 6173–6176.

Sojar, H.T., Lee, J.-Y., and Genco, R.J. (1995). Fibronectin binding domain of *P. gingivalis* fimbriae. *Biochem. Biophys. Res. Commun.* **216**, 785–792.

Sojar, H.T., Sharma, A., and Genco, R.J. (2002). *Porphyromonas gingivalis* fimbriae bind to cytokeratin of epithelial cells. *Infect. Immun.* **70**, 96–101.

Tokuda, M., Duncan, M., Cho, M.I., and Kuramitsu, H.K. (1996). Role of *Porphyromonas gingivalis* protease activity in colonization of oral surfaces. *Infect. Immun.* **64**, 4067–4073.

Tonetti, M.S., Imboden, M.A., Gerber, L., Lang, N.P., Laissue, J., and Mueller, C. (1994). Localized expression of mRNA for phagocyte-specific chemotactic cytokines in human periodontal infections. *Infect. Immun.* **62**, 4005–4014.

Turner, C.E. (2000). Paxillin and focal adhesion signaling. *Nat. Cell Biol.* **2**, E231–E236.

Wang, T., Zhang, Y., Chen, W., Park, Y., Lamont, R.J., and Hackett, M. (2002). Reconstructed protein arrays from 3D HPLC/tandem mass spectrometry and 2D gels: complementary approaches to *Porphyromonas gingivalis* protein expression. *The Analyst* **127**, 1450–1456.

Watanabe, K., Onoe, T., Ozeki, M., Shimizu, Y., Sakayori, T., Nakamura, H., and Yoshimura, F. (1996). Sequence and product analyses of the four genes downstream from the fimbrillin gene (*fimA*) of the oral anaerobe *Porphyromonas gingivalis*. *Microbiol. Immunol.* **40**, 725–734.

Watanabe, K., Yilmaz, O., Nakhjiri, S.F., Belton, C.M., and Lamont, R.J. (2001). Association of mitogen-activated protein kinase pathways with gingival epithelial cell responses to *Porphyromonas gingivalis* infection. *Infect. Immun.* **69**, 6731–6737.

Weinberg, A., Belton, C.A., Park, Y., and Lamont, R.J. (1997). Role of fimbriae in *Porphyromonas gingivalis* invasion of gingival epithelial cells. *Infect. Immun.* **65**, 313–316.

Xie, H., Cai, S., and Lamont, R.J. (1997). Environmental regulation of fimbrial gene expression in *Porphyromonas gingivalis*. *Infect. Immun.* **65**, 2265–2271.

Xie, H., Chung, W., Park, Y., and Lamont, R.J. (2000). Regulation of the *Porphyromonas gingivalis fimA* (fimbrillin) gene. *Infect Immun.* **68**, 6574–6579.

Xie, H. and Lamont, R.J. (1999). Promoter architecture of the *Porphyromonas gingivalis* fimbrillin gene. *Infect. Immun.* **67**, 3227–3235.

Yilmaz, Ö., Watanabe, K., and Lamont, R.J. (2002). Involvement of integrins in fimbriae-mediated binding and invasion by *Porphyromonas gingivalis*. *Cell. Microbiol.* **4**, 305–314.

Yilmaz, Ö., Young, P.A., Lamont, R.J., and Kenny, G.E. (2003). Gingival epithelial cell signaling and cytoskeletal responses to *Porphyromonas gingivalis* invasion. *Microbiology* **149**, 2417–2426.

Yun, P.L., DeCarlo, A.A., and Hunter, N. (1999). Modulation of major histocompatibility complex protein expression by human gamma interferon mediated by cysteine proteinase-adhesin polyproteins of *Porphyromonas gingivalis*. *Infect. Immun.* **67**, 2986–2995.

Zhang, J., Dong, H., Kashket, S., and Duncan, M.J. (1999). IL-8 degradation by *Porphyromonas gingivalis* proteases. *Microb. Pathog.* **26**, 275–280.

Index

INDEX